INTRODUCTION TO LOGIC AND SWITCHING THEORY

INTRODUCTION TO LOGIC AND SWITCHING THEORY

Nripendra N. Biswas

Professor of Electrical Communication Engineering and Computer Science, Indian Institute of Science, Bangalore, India. Formerly, Professor of Electrical Engineering, St. Louis University, Missouri, U.S.A.

GORDON AND BREACH SCIENCE PUBLISHERS
New York London Paris

Gordon and Breach, Science Publishers Ltd.
42 William IV Street
London W.C.2.

Editorial office for the United States of America

Gordon and Breach, Science Publishers, Inc.
1 Park Avenue
New York, N.Y. 10016

Editorial office for France

Gordon & Breach
7–9 rue Emile Dubois
Paris 75014

Library of Congress catalog card number 73-84337. ISBN 0 677 02860 1.
 Printed in Great Britain.

TO MY WIFE REBA

PREFACE

The revolutionary progress of this century's science and technology has introduced many new theories in the disciplines of electrical and electronic engineering. Among these, switching theory is comparatively a newcomer. But unlike others, switching theory is not only a theory, but also a philosophy – a design philosophy – which played, and will continue to play, a vital role in the design and evolution of the various generations of computers. Although this philosophy was nurtured and developed by computer scientists, it has found wide acceptance among communication and control engineers engaged in the design and development of digital systems. By now the subject of logical design and switching theory has firmly established itself as a basic course in any curriculum leading to a degree in computer science or electrical engineering.

In this book I have tried to present the fundamentals of logic and switching theory in a clear and lucid manner. Although the book is intended primarily as a textbook, I have also kept in mind a substantial number of practising engineers who could not study this subject in their student career but whose current professional responsibility demands a thorough knowledge of this subject. It is hoped that this book will be an important aid to them in achieving the objective by self-study.

The material of this book can be conveniently covered either as a two-semester or as a one-semester course depending on the situation. If the course is offered to an undergraduate class who would be exposed to the number system and Boolean algebra for the first time, it is better to offer it as a two-semester course. On the other hand, if the students have some prior knowledge of these, then it can be a one-semester course. At the Indian Institute of Science, I have covered the material in a one-semester course to a combined class attended by final year undergraduate and first year graduate students. In this situation class time can be saved by covering Chapters 0, 1, 2 and 8 at a faster rate with the help of home assignments.

Perhaps it is very difficult for an author to resist the temptation of including in his book some of the methods or algorithms developed by him. The adjacency method of minimizing a Boolean function in Chapter 4, the method of ordered partition for the detection and identification of totally symmetric functions in Chapter 5, virtually the whole of Chapter 8 on logic of flip-flops, the compatibility graph and the implication tree method of minimizing the states of an incompletely specified sequential machine in Chapter 9, appear in this book as a result of the author's falling a victim to this temptation. Nothing can be more rewarding if these methods find favor with the students, the instructors and the practising engineers. To be on the safer side, these methods have been included not by replacing the comparatively well established methods but in addition to them. Hence such of these methods which are not found suitable can be skipped.

While collecting material for writing this book, I have freely consulted many books and papers published in various journals. I take this opportunity to acknowledge my indebtedness to the authors of all these valuable publications. I must specially mention the Institute of Electrical and Electronics Engineers, New York, for their kind permission to reproduce many papers published in their Transactions. I am also thankful to Dr. Leo Hellerman of the IBM and Professor M. L. Dertouzos of the Massachusetts Institute of Technology, and the MIT Press, Cambridge, Mass., U.S.A., for their kind permission to reproduce their works as appendices of this book.

A part of the manuscript of this book was written at the St. Louis University. I would like to express my indebtedness to all my colleagues and specially to Dr. Gerald E. Dreifke, Chairman, Electrical Engineering Department, St. Louis University, for their encouragement in writing this book. I am also grateful to many of my students both in the U.S.A. and in India who contributed a great deal in improving the book by their criticism and suggestions. I am specially thankful to Mr. S. K. Srivatsa and Mr. A. K. Sarje for going through the entire manuscript and suggesting many improvements. I would also like to express my appreciation to Mr. R. Vijayendra for doing an excellent job in preparing the art work for this book. I am also grateful to Professor B. S. Ramakrishna, Chairman of the Department of Electrical Communication Engineering, and Dr. S. Dhawan, Director, Indian Institute of Science, Bangalore, for providing all facilities and encouragement in the preparation and publication of this book. Finally I would like to express my gratitude to Mr. W. Y. Lang, Member (Retd), Technical Staff, Bell Telephone Laboratories, but for whose inspiration, help, and encouragement this book would never have been published.

Nripendra N. Biswas
Indian Institute of Science

TABLE OF CONTENTS

Chapter 0

MATHEMATICAL BACKGROUND

0.1 INTRODUCTION

The purpose of this chapter is to introduce the reader to some basic mathematical concepts which will be necessary to pursue the subject of logic and switching theory. For the sake of clarity important results are presented in the form of theorems and corrollaries. In most cases theorems have been stated without any proof. In some cases proofs have been provided where they help in greater understanding of the problem. This approach has been found adequate for the material of this introductory chapter. For more rigorous or formal proofs the reader may consult textbooks of mathematics.

0.2 SETS

The word 'set' is used by us in everyday life to mean a collection, or a class. It conveys the same meaning in mathematics and we can define it as such.

Definition 0.2.1 A *set* is a collection of certain things. Each object belonging to the set is known as an *element* or a *member* of the set.

Thus, a set A can be the collection of the six things; pen, man, moon, 9, $\vee$ and *. It is written as

$$A = \{\text{pen, man, moon}, 9, \vee, *\}.$$

It can be seen that the six elements of the set A have no common property except that all of them belong to set A. In another example a set X can be as follows

$$X = \{1, 3, 5, 7, 9\}.$$

In this case, the five elements can be defined by a common property, inasmuch as each of them is a positive odd number and less than 10. Thus X can also be written as follows:

$$X = \{x | x \text{ is positive}, x \text{ is odd}, x < 10\}.$$

The fact that x is an element of the set X is written as $x \in X$, and is read as x belongs to X, or x is in X.

Definition 0.2.2 If all the elements of a set also belong to another set, then the former is a *subset* of the latter.

As an example if $X = \{1, 3, 5, 7, 9\}$

and $Y = \{1, 2, 3, 4, 5, 6, 7, 8, 9, 10\}$

then X is a subset of Y, which is written as $X \subset Y$.

It can be easily seen that according to the above definition, $Y \subset Y$. This means that every set is a subset of itself.

Definition 0.2.3 Two sets are equal if and only if they contain exactly the same elements.

Theorem 0.2.1 If two sets are subsets of each other, then they are equal. That is, if $X \subset Y$ and $Y \subset X$, then $X = Y$.

The validity of this theorem can be easily verified.

Definition 0.2.4 A special set which does not contain any element is defined as an *empty set* or a *null set.* It will be denoted by Φ.

It is interesting to note that since no element belongs to the empty set, its elements can be considered to be belonging to every set. Hence the *empty set is a subset of every set.*

Another special set of importance is the universal set. Suppose we are dealing with a number of sets W, X, Y, Z etc. Then we can specify a set U, such that each of the sets W, X, Y, Z, etc. and any new set that may result during any manipulation between any two or more of these sets, will be a subset of U. Then U is the universal set. It must be observed that unlike the empty set, the universal set is not a definite or a fixed set, but can be specified to suit every situation.. It can, therefore, be defined as follows.

Definition 0.2.5 Let the elements of two or more sets and all the new elements that may be obtained by certain manipulations between these sets be defined as the universe of elements in the given situation. Then the set containing all these elements is known as the *universal set* for the given situation.

It is obvious that all other sets will form subsets of the universal set.

Once the universal set has been specified, the complement of a set can be defined.

Definition 0.2.6 The *complement* of a set X, written as X' is defined as the set of all elements of the universal set which are not elements of X. That is,

$$X' = \{x | x \notin X, \text{ and } x \in U\} .$$

Example 0.2.1 Let $A = \{1, 2, 5, 6, 8\}$
$B = \{2, 3, 4, 7, 9, 10\}$
$U = \{1, 2, 3, 4, 5, 6, 7, 8, 9, 10\}$.

Find A' and B'.

Solution: $A' = \{3, 4, 7, 9, 10\}$ $B' = \{1, 5, 6, 8\}$.

Theorem 0.2.2. The complement of the complementary set is the set itself. That is, $(X')' = X$.

The validity of this theorem follows from the definition of the complement of a set.

A very convenient way to visualize sets is due to Venn. In this method a closed area usually a circle is used to show a set. All elements of the set are considered to be inside the closed area, and the elements not belonging to the set are outside. The universal set is shown by a rectangle. Figure 0-1 depicts how the sets X, X' and the relation $Y \subset X$ can be shown on the Venn diagram.

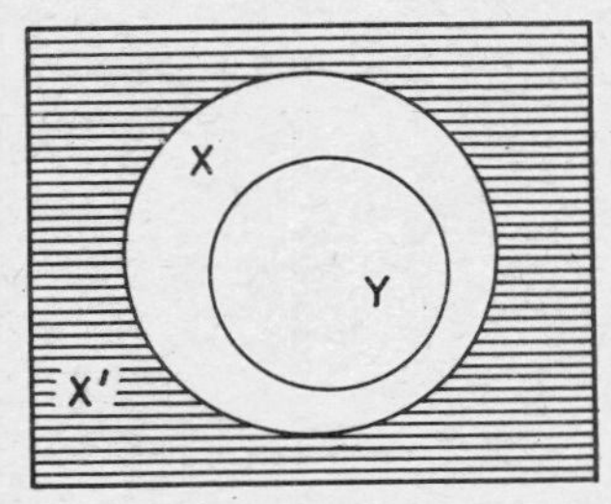

Figure 0-1 Venn diagram representing the sets X and X', and the relation $Y \subset X$.

0.3 ALGEBRA OF SETS

There are many algebraic operations which can be performed between two or more sets. Two of these are the union, denoted by the cup symbol, $\cup$, and the intersection denoted by the cap symbol $\cap$.

Definition 0.3.1 The *union* of two sets is a set containing all the elements of both the sets. Thus,

$$\{1, 2, 5, 6, 8\} \cup \{0, 4, 5, 6\} = \{0, 1, 2, 4, 5, 6, 8\}.$$

The union of two or more sets can be depicted on the Venn diagram as shown in Fig.0-2.

The following theorems regarding the union operation are obvious and are being stated below without any proof. The reader may verify them by taking some examples or on the Venn diagram.

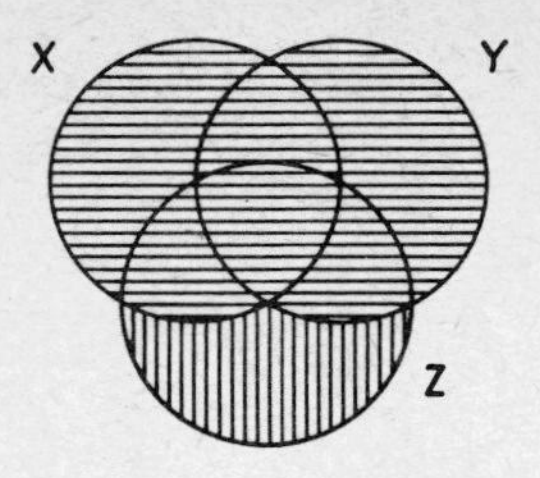

Figure 0-2 Venn diagram depicting union of sets. Horizontally hatched portion is $X \cup Y$. The entire shaded portion is $X \cup Y \cup Z$.

Theorem 0.3.1 The union of a set with itself is the set itself. That is,

$X \cup X = X.$

Theorem 0.3.2 The union of a set with the empty set is the set itself. That is,

$X \cup \Phi = X.$

Theorem 0.3.3 The union of a set with the universal set is the universal set. That is,

$X \cup U = U.$

Theorem 0.3.4 The union of a set with its complement is the universal set. That is,

$X \cup X' = U.$

Theorem 0.3.5 If the set X is in the set Y, then the union of X' and Y is the universal set. That is, if $X \subset Y$, then $X' \cup Y = U$.

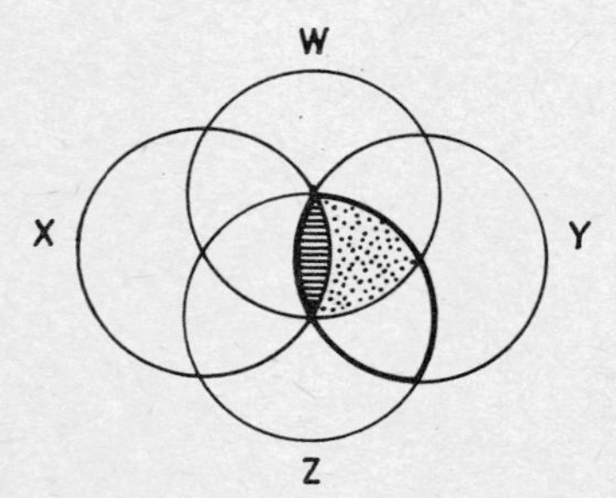

Figure 0-3 Venn diagram showing intersection of sets. The portion within the thick boundary is $Y \cap Z$. The dotted portion is $W \cap X' \cap Y \cap Z$. The horizontal shaded portion is $W \cap X \cap Y \cap Z$.

Definition 0.3.2 The *intersection* of two sets is a set which contains only those elements which are common to both the sets. Thus,

$$\{1, 2, 5, 6, 7\} \cap \{0, 4, 5, 6\} = \{5, 6\}.$$

The intersection of two or more sets can be represented on the Venn diagram as shown in Fig. 0-3.

The following theorems regarding the intersection operation are very useful. These can also be easily verified by examples or on the Venn diagram.

Theorem 0.3.6 The intersection of a set with itself is the set itself. That is,

$$X \cap X = X.$$

Theorem 0.3.7 The intersection of a set with the universal set is the set itself. That is,

$$X \cap U = X.$$

Theorem 0.3.8 The intersection of a set with its complement is the empty set. That is,

$$X \cap X' = \Phi.$$

Theorem 0.3.9 If the set X is in the set Y, then the intersection of X with Y' is the empty set. That is, if $X \subset Y$, then $X \cap Y' = \Phi$.

0.4 PARTITIONS

The idea of partition is a familiar one. It is used in the same sense in the case of elements of a set. If the elements of a set are divided into a number of subsets so that no element belongs to more than one subset, then the subsets constitute a partition of the set. A formal definition of a partition with the help of union and intersection operations can be given as follows.

Definition 0.4.1 If the elements of a set S are divided into n subsets, $s_1, s_2, s_3 \ldots s_n$, so that

$$s_1 \cup s_2 \cup \ldots \cup s_n = S$$

and

$$s_i \cap s_j = \Phi \quad i \neq j,$$

then the set of sets s_1 through s_n is defined as a *partition* of the set S.

For example, for the set, $S = \{1, 2, 4, *, \$\}$ the subsets, $\{1,4\}$ $\{*\}$ and $\{2,\$\}$ form a partition. This partition is written as follows:

$$S = \{\overline{1,4}, \overline{*}, \overline{2, \$}\}.$$

It can be seen that a subset may contain even a single element. Thus the partition which has the largest number of subsets is the one where each subset is the set of one element only. On the other hand the partition with the smallest number of subsets is the set itself. These two extreme cases of partitions for the set S can be written as follows:

$$S = \left\{\overline{1}, \overline{2}, \overline{4}, \overline{*}, \overline{\$}\right\},$$

and

$$S = \left\{\overline{1, 2, 4, *, \$}\right\}.$$

0.5 BINARY OPERATIONS

We are quite familiar with the arithmetic operations of addition, subtraction, multiplication and division. Each of these operations needs two numbers and produces a number by following a definite rule governed by the particular operation. In order to define a binary operation a set is also to be specified in addition to the rule of the operation.

Definition 0.5.1 If the operation * in a set S of elements $(a, b, c \ldots)$ is such that $a * b = c$, where for all $a, b, \in S$, c is also $\in S$, then * is a binary operation.

Thus the important point about a binary operator in a set S is that when it operates between any two elements of the set, the element it produces also belongs to the set S. For example, the arithmetic operation of addition in the set of all integral numbers is a binary operation. But the division operation is not a binary operation as it may produce a fractional number which is not within the set. If now the set is enlarged to contain all fractional numbers as well, then division also will be a binary operation. Again if the set of integral numbers is so restricted that $S = \{0, 1, \ldots 5\}$ then even the arithmetic addition and multiplication operations fail to be binary operations. But the operation sum Mod-6 is binary, since its result is always any one of the 6 digits, 0 through 5. Mod-6 sum is calculated by first finding the ordinary sum and then subtracting 6 from it, if the ordinary sum is more than 5. Thus

$$\text{Mod-6 sum of 3 and 4} = 3 + 4 - 6 = 1.$$

Similarly the Mod-6 product is calculated by first finding the ordinary product and then dividing by 6 if the ordinary product is greater than 5. The resulting remainder is the product Mod-6. Thus the Mod-6 product between 4 and 5 is 2, since

```
        4
        5
  6   |20
        3-2
```

These are shown in the tables of Fig.0-4.

Sum Mod-6

+	0	1	2	3	4	5
0	0	1	2	3	4	5
1	1	2	3	4	5	0
2	2	3	4	5	0	1
3	3	4	5	0	1	2
4	4	5	0	1	2	3
5	5	0	1	2	3	4

Product Mod-6

.	0	1	2	3	4	5
0	0	0	0	0	0	0
1	0	1	2	3	4	5
2	0	2	4	0	2	4
3	0	3	0	3	0	3
4	0	4	2	0	4	2
5	0	5	4	3	2	1

Figure 0-4 Mod-6 addition and multiplication tables.

Sum or product for any other modulus m can be found by following the same procedure replacing 6 by m.

The various laws for binary operation are defined below.

Definition 0.5.2 The binary operation * is *idempotent* if $a * a = a$ for all $a \in S$.

Definition 0.5.3 The binary operation * is *commutative* if $a * b = b * a$ for all $a, b \in S$.

Definition 0.5.4 The binary operation * is *associative* if $a * (b * c) = (a * b) * c$ for all $a, b, c \in S$.

Definition 0.5.5 The binary operation * is *distributive* over another binary operation o if $a * (b \circ c) = (a * b) \circ (a * c)$ for all $a, b, c \in S$.

Definition 0.5.6 There exists an *identity element* $e_* \in S$ for the binary operation * if for all $a \in S$, $a * e_* = a = e_* * a$.

Definition 0.5.7 There exists an *inverse* element for the binary operation *, if for any $a \in S$, there exists an element $b \in S$, so that $a * b = e_* = b * a$. b is then called the *-ive inverse of a.

Example 0.5.1 Show the validity of the following statements about the binary operations of ordinary addition and multiplication in the set Z containing all integers and zero.

a) The multiplication operation is distributive over addition, but addition is not distributive over multiplication.

b) 0 is the identity element for addition, and 1 is the identity element for multiplication.

c) Neither addition nor multiplication has an inverse.

d) The idempotence law does not hold good for both addition and multiplication.

Solution:

a) Let $x, y, z, \in Z$. Now, since $x(y + z) = xy + xz$ multiplication is distributive over addition. But $x + yz \neq (x + y)(x + z)$. $\therefore$ addition is not distributive over multiplication.

b) Since $x + 0 = x = 0 + x$, 0 is the identity element for addition. Again $x1 = x = 1x$, $\therefore$ 1 is the identity element for multiplication.

c) If y is the additive inverse of x, then $x + y = 0$, $\therefore y = -x$. Since x is a positive integer, y must be a negative integer. Hence y is not $\in Z$. $\therefore$ Additive inverse does not exist. Again, if y is the multiplicative inverse, then $xy = 1$. $\therefore y = 1/x$ $\therefore y$ is a fractional number, and therefore y is not $\in Z$. Hence the multiplicative inverse also does not exist.

d) Since $x + x = 2x \neq x$ and $xx = x^2 \neq x$, the law of idempotence does not hold good for either addition or multiplication.

0.6 UNARY OPERATION

As the binary operations operate on two elements, so the unary operation operates on a single element of the set. As an example let a set consist of the 4 pictures as shown in Fig.0-5, and let these pictures be denoted by a, b, c, d respectively. Now, let the unary operation be so defined as to mean rotating the picture by 90° in the clockwise direction. Then $a^\theta = b$, and $b^\theta = c$, and so on.

A property of interest for a unary operation can be defined as follows:

Definition 0.6.1 The *law of involution* for a unary operation * holds good, if for all $a \in S$, $(a\,^*)^* = a$.

The unary operation θ as defined in the above set (rotation by 90° in the clockwise direction) does not obey the law of involution. On the other hand, if a unary operation π is so defined as to mean rotating a picture by 180° clockwise, then $a^\pi = c$, and $c^\pi = a$. Hence the law of involution holds good for the unary operation of π. It can be easily verified that the complementation operation in the algebra of sets is a unary operation for which the law of involution holds good.

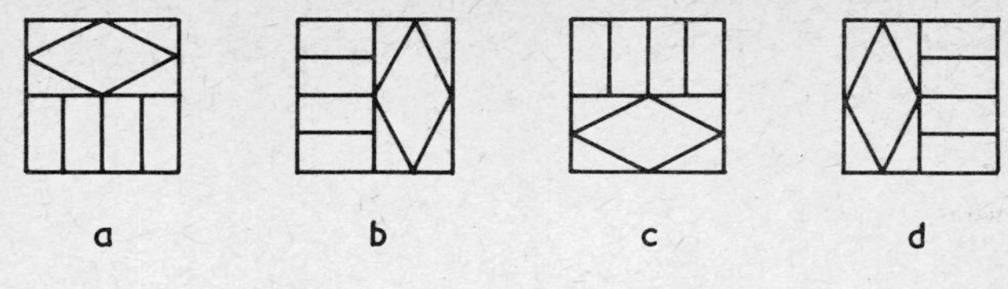

Figure 0-5 A set of 4 pictures.

0.7 NUMBER SYSTEMS

The number system which is so extensively used in our daily life and with which we have been trained to carry out the four basic arithmetic operations of addition, subtraction, multiplication and division is commonly known as the decimal system. In this system any number however big or small can be represented by the use of only ten digits 0, 1, 2, 3, 4, 5, 6, 7, 8 and 9. Obviously any of these digits are used over and over again as the need may be. Thus the same digit may indicate different counts. For example, the digit 7 in 7962 indicates a count of seven thousand, whereas in the number 2736 it indicates a count of seven hundred only. It can be observed that the 7 in 7962 is the 4th digit from the extreme right, and the 7 in 2736 is the 3rd digit from the extreme right. Hence this ability of the same digit to indicate different counts is by virtue of the exact position occupied by the particular digit. The digit 0 which does not count any number plays the vital role of assigning a position to the counting digits 1 through 9. Thus as soon as we see the number 2000, we know that 2 is in the 4th position and the number we are seeing is two thousand. This brings out the tremendous importance of the digit 0. In fact without 0 none of the modern numbering systems will work. The various positions occupied by the digits in a number can be designated serially. Let the position which is at the extreme right be the 0th position, the positions as we move to the left from this position be the 1st, 2nd, 3rd, 4th, . . . nth position. Then a digit a which is at the pth position will indicate a count of $a \times 10^p$. Hence a decimal number, $a_n \ldots a_2a_1a_0$ can be written as a sum as follows:

$$(a_n \ldots a_2a_1a_0) = a_n \times 10^n + \ldots + a_2 \times 10^2 + a_1 \times 10^1 + a_0 \times 10^0. \qquad (0\text{-}1)$$

Thus, strictly speaking the decimal system is a *positional* number system with a *base* or *radix* 10. The principle can be extended to provide a generalized positional number system whose radix is r, so that

$$(a_n \ldots a_2a_1a_0)_r = (a_n \times r^n + \ldots + a_2 \times r^2 + a_1 \times r^1 + a_0 \times r^0)_{10}. \qquad (0\text{-}2)$$

It is imperative that the system with a base r must use only r digits 0, 1, . . , $r - 1$. Thus 0 must be included in all systems, and hence the smallest value that r can have is 2. Such a system which uses only two digits, 0 and 1, is known as the *binary* system. It is interesting to note that most of the present day digital computers find it more convenient to count by this binary number system rather than the decimal system. Another number system we shall encounter later is the system with base 8, known as the *octal* system.

Definition 0.7.1 The *radix* or *base* of a positional number system is defined as the number of digits (including 0) used in the system. A system with radix r is generally known as an *r-ary* system. However, as has been mentioned above the 10-ary, 8-ary and 2-ary systems are more popularly known as the decimal, octal and binary systems respectively.

Definition 0.7.2 Each of the ten numerical symbols (0, 1, . . . 9) used in the decimal system is known as a *digit* and each of the two symbols (0, 1) of the binary system is known as a *bit*.

Any number in an r-ary system can be converted to its decimal equivalent by Eqn.(0-2). The reverse can be achieved by the following theorem.

Theorem 0.7.1 If a decimal number is successively divided by r, then the required pth numeral of the r-ary system is given by the remainder of the $(p + 1)$th division.

Proof: Let the decimal number N be equivalent to the r-ary number, $(a_n \ldots a_2a_1a_0)_r$. Then the values of $a_0, a_1, a_2 \ldots a_n$ are to be determined. Now, $(a_n \ldots a_2a_1a_0)_r = (a_n \times r^n + \ldots + a_2 \times r^2 + a_1 \times r^1 + a_0)_{10}$. Hence the number on the right hand side has been given. Let this be successively divided by r, as follows:

$$
\begin{array}{r|ll}
r & a_n \times r^n + \ldots + a_2 \times r^2 + a_1 \times r + a_0 & \\
\hline
r & a_n \times r^{n-1} + \ldots + a_2 \times r + a_1 & -a_0 \\
\hline
r & a_n \times r^{n-2} + \ldots + a_2 & -a_1 \\
\hline
 & \cdot \quad \cdot \quad \cdot \quad \cdot \quad \cdot \quad \cdot & \\
r & a_n \times r + a_{n-1} & -a_{n-2} \\
\hline
r & a_n & -a_{n-1} \\
\hline
 & 0 & -a_n.
\end{array}
$$

Thus, the required numerals $a_0, a_1 \ldots a_n$ are the remainders of the 1st, 2nd, . . and $(n+1)$th divisions. The least significant digit (LSD) is obtained first, and the most significant digit (MSD) is obtained last. ▲

Example 0.7.1 Express the octal number $(776)_8$ and the binary number $(101011)_2$ as decimal numbers.

Solution: $(776)_8 = (7 \times 8^2 + 7 \times 8 + 6)_{10}$

$$= 7 \times 64 + 7 \times 8 + 6$$

$$= (510)_{10}$$

$$(101011) = 1 \times 2^5 + 1 \times 2^3 + 1 \times 2 + 1$$
$$= 1 \times 32 + 1 \times 8 + 1 \times 2 + 1$$
$$= (43)_{10}.$$

Example 0.7.2 Express the decimal number 98 in octal and binary number systems.

Solution:

Divisor	Quotient	Remainder
8	98	
8	12	– 2 (LSD)
8	1	– 4
	0	– 1 (MSD)

$\therefore (98)_{10} = (142)_8$

Divisor	Quotient	Remainder
2	98	
2	49	– 0 (LSD)
2	24	– 1
2	12	– 0
2	6	– 0
2	3	– 0
2	1	– 1
	0	– 1 (MSD)

$\therefore (98)_{10} = (1100010)_2.$

So far we have discussed how whole numbers can be converted from one radix to another. It will now be seen how these principles can be suitably extended to cover the case of fractional numbers as well. A decimal fractional number $0 \,.\, a_{-1}a_{-2} \ldots a_{-n}$ can be written as

$$(0 \,.\, a_{-1}a_{-2} \ldots a_{-n}) = (a_{-1} \times 10^{-1} + a_{-2} \times 10^{-2} + \ldots a_{-n} \times 10^{-n})_{10}.$$

Similarly a fractional number with radix r can be converted into a decimal fraction with the help of the following equation

$$(0 \,.\, a_{-1}a_{-2} \ldots a_{-n})_r = (a_{-1} \times r^{-1} + a_{-2} \times r^{-2} + \ldots a_{-n} \times r^{-n})_{10}.$$

The procedure for the reverse operation of expressing a decimal fraction into an r-ary fraction is given by the following theorem.

Theorem 0.7.2 If the fractional part (the part on the right side of the decimal point) only is successively multiplied by r then the required pth numeral of the r-ary system is given by the whole number portion (the part on the left side of the decimal point) of the pth product.

Proof: The proof can be carried out in the same way as was done for Theorem 0.7.1 ▲

Conversion of a number given in one radix to a number in another radix can be performed by first converting the number into decimal, and then converting the decimal number to one with the required radix.

Example 0.7.3 Solve for x, in $(327)_9 = (x)_5$.

Solution: $(327)_9 = (3 \times 9^2 + 2 \times 9 + 7)_{10}$

$$= (268)_{10}$$

$$\begin{array}{r|rl} 5 & 268 & \\ 5 & 53 & -\ 3\ \text{(LSD)} \\ 5 & 10 & -\ 3 \\ 5 & 2 & -\ 0 \\ & 0 & -\ 2\ \text{(MSD)} \end{array}$$

$\therefore (327)_9 = (268)_{10} = (2033)_5$

$\therefore x = 2033$

In certain cases conversion from one radix to another can be carried out directly without going *via* the decimal number, although the decimal equivalents of the digits need to be known. Consider the conversion of the binary number $(11101101)_2$ into an octal number.

$$\begin{aligned} (11101101)_2 &= 1 \times 2^7 + 1 \times 2^6 + 1 \times 2^5 + 0 \times 2^4 + 1 \times 2^3 \\ &\quad + 1 \times 2^2 + 0 \times 2^1 + 1 \\ &= \underset{\text{III}}{(0 \times 2^2 + 1 \times 2^1 + 1)} \times 2^6 + \underset{\text{II}}{(1 \times 2^2 + 0 \times 2^1 + 1)} \times 2^3 \\ &\quad + \underset{\text{I}}{(1 \times 2^2 + 0 \times 2^1 + 1)} \\ &= b_2 \times 8^2 + b_1 \times 8^1 + b_0 \times 8^0, \text{ say,} \end{aligned}$$

where b_0, b_1 and b_2 are given by

$(a_2 a_1 a_0) = (b_0)_{10}$

$(a_5 a_4 a_3) = (b_1)_{10}$

$(a_8 a_7 a_6) = (b_2)_{10}$.

But

$$(b_2 \times 8^2 + b_1 \times 8^1 + b_0\, 8^0) = (b_2 b_1 b_0)_8.$$

Thus a binary number can be written directly into its octal equivalent by bunching the bits in groups of three starting from the 0th position, and writing the decimal equivalent of each bunch of three. *Similarly a binary number can be converted directly to 4-ary, 16-ary and in general 2^m-ary system by bunching the bits in groups of 2, 4, and m respectively.*

Example 0.7.4 Convert $(1110101)_2$ into an octal number and 4-ary number.

Solution:

$$(1110101)_2 = (001, 110, 101)_2 = (165)_8$$

$$(1110101)_2 = (01, 11, 01, 01)_2 = (1311)_4$$

Direct conversion for a fractional number can be done by starting the bunching from the -1th position. Octal numbers can be directly converted to binary number by writing each octal digit in its equivalent 3-bit binary number.

Thus, in general, an r-ary number can be written directly into its equivalent in r^m-ary system by bunching them in groups of m digits. Bunching is to be from right to left starting from the 0th position for a whole number, and from left to right starting from the -1th position for a fractional number. The decimal equivalent of each bunch of the r-ary number is the required digit in the r^m-ary system.

Example 0.7.5 Convert directly the tertiary number 1021121.121 into a number with base 9 and convert $(82.76)_9$ to a tertiary number.

Solution: Since $9 = 3^2$, direct conversion can be done by bunching in two. But the decimal of tertiary number must be known before-hand. These are shown on the next page.

$$(1\ 02\ 11\ 21\ .\ 12\ 1)_3 = (01, 02, 11, 21, .\ 12, 10)_3$$

$$= (1247.53)_9$$

$$(82.76)_9 = (2202.2120)_3;$$

3	10
00	0
01	1
02	2
10	3
11	4
12	5
20	6
21	7
22	8

0.8 BINARY ARITHMETIC

The operations of addition, subtraction, multiplication, and division can be carried out following procedures similar to those of the decimal system. The addition and multiplication tables of the binary system are shown in Fig.0-6.

+	0	1
0	0	1
1	1	10

.	0	1
0	0	0
1	0	1

Figure 0-6 Addition and multiplication tables of binary arithmetic.

It is to be noted that 1 + 1 = 10. The 1 on the left is the carry as we know in the decimal system. Just as is done in the decimal system, the carry must be added to the sum of the bits in the next position. In subtraction, when 1 is subtracted from 0, a borrow of 1 must be considered. The following examples will illustrate the four basic operations of addition, subtraction, multiplication, and division.

Example 0.8.1 Perform the following operations in binary arithmetic

a) 43 + 19 b) 43 – 19 c) 43 × 19 d) 43 ÷ 19.

Solution: $(43)_{10} = (101011)_2$

$(19)_{10} = (10011)_2$

a)

101011	Augend	43
10011	Addend	19
111110	Sum	62

b)

101011	Minuend	43
10011	Subtrahend	19
011000	Difference	24

c)

101011	Multiplicand	43
10011	Multiplier	19
101011	Partial	387
101011	Products	43
101011		
1100110001	Product	817

d)

		10	Quotient		2
Divisor	10011	101011	Dividend	19	43
		10011			38
		101	Remainder		5

Although the above procedures yield correct results, the binary arithmetic as performed by the digital computers does not necessarily follow the same general rules with which we are familiar in the decimal system – except to some extent in the addition and multiplication operations. For details of the computer's way of performing binary arithmetic, the reader is referred to texts dealing with the arithmetic operations in digital computers.

0.9 COMPLEMENTS

Most computers perform the subtraction operation by adding what is called the complement of the number. Let the complement of a decimal digit N be defined as $10 - N$. Thus the complements of 2, 3, and 5 are 8, 7, and 5 respectively. Now, if 3 is to be subtracted from 8, then the complement of 3 is added to 8, giving 15, $(8 + 7)$. If the carry digit of the sum is discarded, the result is 5. In this example, 8, 7 and 5 are called the 10's complements of the decimal numbers 2, 3, and 5 respectively. In general it is possible to define two types of complements in any number system.

Definition 0.9.1 Let $N = r_n r_{n-1} \ldots r_2 r_1$ be an n-digit number with radix r, the the *radix' complement* of N, ${}^rN$ is given by ${}^rN = L - N$, where L is the least number in the system with $n + 1$ digits.

It is apparent that in any number system L is the number with 1 followed by n number of 0's.

Example 0.9.1 Find the radix' complement of the following numbers:

a) $(462)_{10}$ b) $(0573)_8$ c) $(0101)_2$.

Solution:

a)
$$\begin{array}{r} 1000 \\ 462 \\ \hline 538 \end{array}$$

$\therefore (538)_{10}$ is the 10's complement of $(462)_{10}$ in the decimal system.

b)
$$\begin{array}{r} 10000 \\ 0573 \\ \hline 7205 \end{array}$$

$\therefore (7205)_8$ is the 8's complement of $(0573)_8$ in the octal system.

c)
$$\begin{array}{r} 10000 \\ 0101 \\ \hline 1011 \end{array}$$

$\therefore (1011)_2$ is the 2's complement of $(0101)_2$ in the binary system.

Definition 0.9.2 Let $N = r_n r_{n-1} \text{---} r_2 r_1$ be an n-digit number with radix r, then the *(radix – 1)'s complement* of N, ${}^{r-1}N$ is given by

$${}^{r-1}N = H - N,$$

where H is the highest number in the system with n digits.

It is evident that H must consist of n numbers of the largest digit of the system which is $r - 1$.

Example 0.9.2 Find the (radix – 1)'s complement of the numbers of the previous example.

Solution:

a)
$$\begin{array}{r} 999 \\ 462 \\ \hline 537 \end{array}$$

$\therefore (537)_{10}$ is the 9's complement of $(462)_{10}$ in the decimal system.

b) 7777
   0573

   7204

$\therefore$ $(7204)_8$ is the 7's complement of $(0573)_8$ in the octal system.

c) 1111
   0101

   1010

$\therefore (1010)_2$ is the 1's complement of $(0101)_2$ in the binary system.

The following theorem establishes the relation between the radix' and (radix − 1)'s complements of a number.

Theorem 0.9.1 In an r-ary number system,

$${}^{r}N = {}^{r-1}N + 1$$

Proof:

$${}^{r}N = L - N$$

and

$${}^{r-1}N = H - N$$

$$\therefore {}^{r}N - {}^{r-1}N = L - H,$$

but since L is the least number of $(n + 1)$ digits and H is the highest number of n digits,

$$L - H = 1.$$

$$\therefore {}^{r}N - {}^{r-1}N = 1$$

or,

$${}^{r}N = {}^{r-1}N + 1. \quad \blacktriangle$$

Theorem 0.9.2 The 1's complement of a binary number can be obtained by simply changing the 0's into 1's and 1's into 0's.

Proof: Let r_p be the pth bit of the binary number. Then by definition 0.9.2 the pth bit of the 1's complement of the binary number is

$$r'_p = 1 - r_p.$$

$\therefore$ If $r_p = 1$, then $r'_p = 0$

and

if $r_p = 0$, then $r'_p = 1$. ▲

A beneficial consequence of this result is that no mathematical calculation is involved to obtain the 1's complement of a binary number. This is specially advantageous in a computer as the 0's and 1's can be interchanged by simple circuit arrangement.

0.10 BINARY CODES

As man uses the decimal system, and the computer uses the binary system, an immediate problem is how to express a decimal number in the binary system. One way will be to convert the decimal number into the binary number by the usual conversion method. However, this method is seldom used. A more convenient method is to write the decimal digits in the binary way. Thus the decimal number 893 can be written as

1000 1001 0011.

Such a system is called a binary coded decimal (BCD) system. The underlying principle of this code is to have 10 code words for the 10 decimal digits 0 through 9. Many times the principle is given effect to by having other codes for the decimal digits than the straight binary code. The various binary codes that can be devised fall broadly into two categories, the weighted codes, and the non-weighted codes. In the former, each position of the number has a definite weight. It can be recalled that the weights of the 4-bit straight binary code are 8, 4, 2, and 1. Some of the weighted codes are shown in the tables of Fig.0-7.

Although only four weighted codes have been shown in Fig.0-7 there are many other possibilities. It may also be seen that a weight may be negative also. The term 'binary code' when used to denote a specific code among the family of all the binary codes, means the 8421 code. All other codes will be called by the proper adjectives such as the 5321 code, the excess-three code etc.

Some of the non-weighted codes are shown in Fig.0-7. In excess-three code the code word for a decimal digit d, is the 8421 binary code word for the digit $(d + 3)$. Thus to obtain the excess-three code word for the digit 6, we find the binary code word for the number (6 + 3) or 9, which is 1001. One interesting property for which excess-three code has been used in many computers is that it is self-complementing.

Definition 0.10.1 Let the decimal digit D have the code word N in a binary code. That is, $(N)_2 = (D)_{10}$. Then if $({}^1N)_2 = ({}^9D)_{10}$, the binary code is called *self-complementing*.

Weighted binary codes

Decimal digits	8421	5321	4311	642(−3)
0	0000	0000	0000	0000
1	0001	0001	0001	0101
2	0010	0010	0011	0010
3	0011	0100	0100	1001
4	0100	0101	0101	0100
5	0101	0110	0111	1011
6	0110	0111	1011	0110
7	0111	1010	1100	1101
8	1000	1011	1101	1010
9	1001	1101	1111	1111

Non-weighted binary codes

Decimal digits	Excess-three	Gray	2-out-of-5
0	0011	0000	00011
1	0100	0001	00101
2	0101	0011	00110
3	0110	0010	01001
4	0111	0110	01010
5	1000	0111	01100
6	1001	0101	10001
7	1010	0100	10010
8	1011	1100	10100
9	1100	1101	11000

Figure 0-7 Weighted and non-weighted binary codes.

The excess-three code for 4 is 0111. 1's complement of 0111 is 1000 which is the code word for 5. But 5 is the 9's complement of 4.

0.11 GRAY CODE

The Gray code has the important feature that while going from one digit to the next, there is change only in one bit. This property is particularly desirable in the design of analog to digital converters. The Gray code has this property as it belongs to the class of *reflected codes.* The formation of the 1-bit, 2-bit, 3-bit and 4-bit Gray codes are shown in the table of Fig.0-8 from which the phenomenon of reflection, and the fact that there is only one change of a bit between adjacent digits will be obvious. The 2-bit code is obtained from the 1-bit code. The two code words of the 1-bit code are reflected. 0's are then prefixed to the incident portion, and 1's to the reflected portion. Following a similar procedure, 3 bit

Decimal numbers	Gray Code 1-bit	2-bit	3-bit	4-bit
0	0	0 0	0 0 0	0 0 0 0
1	1	0 1	0 0 1	0 0 0 1
2		1 1	0 1 1	0 0 1 1
3		1 0	0 1 0	0 0 1 0
4			1 1 0	0 1 1 0
5			1 1 1	0 1 1 1
6			1 0 1	0 1 0 1
7			1 0 0	0 1 0 0
8				1 1 0 0
9				1 1 0 1
10				1 1 1 1
11				1 1 1 0
12				1 0 1 0
13				1 0 1 1
14				1 0 0 1
15				1 0 0 0

Figure 0-8 Gray code as a reflected code.

code is generated from the 2 bit code, and so on. It may also be observed that in an n-bit code, the code words of the numbers 0 and $(2^n - 1)$ differ by one bit only.

Due to the property of reflection Gray code can be very easily derived from the binary code. Let $g_n \ldots g_2 g_1 g_0$ be the code word in Gray code of a number whose code word in binary code is $b_n \ldots b_2 b_1 b_0$. Then it can be easily verified that

$$g_0 = b_0 \oplus b_1$$

$$g_1 = b_1 \oplus b_2$$

$$- - - - - -$$

$$g_i = b_i \oplus b_{i+1}$$

$$- - - - - -$$

$$g_n = b_n.$$

The operation $\oplus$ denotes the mod-2 sum, so that

$$0 \oplus 0 = 0$$

$$0 \oplus 1 = 1$$

$$1 \oplus 0 = 1$$

$$1 \oplus 1 = 0.$$

Example 0.11.1 Determine the Gray code word for the binary number 101011, and the binary number for the Gray code word 111010.

Solution:

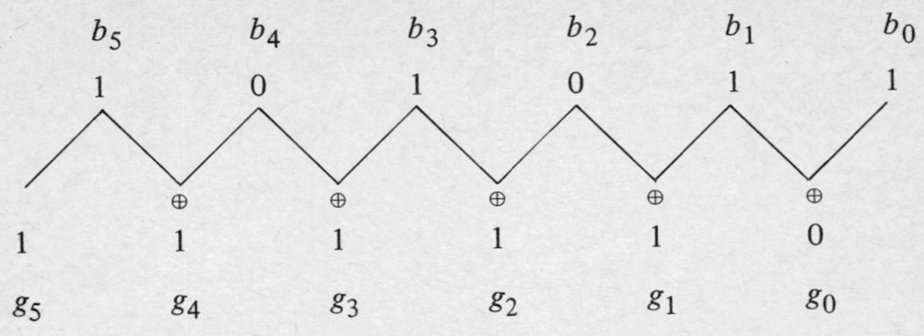

To convert the Gray code word into the corresponding binary word the procedure is reversed. Hence b_n is first determined, and then b_{n-1} and so on, so that the relation, $g_i = b_i \oplus b_{i+1}$ is always satisfied.

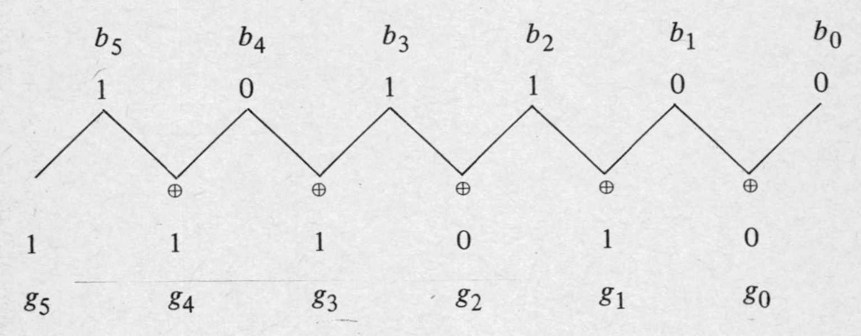

0.12 ERROR DETECTING AND CORRECTING CODES

The 4-bit binary codes as discussed above will be adequate if during the transmission of the message the probability of a 0 being changed into a 1 (or *vice-versa*) is very small. On the other hand, if the probability of such an error is appreciable then measures must be taken to detect the error and then rectify it. In order to detect the error, every code word must possess some distinctive feature which will be disturbed as soon as a single error occurs. The 2-out-of-5 code as given in Fig.0-7 has this features. As soon as a single error occurs, a code word has either one 1 or three 1's. Thus this code is an *error detecting code* (EDC). Another way of converting any of the above 4-bit binary code to an EDC is to add an extra bit called a *parity bit*. The parity bit is so chosen that the number of 1's in a code word is always even or odd. Thus if an *even parity* is decided upon then the code word for 6, 0110 becomes 01100 the parity bit being 0. On the other hand for an *odd parity*, the code word becomes 01101, the parity bit being a 1. It can be seen that such a code fails to detect double errors. However, it is assumed that the probability of double, triple error etc. is extremely small. An EDC also does not say as to which bit has gone wrong. So if we want to correct the error, we need additional information. Obviously we will have to use more than one parity bit. Let us assume that we use k parity check bits for a code having m information bits. It may be mentioned here that the k check positions are not necessarily the last k positions but may be anywhere. In fact they are intermixed with the information bits.

With k parity checks the total number of bits in the code is $m + k$. There may be errors in any of these $m + k$ positions. Therefore the parity check should be capable of spelling out any of these positions. In addition it should also be capable of saying that no error has taken place. Thus the total number of values that the k parity checks should assume is $m + k + 1$. Now let us have a scheme in which each parity check writes a 0, when it finds no error, and it writes a 1, when it finds an error. In this way the k parity checks will write a binary number from right to left, and this number, called the checking number, will be the position of error. Thus if k is 4, 0000 means no error, but 1011 means the error is at the 11th position. Hence k must satisfy the inequality,

$$2^k \geqslant m + k + 1.$$

Let the checking number in a particular case be 110 for $m = 4$ and $k = 3$. This means the error is in the 6th position. Here both the second and third pari checks have found the checking incorrect. This can be achieved by making both the second and third parity checks to check the position 6. In this way we can find out the different positions that a certain parity check should check, and obtain the following table.

Check number 3	2	1	Error position	Checking number
.	.	x	1	001
.	x	.	2	010
.	x	x	3	011
x	.	.	4	100
x	.	x	5	101
x	x	.	6	110
x	x	x	7	111

Figure 0-9 Table showing the various positions that must be checked by a particular parity check. ($m = 4$, and $k = 3$).

The table of Fig.0-9 tells us at once that in this example of a 7 place code the 1s check number should check the 1st, 3rd, 5th and 7th positions; the 2nd check number should check the 2nd, 3rd, 6th, and 7th positions; and the 3rd check number should check the 4th, 5th, 6th, and 7th positions. The next step is to decide as to what positions are to be assigned to the information bits and what t the parity bits. Although any four positions will do, it is better to have the posi tions 3, 5, 6, 7 for the information bits as they would be checked more frequent

The results of writing down the binary equivalents of the decimal digits 0 through 9 using positions 3, 5, 6, 7, and then calculating the values in the check positions, are given in the table of Fig.0-10. The positions 3, 5, 6 and 7 are for the information bits. The positions 1, 2 and 4 are for the parity bits p_1, p_2 and p_3 respectively. The type of parity check decided is the even parity check. Hence, the values of p_1, p_2 and p_3 are given by

	Positions						
	1	2	3	4	5	6	7
Decimal digits	p_1	p_2	m_1	p_3	m_2	m_3	m_4
0	0	0	0	0	0	0	0
1	1	1	0	1	0	0	1
2	0	1	0	1	0	1	0
3	1	0	0	0	0	1	1
4	1	0	0	1	1	0	0
5	0	1	0	0	1	0	1
6	1	1	0	0	1	1	0
7	0	0	0	1	1	1	1
8	1	1	1	0	0	0	0
9	0	0	1	1	0	0	1

Figure 0-10 The Hamming code capable of correcting single error.

$$p_1 = 3 \oplus 5 \oplus 7$$

$$p_2 = 3 \oplus 6 \oplus 7,$$

and

$$p_3 = 5 \oplus 6 \oplus 7.$$

As an illustration of how this code works let us take the symbol 0001111 corresponding to the decimal value 7 and change the 1 in the fifth position to a 0. We now examine the new symbol

0001011

by the method outlined above. The first parity check will check the positions 1, 3, 5, 7, and will find the checking wrong. Thus we write

1.

The second parity check is over positions 2, 3, 6, 7, and this will come out alright. Hence we write a 0 to the left of 1, and obtain,

01.

The third parity check is over positions 4, 5, 6, 7, and will go wrong, yielding finally,

101.

This is the binary way of writing the decimal number 5. Hence the checking number has pointed to the fact that the error is in the 5th position, and when we change the digit in the 5th position, we get the correct number

0001111.

The above scheme is due to Hamming (1950).

REFERENCES

Bartee, T. C.: *Digital Computer Fundamentals,* McGraw-Hill Book Co., New York, 1960.

Flores, I.: *The Logic of Computer Arithmetic,* Prentice-Hall, Englewood, N.J., 1963.

Hamming, R. W.: 'Error detecting and correcting codes', *Bell System Tech. J.* **29**, pp.147–160, April 1950.

McCluskey, E. J.: *Introduction to the Theory of Switching Circuits,* McGraw-Hill Book Co., New York, 1965.

Peterson, W. W.: *Error Correcting Codes,* John Wiley and Sons, Inc., New York, 1960.

Pfeiffer, P. E.: *Sets, Events, and Switching,* McGraw-Hill Book Co., New York, 1964.

Richards, R. K.: *Arithmetic Operations in Digital Computers,* D. Van Nostrand Co, Princeton, N.J., 1955.

Whitesitt, J. E.: *Principles of Modern Algebra,* Addison-Wesley, Reading, Mass., 1964.

PROBLEMS

0.1 Let $W = \{ *, \#, 4 \}$

$$X = \{0, 2, 3, 5, 10, \#, *\}$$

$$Y = \{2, *, 5, @\}$$

$$Z = \{1, 2, 6\},$$

and the universal set,

$$U = \{0, 1, 2, 3, 4, 5, 6, 7, 8, 9, 10, \$, \#, *, @\}$$

Calculate: a) W'; b) Y'; c) $W \cup Z$; d) $X \cap Y$; e) $X \cap Z$; f) $X \cup Y$; g) $W' \cap Z$; h) $X \cup Y \cup Z$; i) $X' \cap Y \cup Z'$; j) $(W \cup X) \cap Y$; k) $W \cup (X \cap Y)$; l) $(W \cup X) \cap (Y \cup Z)$; m) $[(W \cup X) \cap Y] \cup Z$.

0.2 Let $X = \{x|x \text{ is integer, and } 1 \leqslant x < 12\}$

$Y = \{y|y \text{ is integer, and } y = n^2 \text{ where } 2 \leqslant n \leqslant 5\}$.

Calculate: a) $X \cup Y$; b) $X \cap Y$; c) $X \cup (X \cap Y)$; d) $X \cap (X' \cup Y)$.

0.3 For the numbers 0 through 4, write the ordinary addition and multiplication tables. Then write the tables for sum and product Mod-5.

0.4 For the numbers 0 through 7, write the tables for sum and product Mod-8.

0.5 Calculate the weights of the 5th, 8th, and 12th positions in the binary and octal number systems.

0.6 Convert the following decimal numbers into binary numbers: a) 49; b) 123; c) 12.032; d) 0.005.

0.7 Convert the following decimal numbers into octal numbers: a) 526; b) 23.692; c) 1259.6; d) 0.298.

0.8 Convert the following binary numbers into decimal and octal numbers: a) 1101; b) 11010110; c) 100.001; d) 101.11; e) 11010.001.

0.9 Convert the following octal numbers into decimal and binary numbers: a) 236; b) 4172; c) 23.67; d) 12.77.

0.10 Solve for x in the following number system equations:

a) $(235)_7 = (x)_{10}$ b) $(125)_6 = (x)_8$ c) $(x)_3 = (333)_4$

d) $(121)_3 = (34)_x$ e) $(231)_x = (111)_9$.

0.11 Let (0, 1, 2, . . . $, £) be the r digits in ascending order in a number system with radix r. Show that

£ × £ = $ 1.

0.12 Express the decimal numbers 631 and 44 in:

a) Straight binary code b) BCD code
c) 5 3 1 1 code d) Excess-three code.

0.13 Find the 10's and 9's complements of the following decimal numbers: a) 962; b) 7268; c) 4962; d) 02364; e) 0042; f) 02645; g) 000392; h) 026400.

0.14 Find the 2's and 1's complements of the following binary numbers: a) 01; b) 1101; c) 110101; d) 0001011; e) 001001.

0.15 Find the radix' and (radix − 1)'s complements of the following numbers: a) $(239)_{10}$; b) $(33247)_8$; c) $(0456)_7$; d) $(045)_6$.

0.16 Encode the 10 decimal digits by the following weighted binary codes: a) 3, 3, 2, 1; b) 5, 2, 1, 1; c) 7, 3, 2, 1; d) 8, 4, −2, −1; e) 7, 3, 1, −2.

0.17 Show that in order for a weighted code to be self-complementing the sum of the weights must be 9.

0.18 Draw the table of the 5-bit Gray code, and from it write the Gray code words for the decimal numbers 4, 18, 20, 26 and 31.

0.19 Express the following numbers in Gray code: a) $(25)_{10}$; b) $(24)_8$; c) $(101011)_2$.

0.20 Find the equivalent decimal number of the following numbers in Gray code: a) 1011011; b) 111111.

0.21 Extend the table of Fig.0-10 by constructing code words for the decimal numbers 10 through 15.

0.22 A 4 digit decimal number transmitted by the Hamming single error correcting code is as follows:

1001111010101011001101100000.

Find the number transmitted. (It is assumed that, if at all, only single error has occurred.)

Chapter 1

BOOLEAN ALGEBRA

1.1 INTRODUCTION

In the last chapter we developed the mathematical background that will be necessary in the study of Logic and Switching Theory. In this chapter a more detailed and comprehensive discussion of Boolean Algebra will be carried out. It will be seen later that the various switching circuits of interest and their manipulations can be handled very effectively from a knowledge of Boolean Algebra. Hence this algebra forms the main foundation of logic and switching theory.

1.2 BOOLEAN ALGEBRA

There are many ways in which a Boolean Algebra can be defined. Starting from abstract algebra one can define a Lattice, and then proceed to show that a Boolean Algebra is a special case of a Lattice. On the other hand, one can define a Boolean Algebra by enumerating its properties. We shall follow the latter method, as it serves the purpose well without going into the abstract mathematics.

Definition 1.2.1 A set B of elements $(a, b, c, \ldots)$ with two binary operations; one of them addition (denoted by v) and the other multiplication (denoted by . or simply by conjunction) is a Boolean Algebra if and only if the following postulates are satisfied.

P.1 The v and . operations are associative

$$(a \vee b) \vee c = a \vee (b \vee c) = a \vee b \vee c$$

$$(ab)c = a(bc) = abc.$$

P.2 The v and . operations are commutative

$$a \vee b = b \vee a, \text{ and}$$

$$ab = ba.$$

P.3 There exists an additive identity element (denoted by 0, called zero), and a multiplicative identity element (denoted by 1, and called one or unity), within B such that

$a \vee 0 = a$, and

$a1 = a$.

P.4 The two operations are distributive over each other

$a \vee bc = (a \vee b)(a \vee c)$

$a(b \vee c) = ab \vee ac$.

P.5 Each element of B has a complement within B such that, if $\bar{a}$ is the complement of a, then,

$a \vee \bar{a} = 1$, and

$a\bar{a} = 0$.

It should be mentioned here that as in an ordinary algebra so also in a Boolean algebra, the . operation is performed before the $\vee$ operation. Thus,

$a \vee bc = a \vee (bc) \neq (a \vee b)c$.

1.3 EQUALITY THEOREMS

As in any other system, so also in Boolean algebra it becomes necessary to find the necessary and sufficient conditions under which two elements of B are equal. The following two theorems give us the conditions and are useful in proving other theorem

Theorem 1.3.1 For all $a, b \in B$, if

$a \vee \bar{b} = 1$, and

$a\bar{b} = 0$, then

$a = b$.

Proof:

$a = a1$	by P.3
$= a(b \vee \bar{b})$	by P.5
$= ab \vee a\bar{b}$	by P.4
$= ab \vee 0$	since $a\bar{b} = 0$
$= ab \vee b\bar{b}$	since $b\bar{b} = 0$ by P.5
$= b(a \vee \bar{b})$	by P.4
$= b1$	since $a \vee \bar{b} = 1$
$= b$	by P.3. ▲

Theorem 1.3.2 For all $a, b, c, \in B$, if

$a \vee b = a \vee c$, and

$ab = ac$, then

$b = c$.

Proof:

$b = 1b$	by P.3
$= (a \vee \overline{a})b$	since $1 = a \vee \overline{a}$ by P.5
$= ab \vee \overline{a}b$	by P.4
$= ac \vee \overline{a}b$	since $ab = ac$
$= ac \vee \overline{a}b \vee 0$	by P.3
$= ac \vee \overline{a}b \vee \overline{a}a$	by P.5
$= ac \vee \overline{a}(b \vee a)$	by P.4
$= ac \vee \overline{a}(a \vee b)$	by P.2
$= ac \vee \overline{a}(a \vee c)$	since $a \vee b = a \vee c$
$= ac \vee \overline{a}c \vee a\overline{a}$	by P.4 and P.2
$= ac \vee \overline{a}c \vee 0$	by P.5
$= ac \vee \overline{a}c$	by P.3
$= c(a \vee \overline{a})$	by P.4
$= c1$	by P.5
$= c$	by P.3. ▲

1.4 COMPLEMENTARITY THEOREM

This theorem tells us when two elements of B are complements of each other.

Theorem 1.4.1 For all $a, b \in B$, if

$a \vee b = 1$, and

$ab = 0$, then

$a = \overline{b}$, and

$b = \overline{a}$.

Proof: $a \vee b = 1 = a \vee \overline{a}$ by P.5.

Also $ab = 0 = a\bar{a}$ by P.5

$\therefore$ We have $a \vee b = a \vee \bar{a}$

and $ab = a\bar{a}$.

$\therefore$ By Theorem 1.3.2 $b = \bar{a}$. ▲

Again, $a \vee b = 1 = b \vee \bar{b}$ by P.5

and $ab = 0 = b\bar{b}$ by P.5.

$\therefore$ We have $b \vee a = b \vee \bar{b}$

and $ba = b\bar{b}$.

$\therefore$ By Theorem 1.3.2 $a = \bar{b}$. ▲

1.5 LAWS OF BOOLEAN ALGEBRA

In this section we prove the basic theorems which establish the laws of Boolean algebra.

Theorem 1.5.1 The law of involution holds in Boolean algebra. That is, for all $a \in B$

$\bar{\bar{a}} = a$.

Proof: Let $\bar{a} = x$, then $\bar{\bar{a}} = \bar{x}$.

Now $a \vee \bar{a} = 1$, by P.5

and $a\bar{a} = 0$ by P.5

$\therefore$ $a \vee x = 1$, since $\bar{a} = x$

and $ax = 0$ since $\bar{a} = x$

$\therefore$ $\bar{x} = a$ by Th.1.4.1.

Or, $\bar{\bar{a}} = a$ since $\bar{x} = \bar{\bar{a}}$. ▲

Theorem 1.5.2 For all $a \in B$.

$a \vee 1 = 1$ and

$a0 = 0$.

Proof: $a \vee 1 = (a \vee 1)1$ by P.3

$= (a \vee 1)(a \vee \bar{a})$ since $a \vee \bar{a} = 1$ by P.5

$= a \vee 1\bar{a}$ by P.4

$= a \vee \bar{a}$ by P.3

$= 1$ by P.5. ▲

Also $a0 = a0 \vee 0$ by P.3

$= a0 \vee a\bar{a}$ by P.5

$= a(0 \vee \bar{a})$ by P.4

$= a\bar{a}$ by P.3

$= 0$ by P.5. ▲

Theorem 1.5.3 The two identity elements of Boolean Algebra are complements of each other. That is,

$\bar{0} = 1$, and

$\bar{1} = 0$.

Proof: Since $0 \in B$, then from Th.1.5.2

$0 \vee 1 = 1$.

Again, since $1 \in B$, then from Th.1.5.2

$10 = 0$.

Hence we have

$0 \vee 1 = 1$

and

$01 = 0$.

$\therefore$ By Th.1.4.1,

$\bar{0} = 1$,

and

$\bar{1} = 0$. ▲

Theorem 1.5.4 The law of idempotence holds in Boolean algebra. That is, for all $a \in B$

$a \vee a = a$, and

$aa = a$.

Proof: $a \vee a = (a1) \vee (a1)$ by P.3

$= a(1 \vee 1)$ by P.4

$= a1$ by Th.1.5.2

$= a$ by P.5. ▲

Also $aa = (a \vee 0)(a \vee 0)$ by P.3

$= a \vee 00$ by P.4

$= a \vee 0$ by Th.1.5.2

$= a$ by P.5. ▲

Theorem 1.5.5 The absorption law holds in Boolean algebra. That is, for all $a, b \in B$,

$a \vee ab = a$ and

$a \vee \bar{a}b = a \vee b$.

Proof: $a \vee ab = a1 \vee ab$ by P.3

$= a(1 \vee b)$ by P.4

$= a1$ by Th.1.5.2

$= a$ by P.5. ▲

Also $a \vee \bar{a}b = (a \vee \bar{a})(a \vee b)$ by P.4

$= 1(a \vee b)$ by P.5

$= a \vee b$ by P.3. ▲

COROLLARY **1.5.5.A** The law of cancellation does not hold good in Boolean algebra. That is, if $a \vee b = a \vee c$, then $b \neq c$ in general.

Proof: $a \vee ab = a = a \vee 0$ by P.3.

If the law of cancellation holds good, then $ab = 0$, which may not be true. Hence the law of cancellation does not hold good. ▲

Theorem 1.5.6 De Morgan's Theorem. For all $a, b \in B$

$\overline{a \vee b} = \bar{a}\,\bar{b}$ and

$\overline{ab} = \bar{a} \vee \bar{b}$

Proof: From the complementarity theorem

$\overline{a \vee b} = \bar{a}\bar{b}$

if $(a \vee b) \vee \bar{a}\,\bar{b} = 1$,

and $(a \vee b)\bar{a}\,\bar{b} = 0$.

Now, $(a \vee b) \vee \bar{a}\,\bar{b} = (a \vee b \vee \bar{a})\,(a \vee b \vee \bar{b})$

$$= [b \vee (a \vee \bar{a})]\ [a \vee (b \vee \bar{b})]$$

$$= (b \vee 1)\,(a \vee 1)$$

$$= 1 \cdot 1$$

$$= 1$$

and $(a \vee b)\bar{a}\,\bar{b} = (a\bar{a}\bar{b}) \vee (b\bar{a}\bar{b})$

$$= (a\bar{a})\bar{b} \vee (b\bar{b})\bar{a}$$

$$= 0 \vee 0$$

$$= 0 \text{ since } a\bar{a} = b\bar{b} = 0.$$

Hence $\overline{a \vee b} = \bar{a}\,\bar{b}$. ▲

Similarly $\overline{ab} = \bar{a} \vee \bar{b}$

if $ab \vee (\bar{a} \vee \bar{b}) = 1$,

and $ab(\bar{a} \vee \bar{b}) = 0$.

Now, $ab \vee (\bar{a} \vee \bar{b}) = (a \vee \bar{a} \vee \bar{b})\,(b \vee \bar{a} \vee \bar{b})$

$$= 1 \cdot 1$$

$$= 1$$

and $ab(\bar{a} \vee \bar{b}) = (ab\bar{a}) \vee (ab\bar{b})$

$$= 0 \vee 0$$

$$= 0.$$

Hence $\overline{ab} = \bar{a} \vee \bar{b}$. ▲

1.6 THE INCLUSION RELATION

The inclusion relation has been discussed in the last chapter while discussing the algebra of sets. An inclusion relation can also be described in Boolean algebra. It is written as $a \leqslant b$, or $b \geqslant a$ which can be read as 'a is in b', or 'b covers a', or 'a implies b'. It is defined as follows:

Definition 1.6.1 For all $a, b \in B$,

$a \leqslant b$, if and only if

$ab = a$.

Another equivalent definition is

$a \leqslant b$ if and only if

$a \vee b = b.$

We shall now prove the following theorems regarding the inclusion relation.

Theorem 1.6.1 The inclusion relation is reflexive, that is for all $a \in B$, $a \leqslant a$.

Proof: $aa = a$ by Th.1.5.4

$\therefore a \leqslant a.$ ▲

Theorem 1.6.2 The inclusion relation is antisymmetric. That is, for all $a, b \in B$, if $a \leqslant b$, and $b \leqslant a$, then $a = b$.

Proof: Since $a \leqslant b$ $ab = a = aa.$

Again, since $b \leqslant a$ $b \vee a = a = a \vee a.$

Hence, $a \vee a = a \vee b$, and

$aa = ab$

$\therefore$ by Th.1.3.2 $a = b.$ ▲

Theorem 1.6.3 The inclusion relation is transitive. That is, for all, $a, b, c \in B$, if $a \leqslant b$, and $b \leqslant c$, then $a \leqslant c$.

Proof: Since $a \leqslant b$ $ab = a$, and

since $b \leqslant c$ $bc = b.$

Now, $ac = (ab)c$ since $a = ab$

$= a(bc)$ by P.1

$= ab$ since $bc = b$

$= a$ since $ab = a$

$\therefore a \leqslant c.$ ▲

Theorem 1.6.4 For all $a \in B$,

$0 \leqslant a \leqslant 1.$

Proof: By Th.1.5.2

$0a = 0$

$\therefore 0 \leqslant a.$

Again, by Th.1.5.2

$a \vee 1 = 1$

$a \leqslant 1$

$\therefore 0 \leqslant a \leqslant 1.$ ▲

1.7 UNIQUENESS THEOREMS

Theorem 1.7.1 The identity elements 0 and 1 are unique.

Proof: If 0 is not unique, then let there be two 0's, 0_1 and 0_2 both in B. Now, since 0_1 is the identity element and $0_2 \in B$, then by Th.1.6.4

$0_1 \leqslant 0_2.$

Again, since 0_2 is the identity element and $0_1 \in B$, then

$0_2 \leqslant 0_1.$

Hence, we have, $0_1 \leqslant 0_2$ and $0_2 \leqslant 0_1$.

$\therefore$ By Th.1.6.2

$0_1 = 0_2.$

Hence, 0 is unique. ▲

Similarly it can be shown that if there are two 1's, 1_1 and 1_2 both in B, then $1_1 \leqslant 1_2$ and $1_2 \leqslant 1_1$. $\therefore 1_1 = 1_2$, proving 1 to be unique. ▲

Theorem 1.7.2 For all $a \in B$, the complement of a, $\bar{a}$ is unique.

Proof: If $\bar{a}$ is not unique, let there be two $\bar{a}$'s, $\bar{a}_1$ and $\bar{a}_2$. Since $\bar{a}_1$ is complement of a,

$a \vee \bar{a}_1 = 1$ and

$a\bar{a}_1 = 0.$

Again since $\bar{a}_2$ is also a complement of a

$a \vee \bar{a}_2 = 1$ and

$a\bar{a}_2 = 0.$

$\therefore$ We have $a \vee \bar{a}_1 = a \vee \bar{a}_2$ and $a\bar{a}_1 = a\bar{a}_2$.

$\therefore$ By Th.1.3.2

$\bar{a}_1 = \bar{a}_2.$

Hence, $\bar{a}$ is unique. ▲

1.8 DUALITY IN BOOLEAN ALGEBRA

Before we discuss duality, let us define a Boolean function.

Definition 1.8.1 When two or more elements of a Boolean algebra are connecte by one or more of the three operators, v, . , and –, the resulting expression is called a Boolean function. Thus f_1, f_2 and f_3 as given below are all examples of Boolean functions.

$$f_1 = a \vee b$$

$$f_2 = (\bar{a} \vee b)(c \vee d)$$

$$f_3 = [(a \vee b)c \vee \bar{d}]e \vee \bar{b}.$$

It should be mentioned here that all the theorems of Boolean algebra as have been developed in the preceding sections, hold good for the Boolean functions also. This is because by the closure property of the binary operations v and . , and of the unary operation –, a Boolean function f is also an element of B. As the theorems are valid for all $a \in B$, they are also valid for f which is also $\in B$.

Now, we can define a dual function.

Definition 1.8.2 If in a Boolean function the binary operations (v, .) and the identity elements (0, 1) are interchanged pairwise, then the resulting function and the original function are duals of each other. That is, if

$$u = f(a, b, c, \ldots, 0, 1, \vee, .)$$

then

$$u^D = f(a, b, c, \ldots, 1, 0, ., \vee).$$

Thus if

$$f = (a \vee b)c$$

and

$$g = ab \vee c,$$

then f and g are duals of each other,

$$f^D = g \text{ and } g^D = f.$$

It should be seen that each of the elements $a, b, c, \ldots$ in the expression $u = f(a, b, c, \ldots, 0, 1, \vee, .)$ cannot be either 0 or 1. While all these elements remain unchanged in the dual expression, the element 0 becomes 1, and the element 1 becomes 0 in the dual expression. This special behavior of 0 and 1 can be accounted for by the following theorem.

Theorem 1.8.1 The two identity elements of Boolean algebra, 0 and 1, are duals of each other.

Proof: $0 = a \, . \, \bar{a} = f$ say

then $f^D = a \vee \bar{a} = 1.$

Again $1 = a \vee \bar{a} = g$, say,

then $g^D = a \, . \, \bar{a} = 0.$

Hence $0^D = 1$ and $1^D = 0.$ ▲

In Th.1.5.6 (De Morgan's Theorem), we have seen how the complement of a sum or a product term can be taken. In the following example, it is shown how by repeated applications of this theorem, the complement of a Boolean function can be found out.

Example 1.8.1. Compute the complement of the following function,

$f = (a \vee \bar{b}\bar{c})\,(b \vee c) \vee a\bar{d}.$

Solution:

$$\bar{f} = \overline{(a \vee \bar{b}\bar{c})\,(b \vee c) \vee a\bar{d}}$$

$$= \overline{(a \vee \bar{b}\bar{c})\,(b \vee c)} \, . \, \overline{a\bar{d}}$$

$$= [\overline{(a \vee \bar{b}\bar{c})} \vee \overline{(b \vee c)}]\,(\bar{a} \vee d)$$

$$= [\bar{a} \, . \, \overline{\bar{b}\bar{c}} \vee \bar{b}\bar{c}]\,(\bar{a} \vee d)$$

$$= [\bar{a}(b \vee c) \vee \bar{b}\bar{c}]\,(\bar{a} \vee d).$$

A study of the ultimate expression for $\bar{f}$ reveals that, it could have been written without going through the intermediate steps, by changing every element by its complement and interchanging the v's and .'s. Thus we can arrive at this generalized definition of a complementary function.

Definition 1.8.3 If in a Boolean function the v's are changed to .'s, and .'s to v's, and every element is replaced by its complement then the resulting function and the original function are complements of each other. That is, if

$u = f(a, b, c, \ldots, \vee, .)$

then

$\bar{u} = f(\bar{a}, \bar{b}, \bar{c}, \ldots, ., \vee).$

Another interesting consequence of De Morgan's theorem is stated in the following theorem.

Theorem 1.8.2 Every identity of Boolean algebra has its dual.

Proof: Let an identity be expressed as follows:

$$f(a, b, c, \ldots, \vee, .) = g(x, y, z, \ldots, \vee, .).$$

Since it is an identity, it will hold good if every element of f and g is replaced by its complement, that is, if,

$$f(\bar{a}, \bar{b}, \bar{c}, \ldots, \vee, .) = u$$

and

$$g(\bar{x}, \bar{y}, \bar{z}, \ldots, \vee, .) = w$$

then

$$u = w.$$

Now taking complements of both sides,

$$\bar{u} = \bar{w},$$

but

$$\bar{u} = \overline{f(\bar{a}, \bar{b}, \bar{c}, \ldots, \vee, .)} = f(a, b, c, \ldots, \cdot, \vee) = f^D \qquad \text{by Def.1.8.2}$$

and

$$\bar{w} = \overline{g(\bar{x}, \bar{y}, \bar{z}, \ldots, \vee, .)} = g(x, y, z, \ldots \cdot, \vee) = g^D \qquad \text{by Def.1.8.2}$$

Hence, $f^D = g^D$. ▲

It is interesting to note here that many of the previous theorems are also Boolean identities; and in some of them, where two parts have been proved, one part is the dual of the other.

A word of caution must be mentioned here. Whereas it is possible to write the dual f^D of every Boolean function f, and also every identity has its dual, every relation may not have its dual. For example, if

$$x \vee y = 1,$$

it does not necessarily follow that the dual relation,

$$xy = 0$$

is true.

1.9 SETS AND BOOLEAN ALGEBRA

It is interesting to note that the sets as discussed in the last chapter form a Boolean algebra.

Theorem 1.9.1 The sets $A, B, C, \ldots$ with their universal set U, the null set Φ and the binary operations of union $\cup$, intersection $\cap$, and the unary operation of complementation $'$ constitute a Boolean algebra.

Proof: The theorem is proved if the validity of the five postulates as given in Def.1.2.1 is demonstrated. That this is so can be seen from the following relations already established in the algebra of sets.

P.1 $(A \cup B) \cup C = A \cup (B \cup C) = A \cup B \cup C$, and

$(A \cap B) \cap C = A \cap (B \cap C) = A \cap B \cap C$

P.2 $A \cup B = B \cup A$, and

$A \cap B = B \cap A$

P.3 $A \cup \Phi = A$, and

$A \cap U = A$

P.4 $A \cup (B \cap C) = (A \cup B) \cap (A \cup C)$

$A \cap (B \cup C) = (A \cap B) \cup (A \cap C)$

P.5 $A \cup A' = U$, and

$A \cap A' = \Phi$. ▲

From these it can be concluded that the operations $\cup$ and $\cap$ are analogous to

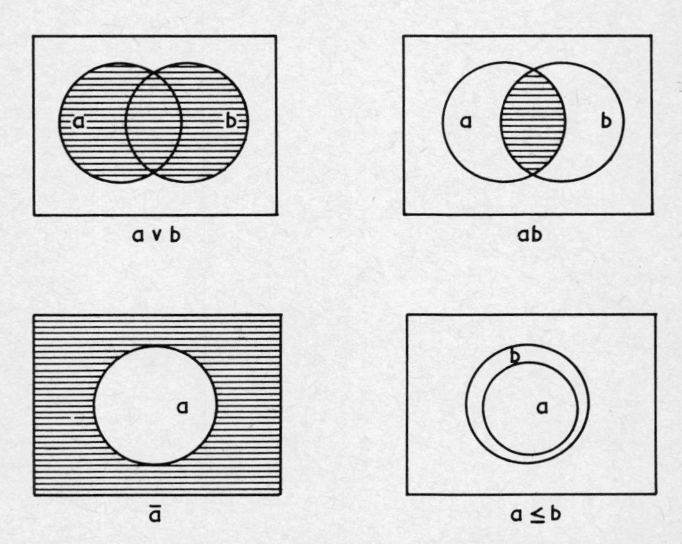

Figure 1-1 Venn diagram representations of Boolean operations and the inclusion relation.

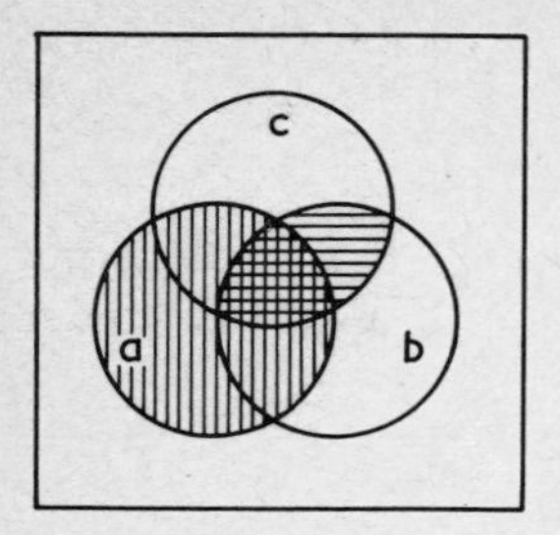

Vertically hatched area is a.
Horizontally hatched area is bc.
Entire hatched area is $a \vee bc$.

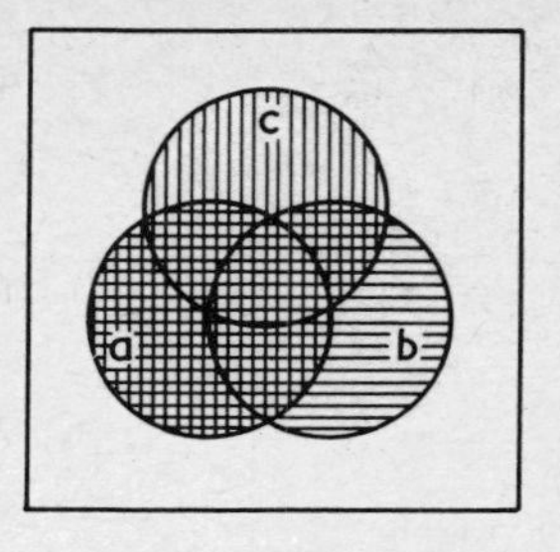

Horizontally hatched area is $a \vee b$.
Vertically hatched area is $a \vee c$.
Doubly hatched area is $(a \vee b)(a \vee c)$

Figure 1-2 $a \vee bc = (a \vee b)(a \vee c)$ verified on Venn diagram.

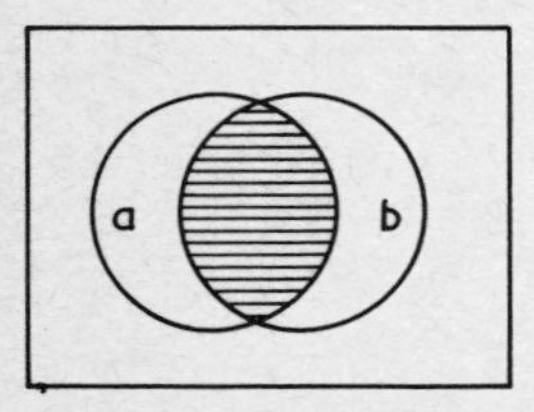

Shaded area is ab. Unshaded area is $\overline{ab}$.

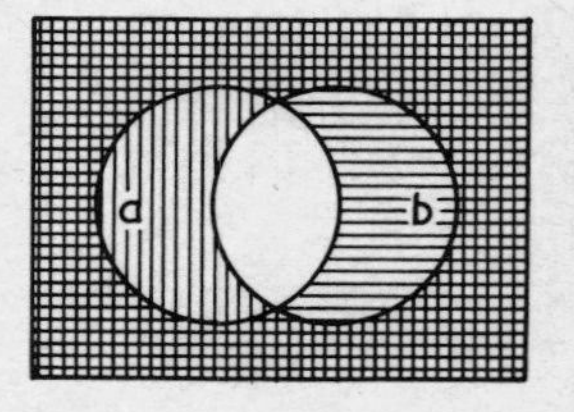

Horizontally hatched area is $\overline{a}$.
Vertically hatched area is $\overline{b}$.
Entire hatched area is $\overline{a} \vee \overline{b}$.

Figure 1-3 $\overline{ab} = \overline{a} \vee \overline{b}$ verified on the Venn diagram.

the operations $\vee$ and . . The universal set U and the null set Φ are the identity elements 1 and 0 respectively. An important result of these analogous relations is that the Venn diagram can be used to produce a pictorial depiction of the operations of Boolean algebra.

Figures 1-1 through 1-3 show how some of the postulates and theorems can also be verified on the Venn diagram.

REFERENCES

Boole, George: *An Investigation of the Laws of Thought,* Dover Publications, New York, 1854.
Harrison, M. A.: *Introduction to Switching and Automata Theory,* McGraw-Hill Book Co., New York, 1965.
Huntington, E. V.: 'The algebra of logic' *Trans. American Mathematical Society,* **5**, pp.288–309, 1904.
Krieger, M.: *Basic Switching Circuit Theory,* McMillan Co., New York, 1967.
Phister, Jr., M.: *Logical Design of Digital Computers,* John Wiley and Sons, Inc. New York, 1959.
Prather, R.: *Introduction to Switching Theory,* Allyn and Bacon, Boston, Mass., 1967.
Whitesitt, J. E.: *Boolean Algebra and its Applications,* Addison-Wesley Publishing Co., Reading, Mass., 1967.

PROBLEMS

1.1 Prove the following identities:

a) $(a \vee b)(\bar{a} \vee \bar{c}) = a\bar{c} \vee \bar{a}b$

b) $(a \vee b)(b \vee c)(c \vee \bar{a}) = (a \vee b)(c \vee \bar{a})$

c) $ab \vee bc \vee c\bar{a} = ab \vee c\bar{a}$ *(Consensus Theorem).*

1.2 Prove the following:

a) If $x \leqslant y$, then $\bar{x} \vee y = 1$, and $x\bar{y} = 0$.

b) If $x \leqslant y$, then $\bar{y} \leqslant \bar{x}$.

1.3 If a is an element of B, and f is a Boolean function, then prove that a covers af, that is, $af \leqslant a$.

1.4 Find the complements of the following functions:

a) $b \vee c(\bar{a} \vee d)$

b) $a \vee \bar{a}b \vee c(d \vee e)$

c) $\overline{a(b \vee c)} \vee b \vee c\bar{d}$

d) $d[a \vee \bar{b}(\bar{c} \vee \overline{ad})]$.

Check for correctness by showing that $f\bar{f} = 0$, and $f \vee \bar{f} = 1$.

1.5 Find the duals of the following functions:

a) $abc \vee \bar{a}b\bar{c}$

b) $a \vee \bar{a}b \vee \bar{b}c$

c) $ab \vee 1$

d) $(a \vee b)(\bar{c} \vee 0)$.

1.6 Show that if in the function

$$u = f(x_1, x_2, \ldots, x_n),$$

none of the elements $x_1, x_2, \ldots, x_n$ is 0 or 1, then the dual of u is given by,

$$u^D = \bar{f}(\bar{x}_1, \bar{x}_2, \ldots, \bar{x}_n)$$

1.7 Verify the identities of problem 1.1 on the Venn diagram.

Chapter 2

GATE AND CONTACT NETWORKS

2.1 INTRODUCTION

In the last chapter a Boolean algebra has been formally defined. There can be many systems which can qualify to be Boolean algebras. In this chapter it will be shown how an electronic system made of unilateral gates, and another system made of bilateral contacts may constitute Boolean algebras. The electronic gates are usually made of semiconductor diodes, transistors, etc. The contacts are usually those of electromechanical relays, or cryotrons. To start with let the electronic system be considered first. For this purpose let us define the following.

Definition 2.1.1 An electronic circuit which has one or more inputs and one output, and where the electrical condition of the output is dependent on those of the inputs, is called an *electronic gate.*

The number of inputs normally varies from one to four. In subsequent sections we shall mostly confine our discussion to electronic gates with two inputs. In switching circuits of interest to us, it is so arranged that the inputs and output of an electronic gate can assume one of only two distinct values. One of these values, we shall call the 0 state and the other 1 state. Depending on the particular design, the 0 state and the 1 state can be respectively a negative voltage (0), and a positive voltage (1), or, a zero voltage (0), and a positive voltage (1), or, a low positive voltage (0) and a high positive voltage (1).

Although we have mentioned only the above combinations for 0 and 1 as examples, it must be mentioned that many other combinations are possible. The main point from the engineering point of view is that the electrical circuit should be capable of distinguishing between the two states, namely, the 0 state and the 1 state.

2.2 BINARY AND UNARY OPERATIONS WITH ELECTRONIC GATES

It is possible to design two types of electronic gates to perform the two binary operations, those of addition and multiplication. We have already specified that the inputs and outputs of the gates will be either 0 or 1. Now, let the addition operation (denoted by v), be defined by the following table:

v	0	1
0	0	1
1	1	1

Written in the form of equations these are:

$$0 \vee 0 = 0 \tag{2.1}$$

$$0 \vee 1 = 1 \tag{2.2}$$

$$1 \vee 0 = 1 \tag{2.3}$$

$$1 \vee 1 = 1. \tag{2.4}$$

An electronic gate which produces an output satisfying the above table or equations is called an *OR gate* and is represented as in Fig.2-1. Its outputs are for the four different combinations of its input as given in the table in Fig.2-1. Such a table is called the *truth table* for the OR gate. The reason why such a gate is called an OR gate will be clear from the truth table. It can be seen from this table that the output of the OR gate is 1, when either *a or b* (or both) are 1.

Similarly the electronic gate which performs the binary operation of multiplication (denoted by .) must satisfy the following table:

.	0	1
0	0	0
1	0	1

In the form of equations, these are

$$0 \cdot 0 = 0 \tag{2.5}$$

$$0 \cdot 1 = 0 \tag{2.6}$$

Gate symbol	Truth table: In voltage	Truth table: In 0 and 1	Boolean operation
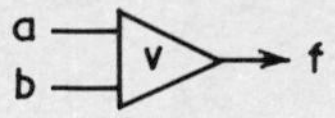 OR Gate (a, b → V → f)	a b | f 0 0 | 0 0 e_h | e_h e_h 0 | e_h e_h e_h | e_h	a b | f 0 0 | 0 0 1 | 1 1 0 | 1 1 1 | 1	$a \vee b$
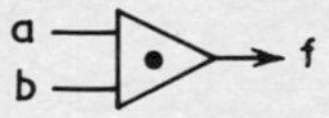 AND Gate (a, b → • → f)	a b | f 0 0 | 0 0 e_h | 0 e_h 0 | 0 e_h e_h | e_h	a b | f 0 0 | 0 0 1 | 0 1 0 | 0 1 1 | 1	ab
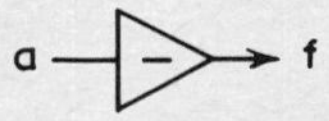 NOT Gate (a → – → f)	a | f 0 | e_h e_h | 0	a | f 0 | 1 1 | 0	$\bar{a}$

Figure 2-1 Boolean operations by electronic gates.

$$1 \cdot 0 = 0 \tag{2.7}$$

$$1 \cdot 1 = 1. \tag{2.8}$$

It is called an *AND gate* because the output is 1, only when both *a and b* are 1. Its symbol and truth table are given in Fig.2-1. It must be remembered that in addition to the two binary operations Boolean algebra also needs the unary operation of complementation. The gate which performs such an operation is called an *INVERTER gate*, as it will be evident from its truth table (Fig.2-1) that it inverts the input. Unlike the AND and OR gates, the inverter gate which is also called a *NOT gate* has only one input.

2.3 GATE NETWORKS

When several of the AND, OR, and NOT gates are interconnected in a certain manner, the resulting circuit is called a gate network. The outputs of these networks are Boolean functions of the input variables. These are illustrated by the following examples:

Example 2.3.1 Calculate the output functions of the gate networks of Fig.2-2.

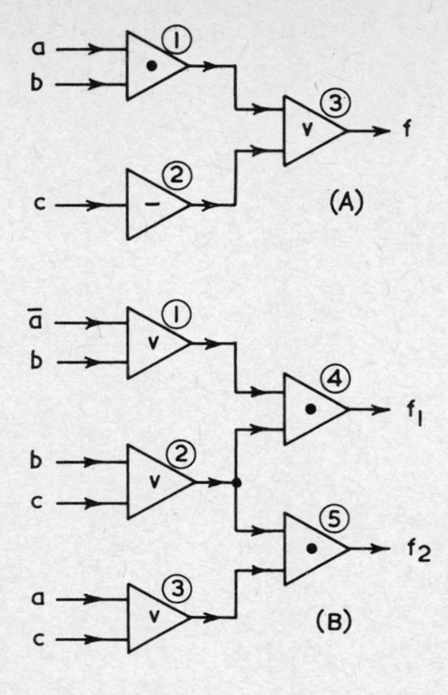

Figure 2-2

Solution: Mark the outputs of individual gates of Fig.2-2(A) as shown. Then,

(1) $= ab$ (3) $=$ (1) $\vee$ (2)

(2) $= \overline{c}$ $= ab \vee \overline{c} = f.$

For Fig.2-2(B)

(1) $= \overline{a} \vee b$ (2) $= b \vee c$

(3) $= a \vee c$

(4) $=$ (1) $\cdot$ (2) $= (\overline{a} \vee b)(b \vee c)$

$= f_1$

(5) $= f_2 =$ (2) $\cdot$ (3)

$= (b \vee c)(a \vee c).$

Definition 2.3.1 An interconnected gate network where the output Boolean function depends on the input variables only is called a *combinational circuit.* Combinational circuits with only one output are called *single output*, and those with more than one output are called *multiple output* circuits.

The other type of gate network is called a sequential circuit. It is defined as follows:

Definition 2.3.2 In some interconnected gate networks the output Boolean function depends not only on the input variables but also on its past history. Such a circuit is called a *sequential circuit.*†

While the analysis and synthesis of both combinational and sequential circuits will be treated in detail in later chapters, the two examples above give a very preliminary idea about the analysis of a combinational circuit. The synthesis can also be carried out by reversing the procedure as can be seen in the following example.

Example 2.3.2 Draw a gate circuit whose output function is $f = (a \vee bc)d$.

Solution: Here the synthesis is started from the extreme right as shown in the Fig. 2-3. In step 1, an AND gate is drawn with two inputs, $a \vee bc$ and d. $a \vee bc$ must be the output of an OR gate with inputs a and bc. This is drawn in the second step. In the third and final step, bc is derived from an AND gate with b and c at the two inputs.

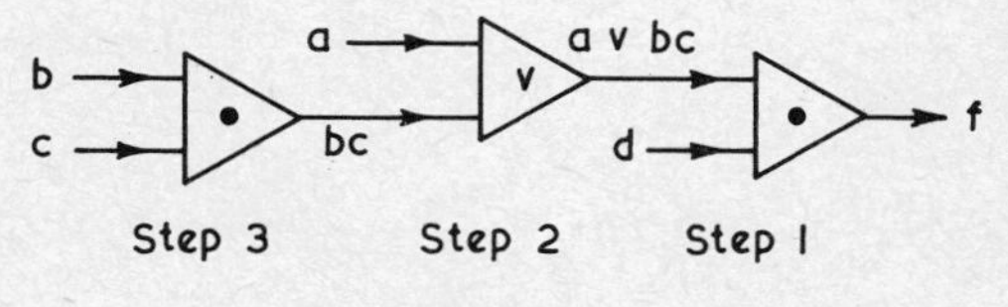

Figure 2-3

It should not be thought, however, from these examples that the analysis and synthesis are so simple. There are many other factors which are to be taken into consideration while designing a practical circuit.

2.4 ELECTRONIC GATE CIRCUITS

There are many electronic devices from which the AND and OR gates may be made. The simplest of these are gates made from semiconductor diodes. In this section we shall study these circuits.

† A formal definition of a sequential circuit will be given in Chapter 9.

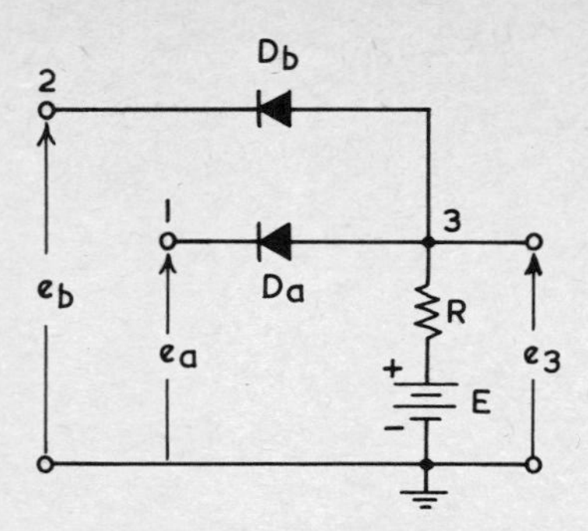

Figure 2-4 Diode AND gate.

Figure 2-4 gives the circuit of a diode AND gate with two inputs. Here the Boolean variable a appears as 1 when the voltage e_a has certain positive value $e_h(0 < e_h < E)$, and it appears as 0 when e_a is zero volts. Similarly the variable b is 1 or 0, depending on whether e_b has the same positive value or zero. Let us assume that the diodes are ideal, that is, their forward resistances are zero, and the reverse resistances are infinite. Now suppose, $e_b = e_h$ and $e_a = 0$. Then, since E is greater than both e_h and 0, initially both the diodes D_a and D_b will tend to conduct. But as soon as diode D_a starts conducting, the output terminal 3 will be at the same potential as the terminal 1. Since $e_a = 0$, 1 is at the ground potential. Therefore, terminal 3 also will be at ground potential, making $e_3 = 0$. Since 3 is at ground potential, and $e_b = e_h$, diode D_b will be reverse biased, and will not conduct. Thus there is no risk of the voltage source e_b being short circuited as a result of the terminal 3 being grounded. The circuit behaves in a similar manner when $e_b = 0$, and $e_a = e_h$. It can be easily seen that when both e_a and e_b are zero volts, e_3 is also zero. But when $e_a = e_b = e_h$, both diodes will conduct (since $E > e_h$), and the terminal 3 will have the same potential as terminals 1 or 2 which is e_h. Thus the behavior of the circuit can be tabulated as in the truth table for the AND operation in Fig.2-1. Here the voltages 0 and e_h are interpreted as the 0 and 1 of the Boolean algebra.

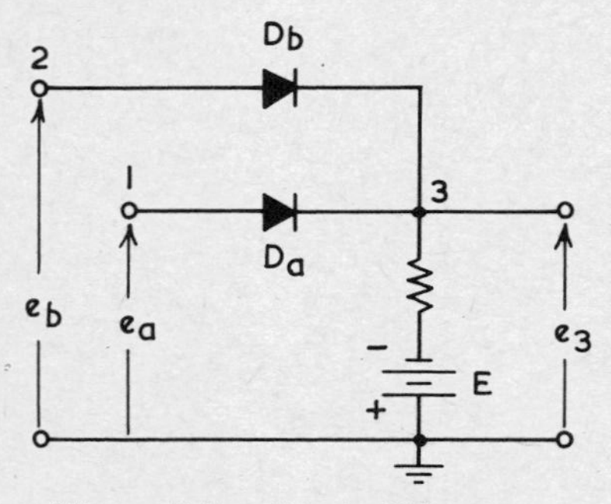

Figure 2-5 Diode OR gate.

Figure 2-5 shows the circuit of a diode OR gate. Here both the diodes and supply voltage E have been reversed. In this circuit, if both e_a and e_b are e_h then both the diodes conduct making $e_3 = e_h$.

When both e_a and e_b are zero, then also both the diodes conduct and $e_3 = 0$. Now, suppose $e_a = e_h$ and $e_b = 0$ then initially both D_a and D_b tend to conduct, and the terminal 3 tends to become equal to zero and e_h simultaneously. However, if e_3 becomes 0, diode D_a will still conduct raising its potential to e_h. At this time diode D_b will be reverse biased, and conduction stops. Hence the circuit will soon stabilize in this state. Similarly e_3 becomes equal to e_h when $e_a = 0$ and $e_b = e_h$. Hence this circuit produces the truth table of OR operation (Fig.2-1).

Although in both the circuits above, only two diodes have been shown, three or more diodes may be connected when the number of input variables is three or more.

When a gate network such as the one in Fig.2-2(B) is implemented by diode AND and OR gates, the circuit is as shown in Fig.2-6.

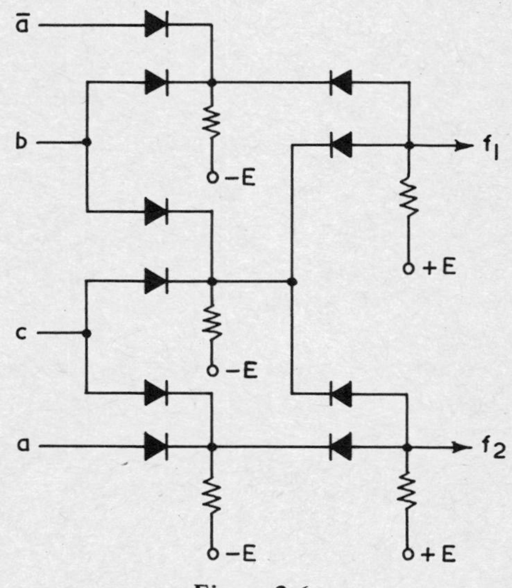

Figure 2-6

While AND and OR gates can be built up from diodes only, an active element such as a transistor is essential to perform the operation of complementation. Figure 2-7 shows such a circuit which is called an INVERTER gate. Here, when the signal at terminal 1 is a logical 0, that is, the terminal is at ground potential, the transistor is in the off condition, that is, it does not conduct. Consequently the potential of terminal 2 is V_{cc+}. This high voltage at 2 is interpreted as a logical 1 output. If now a positive voltage (a logical 1) is applied at 1, the transistor is

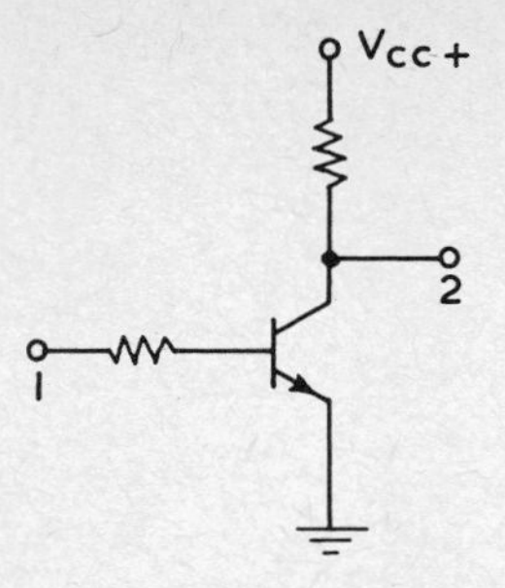

Figure 2-7 A transistor INVERTER gate.

driven into saturation. Voltage drop across the collector and emitter is now almost zero, and terminal 2 can be considered to be virtually at ground potential. This is interpreted as a logical 0 output. Hence the circuit produces a 1 output for a 0 input, and a 0 output for a 1 input, and thus performs the unary operation of complementation.

It must be mentioned once again that there are many devices as well as many varieties of circuits to perform the function of various electronic gates. The above circuits are described only to provide a basic idea as to their modes of operation. More detailed circuits can be studied from the appropriate references listed at the end of this chapter.

2.5 CONTACT NETWORKS

It will now be shown that the familiar mechanical contacts also form a Boolean algebra when certain conventions are observed. For the purpose of this development we shall, for the time being, confine our attention to only two simple types of contacts, namely a make contact, and a break contact, which are defined as follows:

Definition 2.5.1 A *make contact* (Fig.2-8) has two terminals, and provides an open path between them when it is in normal, that is, unoperated condition, and provides a closed path in the operated condition.

Following the usage in electrical circuits, an open path will also be called an open-circuit, and a closed path a short-circuit.

Definition 2.5.2 A *break contact* (Fig.2-8) has two terminals, and provides a closed path between them when it is in the normal, that is, unoperated condition whereas in the operated condition it provides an open path.

The conventions to be followed are that when a contact is in normal condition it is considered to carry the logical 0 signal, and when in the operated condition it is considered to carry the logical 1 signal.

Contact	Symbol	Truth table				Transmission function
		In words		In 0 & 1		
		Condition of x	Nature of path	x	t_{12}	
Make	1 x 2 or 1 — x — 2	Normal	Open circuit	0	0	$t_{12} = x$
		Operated	Short circuit	1	1	
		Condition of y	Nature of path	y	t_{12}	
Break	1 y 2 or 1 — $\bar{y}$ — 2	Normal	Short circuit	0	1	$t_{12} = \bar{y}$
		Operated	Open circuit	1	0	

Figure 2-8 Make and break contacts.

Definition 2.5.3 When the path between two points is open (an open-circuit), it is interpreted to have a *transmission function* of 0, and when the path is closed (a short-circuit), it is said to have a *transmission function* of 1.

The transmission function as defined above is a Boolean function, and its two probable values, 0 and 1, are the 0 and 1 of Boolean algebra.

Following these conventions the truth table of a single make contact in words and in terms of 0 and 1 are given in the tables of Fig.2-8. It will be clear from the truth tables that the transmission function of the make contact is x, and that of the break contact is $\bar{y}$.

Let us now study the condition of the path between 1 and 2 of Fig.2-9, where two make contacts x_1 and x_2 have been connected in parallel. The different conditions of the path for various conditions of the two circuits are given in the adjoining table. The transmission function of the series circuit is a function of the transmission functions of the individual contacts. It can be easily verified that the equation

$$t_{12} = t_{x_1} \vee t_{x_2} = x_1 \vee x_2$$

satisfies the truth table. Thus this particular type of connection performs the binary operation of addition in contact networks.

Definition 2.5.4 The *parallel connection* between two contacts is equivalent to $\vee$ *operation* of Boolean algebra.

In a similar manner the definition of . operation as given below can also be verified (Fig.2-9).

Connection	Circuit	Truth table						Boolean operation
		In words			In 0 & 1			
		Condition of x_1	Condition of x_2	Nature of path	x_1	x_2	t_{12}	
Parallel	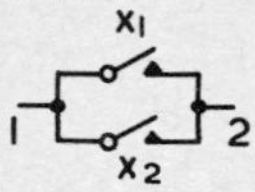	Norm	Norm	OC	0	0	0	$x_1 \vee x_2$
		Norm	Optd	SC	0	1	1	
		Optd	Norm	SC	1	0	1	
		Optd	Optd	SC	1	1	1	
Series	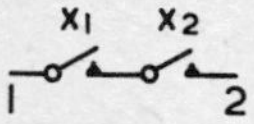	Norm	Norm	OC	0	0	0	$x_1 x_2$
		Norm	Optd	OC	0	1	0	
		Optd	Norm	OC	1	0	0	
		Optd	Optd	SC	1	1	1	

Figure 2-9 Boolean operations by contact networks. Norm: normal; Optd: operated; OC: open-circuit; SC: short-circuit.

Definition 2.5.5 The *series connection* between two contacts is equivalent to . *operation* of Boolean algebra.

Thus the parallel and series connections are contact equivalents of OR and AND gates respectively. There is, however, one important difference. The gate circuits are unidirectional or unilateral, that is, signal always flows from the input to the output terminals, which cannot be interchanged. On the other hand contact networks are bilateral, so that a terminal which is acting as an input in one circuit may act as an output in another circuit.

As regards the complementation operation, let x be a make contact and y be a break contact, and let them be so coupled mechanically that they operate simultaneously, then x and y are complements of each other. This is because (as will be evident from Fig.2-10)

$$x \vee \bar{y} = 1$$

and

$$x\bar{y} = 0.$$

Thus we have the following definition.

Definition 2.5.6 If a make and a break contact are so coupled that they operate simultaneously, then they are *complements* of each other.

As the output function of a gate depends on its inputs, so also the transmission function of a path is dependent on the states of the contacts constituting the path. It can be easily verified that following the above definitions contact

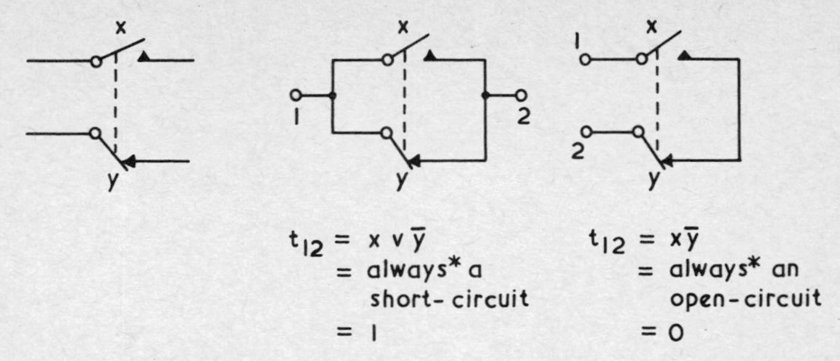

Figure 2-10 Complementation operation in contact network.

networks satisfy the basic postulates of Boolean Algebra, and therefore, they can be analyzed and synthesized by the application of the theorems of Boolean algebra In fact as all the theorems of the Boolean algebra can be verified on the Venn diagram, they can also be verified by the contact networks. As an example, Fig.2-11 shows that so far as the transmission between the points 1 and 2 is concerned, the network of Fig.2-11(A) serves the same purpose as the single contact of Fig.2-11(B). Therefore the transmission functions of these two networks are equal. This proves the familiar absorption theorem.

The transmission function of a series–parallel network can be calculated as illustrated by the following example.

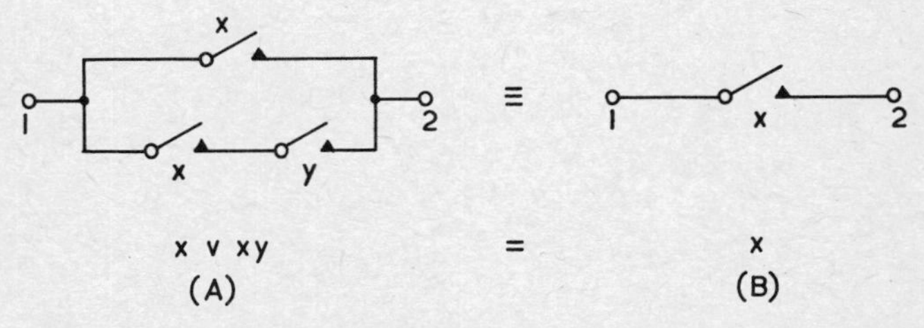

Figure 2-11 Absorption theorem verified by contact network.

* The description "always" is not exactly correct. During the interval in which the movable contact spring of x has not yet touched the other contact spring,but the movable contact spring of y has moved away from its fixed contact spring, the paths between 1 and 2 in the parallel connection are open. Similarly, there is an interval during the transition of spring contacts when the path between 1 and 2 in the series connection is closed. However, for most practical applications, this discrepancy during the transit time of the contacts can be ignored, and the contacts are idealized assuming that they change state "instantaneously".

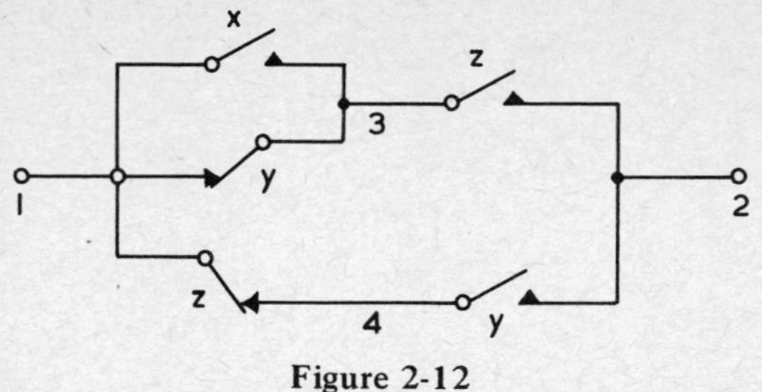

Figure 2-12

Example 2.5.1 Find the transmission function of the terminal network of Fig.2-12.

Solution: Mark the points 1, 2, 3 and 4 as shown in the figure. Now, the net work between points 1 and 3 is the parallel connection of the make contact x and the break contact y. Therefore, the transmission function,

$t_{13} = x \vee \bar{y}.$

Similarly,

$t_{32} = z,$

but paths 13 and 32 are in series, therefore

$t_{132} = t_{13}t_{32} = (x \vee \bar{y})z$

and

$t_{142} = \bar{z}y.$

But the paths 132, and 142 are in parallel, therefore

$t_{12} = t_{132} \vee t_{142} = (x \vee \bar{y})z \vee \bar{z}y = xz \vee \bar{y}z \vee \bar{z}y.$

Alternatively, the transmission function between 1 and 2 can be computed by finding the transmission functions of all independent paths from 1 to 2, and then considering them in parallel. Thus in the above problem, there are three paths from 1 to 2, and their transmission functions are as follows

t_{132} (*via* x contact) $= xz$

t_{132} (*via* $\bar{y}$ contact) $= \bar{y}z$

$t_{142} = \bar{z}y.$

Therefore

$t_{12} = xz \vee \bar{y}z \vee \bar{z}y.$

The second approach is more versatile, as it can be applied to any network, whereas the first approach is valid only for a series–parallel network.

While dealing with multiterminal networks, the concept of a connection function as defined below is very useful in calculating the transmission function between any two terminals or nodes.

Definition 2.5.7 The *connection function* c_{ij} between any two nodes of a bilateral contact network is the transmission function of the contact or group of contacts connected directly (that is, without going through any other node) between these two nodes.

Example 2.5.2 Compute the connection and transmission functions of the network of Fig.2-13 between all pairs of nodes.

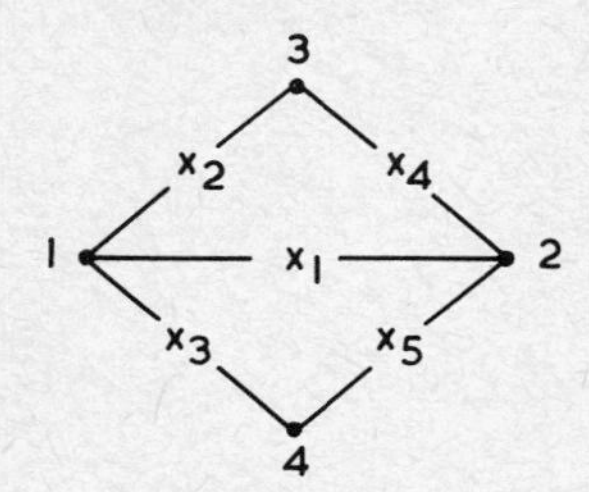

Figure 2-13

Solution:

$c_{12} = x_1,\ c_{13} = x_2, \quad c_{14} = x_3$

$c_{23} = x_4, \quad c_{24} = x_5, \quad c_{34} = 0$

$t_{12} = c_{12} \vee c_{13}c_{32} \vee c_{14}c_{42}$

$\quad = x_1 \vee x_2x_4 \vee x_3x_5$

$t_{13} = c_{13} \vee c_{12}c_{23} \vee c_{14}c_{42}c_{23}$

$\quad = x_2 \vee x_1x_4 \vee x_3x_5x_4$

$t_{14} = c_{14} \vee c_{12}c_{24} \vee c_{13}c_{32}c_{24}$

$\quad = x_3 \vee x_1x_5 \vee x_2x_4x_5$

$t_{23} = c_{23} \vee c_{21}c_{13} \vee c_{24}c_{41}c_{13}$

$\quad = x_4 \vee x_1x_2 \vee x_5x_3x_2$

$t_{24} = c_{24} \vee c_{21}c_{14} \vee c_{23}c_{31}c_{14}$

$\quad = x_5 \vee x_1x_3 \vee x_4x_2x_3$

$t_{34} = c_{31}c_{14} \vee c_{32}c_{24} \vee c_{31}c_{12}c_{24} \vee c_{32}c_{21}c_{14}$

$\quad = x_2x_3 \vee x_4x_5 \vee x_2x_1x_5 \vee x_4x_1x_3.$

2.6 BOOLEAN MATRICES

In many occasions Boolean matrices prove to be powerful tool in manipulating switching networks. Here we shall develop certain basic properties of such matrices which will be helpful in dealing with the bilateral contact networks.

Definition 2.6.1 A *Boolean matrix* is a rectangular array of elements which are either Boolean elements or functions.

If $A = [a_{ij}]$ and $B = [b_{ij}]$ are both $(m \times n)$ Boolean matrices, that is, both have m rows and n columns, then we define the following operations between A and B.

Definition 2.6.2 $A = B$, if and only if $a_{ij} = b_{ij}$ for $1 \leqslant i \leqslant m$, and $1 \leqslant j \leqslant n$.

Definition 2.6.3 $A \vee B = D$, where $D = [d_{ij}]$ is also an $m \times n$ matrix such that $d_{ij} = a_{ij} \vee b_{ij}$.

It can be easily verified that

$$A \vee A = A. \tag{2.9}$$

Definition 2.6.4 $AB = E$, where $E = [e_{ij}]$ is also an $m \times n$ matrix such that $e_{ij} = a_{ij}b_{ij}$. Here also it can be verified that

$$AA = A. \tag{2.10}$$

The results of the above two operations are known as the Boolean sum and Boolean product or sometimes as the logical sum and logical product of two Boolean matrices.

However, we can also define the product of two Boolean matrices as in the ordinary algebra.

Definition 2.6.5 $A \times B = H$, where $A = [a_{ij}]$ is an $m \times r$ matrix, $B = [b_{ij}]$ is an $r \times n$ matrix, and $H = [h_{ij}]$ is an $m \times n$ matrix such that

$$h_{ij} = \bigvee_{k=1}^{r} a_{ik}b_{kj}.$$

The result of this operation is called the matrix product or sometimes cross-product of two Boolean matrices. It should be noted that the product elements of the product matrix are computed in the same way as in ordinary matrix algebra except that the Boolean operations $\vee$ and . replace the arithmetic operations of addition and multiplication respectively. We will use the following notation

$$A \times A = A^2 \qquad A^2 \times A = A^3 \text{ etc.} \tag{2.11}$$

If for the network of Fig.2-13 we write the connection functions between all pairs of nodes in the form of a matrix, we get the following matrix.

$$C = \begin{array}{c} \\ 1 \\ 2 \\ 3 \\ 4 \end{array} \begin{array}{c} \begin{array}{cccc} 1 & 2 & 3 & 4 \end{array} \\ \left[\begin{array}{cccc} 1 & x_1 & x_2 & x_3 \\ x_1 & 1 & x_4 & x_5 \\ x_2 & x_4 & 1 & 0 \\ x_3 & x_5 & 0 & 1 \end{array}\right] \end{array}$$

Such a matrix called a connection matrix may be defined as follows:

)efinition 2.6.6 A *connection matrix* $C = [c_{ij}]$ for a network with p nodes is ı $p \times p$ matrix, where c_{ij} is the connection function between the nodes i and j.

A connection matrix is a member of the subclass of Boolean matrices known .s switching matrices.

)efinition 2.6.7 A *switching matrix* is a symmetric Boolean matrix, whose all lements of the principal diagonal are 1's.

Another operation which is defined only for switching matrices is as follows:

)efinition 2.6.8 If $A = [a_{ij}]$ is a switching matrix, then

$$\bar{A} = [b_{ij}],$$

vhere

$$b_{ij} = \begin{cases} \bar{a}_{ij} & \text{if } i \neq j \\ 1 & \text{if } i = j. \end{cases}$$

It can be easily verified that a connection matrix is also a switching matrix. \ transmission matrix which is defined as follows is also a switching matrix.

)efinition 2.6.9 A *transmission matrix* $T = [t_{ij}]$ for a network with p nodes s a $p \times p$ matrix, where t_{ij} is the transmission function between the nodes i .nd j.

For the network of Fig.2-13 the transmission matrix will be as follows:

$$T = \begin{array}{c} \\ 1 \\ \\ 2 \\ \\ 3 \\ \\ 4 \\ \\ \end{array} \begin{array}{c} \begin{array}{cccc} 1 & 2 & 3 & 4 \end{array} \\ \left[\begin{array}{cccc} 1 & x_1 \vee x_2x_4 & x_2 \vee x_1x_4 & x_3 \vee x_1x_5 \\ & \vee x_3x_5 & \vee x_3x_5x_4 & \vee x_2x_4x_5 \\ x_1 \vee x_2x_4 & 1 & x_4 \vee x_1x_2 & x_5 \vee x_1x_3 \\ \vee x_3x_5 & & \vee x_5x_3x_2 & \vee x_4x_2x_3 \\ x_2 \vee x_1x_4 & x_4 \vee x_1x_2 & 1 & x_2x_3 \vee x_4x_5 \\ \vee x_3x_5x_4 & \vee x_5x_3x_2 & & \vee x_2x_1x_5 \vee x_4x_1x_3 \\ x_3 \vee x_1x_5 & x_5 \vee x_1x_3 & x_2x_3 \vee x_4x_5 & 1 \\ \vee x_2x_4x_5 & \vee x_4x_2x_3 & \vee x_2x_1x_5 \vee x_4x_1x_3 & \end{array}\right] \end{array}$$

If we study the transmission function of the network between any two nodes say 1 and 3, then we find that the transmission function $x_2 \vee x_1x_4 \vee x_3x_5x_4$ is the computation of transmission functions for three parallel paths. For each of these parallel paths a concept of path length can be introduced. The direct path between any two nodes has a path length of 1, the path going through any other node has a length of 2. If the path goes through two other nodes, then the length is 3.

Definition 2.6.10 The *path length* of a particular path between any two nodes of a multinode network, is given by

$$l = n_i + 1,$$

when the path is going through n_i number of intermediate nodes.

If the network has p nodes, then the minimum and the maximum values of n_i are 0 and $p - 2$. Hence

$$l_{min} = 1 \qquad (2.12)$$

and

$$l_{max} = p - 1. \qquad (2.13)$$

Definition 2.6.11 If the transmission function between any two nodes of a network is computed by excluding all paths of length greater than l, then the transmission function so computed is called the *partial transmission function* for path length l.

Now, each entry in the connection matrix of a network gives the partial transmission function of path length l between the two nodes. It is interesting to note that each entry in the matrix $C^2 = C \times C$ is the partial transmission function of path length 2, and similarly the matrix $C^3 = C^2 \times C$ gives the partial transmission function of path length 3. Thus for a p node network, $l_{max} = p - 1$ and the matrix C^{p-1} gives the complete transmission function of the network. These are so because of the following theorem.

Theorem 2.6.1 For a p node network having the connection matrix C, if $C^q(1 \leqslant q \leqslant p - 1) = [x_{ij}]$, then x_{ij} is the partial transmission function of path length q between the nodes i and j.

Proof: $C = [c_{ij}]$. Now

$$C^2 = C \times C = \bigvee_{k=1}^{p} c_{ik}c_{kj},$$

but

$$\bigvee_{k=1}^{p} c_{ik}c_{kj}$$

is equal to the sum of transmission functions of all paths between nodes i and j going through the node $k (1 \leqslant k \leqslant p)$ = partial transmission function of path length 2, between nodes i and j.

Again, if $C^2 = [\beta_{ij}]$, then

$$C^3 = C^2 \times C = \bigvee_{k=1}^{p} \beta_{ik} c_{kj}.$$

Now, each $\beta_{ik} c_{kj}$ is the transmission function of a path of length 2 between i and k in series with a path of length 1 between k and j.

Therefore

$$\bigvee_{k=1}^{p} \beta_{ik} c_{kj}$$

is equal to partial transmission function of path length 3 between i and j nodes.

Repeating this procedure the theorem is proved. ▲

COROLLARY **2.6.1A** For a p-node network if C is the connection matrix, and T is the transmission matrix, then,

$$C^{p-1} = T.$$

Proof: All entries of C^{p-1} are partial transmission functions of path length $p - 1$. But l_{max} for the p-node network is $p - 1$. Hence, each of these transmission functions is also the complete transmission function. Therefore $C^{p-1} = T$. ▲

This shows the importance of Boolean matrices in the computation of transmission function of a bilateral switching network. Some more operations of Boolean matrices will be discussed in Chapters 3 and 5.

2.7 RELAYS: EXCITATION FUNCTIONS

A relay has a number of contact springs which are mechanically coupled. Its coil acts as an electromagnet. When current flows through the coil it attracts an armature which operates all the contacts simultaneously. Usually a relay has four types of contacts as shown in Fig.2-14. The change-over or the transfer contact is a combination of a make and break contact. The make-before-break contact is also a change-over contact with the difference that here the make contact closes earlier than the opening of the break contact. For this reason this is also called the continuity transfer contact. A relay is normally designated by a capital letter under its coil. The contacts are then designated by the corresponding small letter. Thus a relay X may have several x contacts. When X

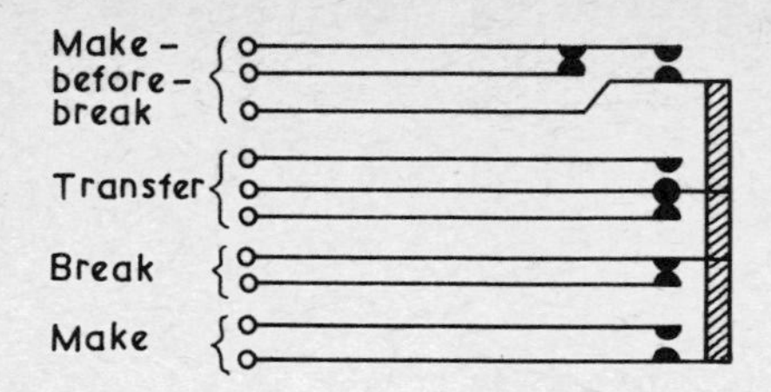

Figure 2-14 Types of relay contacts.

operates all x contacts change state simultaneously. Strictly speaking, because of the differences in air gaps between the contacts and the mechanical coupling arrangement by which the movement of the armature is transferred to various movable contact springs, all contacts do not open or close exactly at the same instant of time. But for many applications this discrepancy can be ignored and the assumption that the contacts operate simultaneously remains valid. On the other hand, there are occasions when these minute differences in operation can be hazardous to the circuit operation, and then remedial measures must be taken. The symbols by which these contacts are represented in the schematic circuit diagram are shown in Fig.2-15.

In a circuit consisting of one or more relays, the energizing path of a relay coil may be in series with a contact network. In such a case the relay will be energized only when the transmission function of the network is 1. If a contact network is in parallel with a relay, then the relay is prevented from operating so long as its transmission function is 1.

	Type	Symbols: Used in this text	Symbols: Others
CONTACTS	Make	or — x —	
	Break	or — $\bar{x}$ —	
	Transfer	or x — / $\bar{x}$ —	
	Make-before-break	or x — / $\bar{x}$ —	
COIL			

Figure 2-15 Symbols of relay contacts and coil.

Definition 2.7.1 The *excitation function* of a relay is the Boolean function of the contact network connected with the coil, such that the relay operates when the excitation function is 1, and the relay releases (comes back to normal condition) when the excitation function is 0.

Theorem 2.7.1 The excitation function of the X relay is given by $X = t_s \overline{t_p}$, where t_s = transmission function of the contact network in series with its coil, and t_p = transmission function of the contact network in parallel with its coil.

Proof: The relay can operate only when the contact network in series with the coil presents a short-circuit, and the network shunting the coil presents an open-circuit, that is, when t_s is 1, and t_p is 0. Under this condition X is 1, and in all other conditions X is 0. Hence $X = t_s \overline{t_p}$. ▲

X can also be computed by separately computing the transmission functions of all the series and parallel paths. If t_{s1}, t_{s2} etc. are transmission functions of series paths and t_{p1}, t_{p2} are those for parallel paths, then,

$$X = (t_{s1} \vee t_{s2} \vee \ldots.) \overline{t_{p1}}\, \overline{t_{p2}}. \tag{2.14}$$

Example 2.7.1 Find the excitation functions of the X and the Y relays of Fig.2-16.

Solution: For the X relay $t_s = z(x \vee \overline{y})$, and $t_p = u \vee \overline{w}$. Therefore

$$X = z(x \vee \overline{y})\, \overline{u \vee \overline{w}}$$

$$= (zx \vee \overline{y}z)\, (\overline{u}w)$$

$$= \overline{u}wxz \vee \overline{u}w\overline{y}z.$$

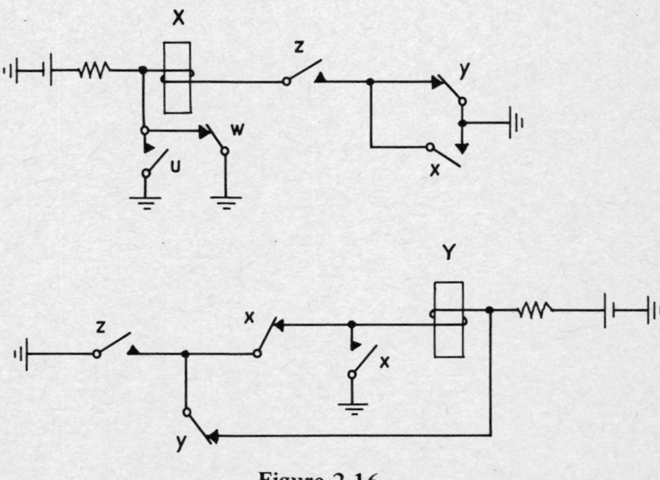

Figure 2-16

For the Y relay, series path 1 is *via* x, therefore $t_{s1} = x$. Series path 2 is *via* $\bar{x}$ and z, therefore $t_{s2} = \bar{x}z$. Parallel path 1 is *via* $\bar{y}$ and z, therefore $t_{p1} = \bar{y}z$. Parallel path 2 is *via* $\bar{y}$, $\bar{x}$ and x therefore $t_{p2} = \bar{y}\,\bar{x}\,x = 0$. Therefore

$$Y = (x \vee \bar{x}z)\,(\overline{\bar{y}z}) \cdot \bar{0}$$

$$= (x \vee z)\,(y \vee \bar{z})$$

$$= xy \vee yz \vee x\bar{z}.$$

REFERENCES

Higonnet, R. and R. Grea: *Logical Design of Electrical Circuits*, McGraw-Hill Book Co., New York 1958.

Hohn, F. E. and R. L. Schissler: "Boolean matrices and the design of combinational relay switching circuits", *BSTJ*, **34**, pp.170–202, Jan. 1955.

—— : "A matrix method for the design of relay circuits", *IRE Trans Circuit Theory*, **CT-2**, pp.154–161, June 1955.

Humphrey, W. S. Jr.: *Switching Circuits with Computer Applications*, McGraw-Hill Book Co., New York, 1958.

Hurley, R. B.: *Transistor Logic Circuits*, John Wiley and Sons, Inc., New York, 1961.

Keister, W., Richie, A. E., and S. H. Washburn: *The Design of Switching Circuits*, D. Van Nostrand Co., Princeton, N.J., 1962.

Miller, R. E.: *Switching Theory*, Vol.I, John Wiley and Sons, Inc., New York, 1966.

Shanon, C. E.: "A symbolic analysis of relay and switching circuits", *AIEE Trans.* **57**, pp.713–723, 1938.

—— : "The synthesis of two-terminal switching circuits", *BSTJ*, pp.59–98, Jan. 1949.

PROBLEMS

2.1 Write the truth tables of the circuits of Fig.2-17.

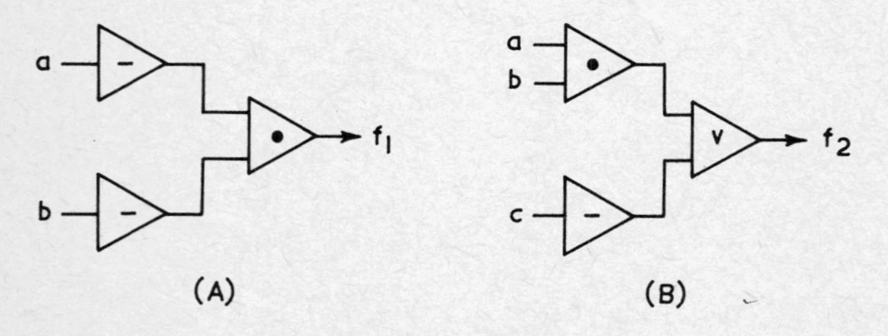

Figure 2-17

2.2 Calculate f_1, f_2 and f_3 in Fig.2-18.

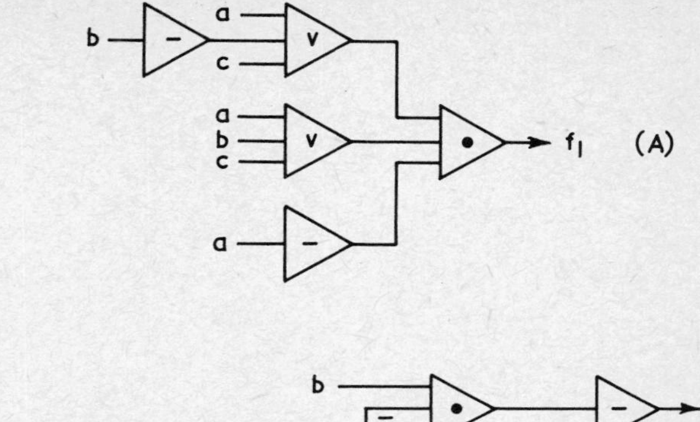

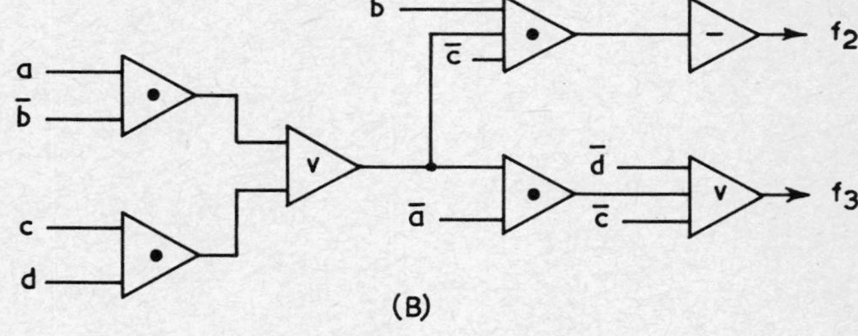

Figure 2-18

2.3 Implement the following functions with diode gates. (All variables are available in both true and complemented forms.)

$f_1 = abc \vee \bar{a}\bar{b} \vee \bar{c}$

$f_2 = (a \vee b)\bar{c} \vee d$

$f_3 = (x_1 \vee x_2 \vee x_3)\,(\bar{x}_1 \vee \bar{x}_2) \vee x_3.$

2.4 Write the truth tables of the circuits of Fig.2-19.

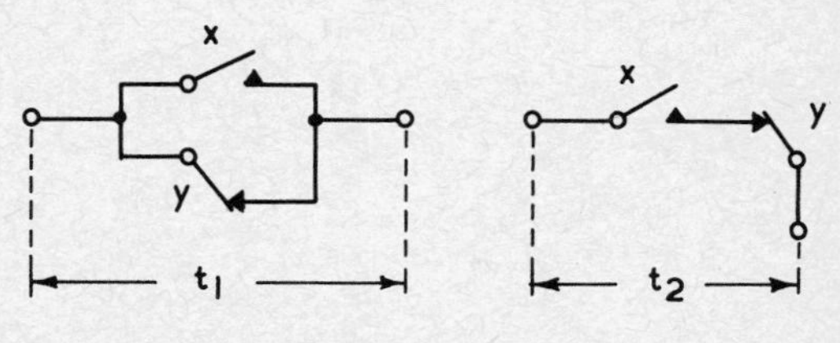

Figure 2-19

2.5 Assuming f_1, f_2 and f_3 of problem 2.3 as transmission functions, draw the contact networks of these functions.

2.6 Calculate the transmission functions t_{12} of the networks of Fig. 2-20.

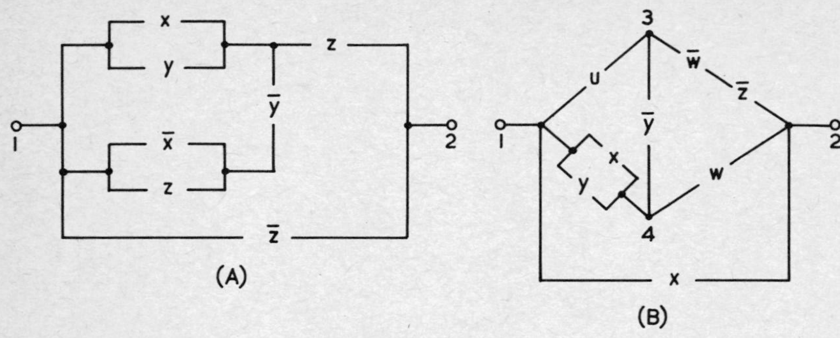

Figure 2-20

2.7 Draw the complement and dual networks of the networks of problem 2.6. (Hint: Find first the complements and duals of the transmission functions.)

2.8 Write the connection matrices of the networks of Fig. 2.20.

2.9 Compute the transmission matrices of the network of Fig. 2.20(B).

2.10 Find the excitation functions of the X and Y relays of Fig. 2-21.

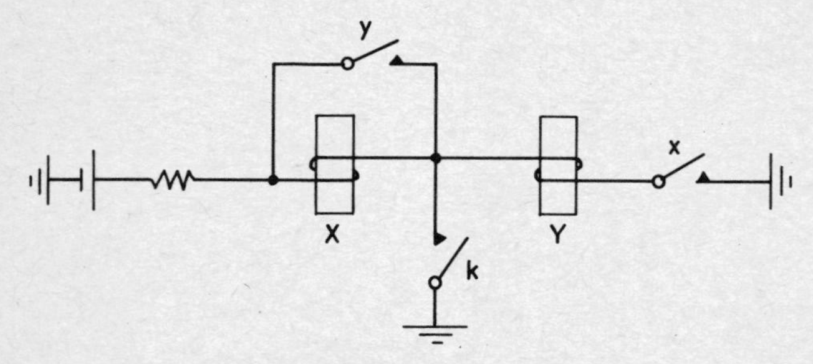

Figure 2-21

2.11 Draw the relay circuits whose excitation functions are as follows:

$X = (\overline{x}y \vee x\overline{y})k$

$Y = (x \vee \overline{y})k.$

Chapter 3

BOOLEAN EXPRESSIONS

3.1 INTRODUCTION

In Chapter 1 (Sec. 1.8) we have defined a Boolean function. In Chapter 2, the transmission functions of contact networks, excitation functions of relays, and the functions expressing the outputs of gate networks, are typical examples of Boolean functions. In this chapter various aspects of these functions as well as those terms comprising such functions will be studied.

3.2 NORMAL FORMS

Definition 3.2.1 A Boolean variable in the true form or in the complemented form is called a *literal.*

Thus a, $\bar{a}$, b, $\bar{b}$ etc. are literals.

Definition 3.2.2 The Boolean product of two or more literals is called a *product term.*

It should be noted that a variable cannot appear more than once in a product term. For example, if there is a term like, aab, then

$$aab = (aa)b = ab$$

and, if there is a term like $a\bar{a}b$, then

$$a\bar{a}b = (a\bar{a})b = 0b = 0.$$

Definition 3.2.3 The Boolean sum of two or more literals is called a *sum term.*

It should be verified that as in the product term so also in the sum term a variable cannot appear more than once.

Definition 3.2.4 When a Boolean function appears as a sum of several product terms, it is said to be expressed as a *sum-of-products (SP).* The SP form is also called the *disjunctive normal form (DNF).*

$$f(abcd) = \bar{a} \vee b\bar{c} \vee cd$$

is an example of the SP form or the DNF.

Definition 3.2.5 When a Boolean function appears as a product of several sum terms, it is said to be expressed as a *product-of-sums (PS)*. The PS form is also called the *conjunctive normal form (CNF).*

$$f(abcd) = (a \vee b)(b \vee c \vee d)$$

is an example of the PS form or the CNF.

A Boolean function which is neither in the disjunctive nor in the conjunctive normal form can be 'simplified' to such a form by applying postulate P.4.

Example 3.2.1 Express the following functions in normal forms.

a) $f_1 = \bar{a}\bar{b} \vee b(c \vee \bar{d})$

b) $f_2 = (\bar{a} \vee b)(a \vee cd)$.

Solution: $f_1 = \bar{a}\bar{b} \vee bc \vee b\bar{d}$

$f_2 = (\bar{a} \vee b)(a \vee c)(a \vee d)$.

3.3 CANONICAL FORMS

Definition 3.3.1 When each of the terms of a Boolean function expressed either in the SP or PS form has all the variables in it, then it is said to be expressed in the *canonical form.*

Here also by the idempotence law of Boolean algebra the canonical form cannot have the same term more than once.

The canonical SP form is called the *disjunctive canonical form (DCF)*, and the canonical PS form is called the *conjunctive canonical form (CCF).*

We now proceed to show that any Boolean function can be expressed in the canonical forms. For this we first prove the following theorems.

Theorem 3.3.1 A Boolean product term of k variables can be expressed as a sum of 2^{n-k} product terms of n variables $(n > k)$.

To prove this theorem, we prove the following lemma.

LEMMA **3.3.1** *A Boolean product term of k variables can be expressed as a sum of 2 product terms of k + 1 variables.*

Proof: Let $(x_1x_2 \ldots x_k)$ be a term of k variables. Now,

$$(x_1x_2 \ldots x_k) = (x_1x_2 \ldots x_k) \cdot 1$$
$$= (x_1x_2 \ldots x_k)(\bar{x}_{k+1} \vee x_{k+1})$$
$$= (x_1x_2 \ldots x_k\bar{x}_{k+1}) \vee (x_1x_2 \ldots x_kx_{k+1}). \quad \blacktriangle$$

Theorem 3.3.1 can now be proved by repeated applications of lemma 3.3.1. The product term of k variables can be expressed as a sum of 2 product terms of $k + 1$ variables. Again each of these 2 terms of $k + 1$ variables can be expressed as a sum of 2 terms of $k + 2$ variables. Hence the product term of k variables can be expressed as a sum of 2^2 product terms of $k + 2$ variables, 2^3 product terms of $k + 3$ variables, and 2^m product terms of $k + m$ variables.

Putting $k + m = n$ or, $m = n - k$, proof of the theorem is completed. ▲

COROLLARY 3.3.1A *The identity element 1 can be expressed as a sum of 2^n product terms of n variables.*

Proof: $1 = x_1 \vee \bar{x}_1$

Now both x_1 and $\bar{x}_1$ can be expressed as a sum of 2^{n-1} product terms of n variables. Hence 1 can be expressed as a sum of $2^{n-1} + 2^{n-1}$ or 2^n number of product terms of n variables. ▲

Theorem 3.3.2 A Boolean sum term of k variables can be expressed as a product of 2^{n-k} $(n > k)$ sum terms of n variables.

COROLLARY 3.3.2A *The identity element 0 can be expressed as a product of 2^n number of sum terms of n variables.*

The proof of this theorem and its corollary can be carried out in the same manner as for the previous theorem and corollary. This is left as an exercise for the reader.

Theorem 3.3.3 Every Boolean function can be expressed in a canonical form.

Proof: A Boolean function of n variables given in any form can be simplified to one of the two normal forms, disjunctive or conjunctive.

Case a) Let the function simplify to disjunctive normal form, that is, it is a sum of a number of product terms. Now, all those product terms which do not contain all the variables can be expanded as a sum of product terms containing all the n variables (Th. 3.3.1). Hence the DNF is expanded into disjunctive canonical form.

Case b) Let the function simplify to conjunctive normal form, that is, as a product of several sum terms. By Th. 3.3.2, all those terms not containing all the n variables can be expanded as a product of sum terms containing all the n variables. Hence the CNF is expanded into the conjunctive canonical form. ▲

Thus every Boolean function can be expressed in a canonical form, either disjunctive or conjunctive.

It will be shown later that the disjunctive canonical form of a Boolean function can be converted into the conjunctive canonical form, and *vice-versa.*

Example 3.3.1 Express each of the following functions in a canonical form:

$f_1 = a\bar{b}c \vee b\bar{c} \vee ac$

$f_2 = (a \vee b)(b \vee \bar{c})$

$f_3 = a \vee \bar{a}(b \vee \bar{c})$.

Solution:

$$f_1 = a\bar{b}c \vee b\bar{c}(\bar{a} \vee a) \vee ac(b \vee \bar{b})$$

$$= a\bar{b}c \vee \bar{a}b\bar{c} \vee ab\bar{c} \vee abc \vee a\bar{b}c$$

$$= a\bar{b}c \vee \bar{a}b\bar{c} \vee ab\bar{c} \vee abc$$

$$f_2 = (a \vee b \vee c\bar{c})(b \vee \bar{c} \vee a\bar{a})$$

$$= (a \vee b \vee c)(a \vee b \vee \bar{c})(a \vee b \vee \bar{c})(\bar{a} \vee b \vee \bar{c})$$

$$= (a \vee b \vee c)(a \vee b \vee \bar{c})(\bar{a} \vee b \vee \bar{c})$$

$$f_3 = a \vee \bar{a}(b \vee \bar{c})$$

$$= a \vee \bar{a}b \vee \bar{a}\bar{c}$$

$$\doteq a(b \vee \bar{b}) \vee \bar{a}b(c \vee \bar{c}) \vee \bar{a}\bar{c}(b \vee \bar{b})$$

$$= ab \vee a\bar{b} \vee \bar{a}bc \vee \bar{a}b\bar{c} \vee \bar{a}b\bar{c} \vee \bar{a}\bar{b}\bar{c}$$

$$= ab(c \vee \bar{c}) \vee a\bar{b}(c \vee \bar{c}) \vee \bar{a}bc \vee \bar{a}b\bar{c} \vee \bar{a}\bar{b}\bar{c}$$

$$= abc \vee ab\bar{c} \vee a\bar{b}c \vee a\bar{b}\bar{c} \vee \bar{a}bc \vee \bar{a}b\bar{c} \vee \bar{a}\bar{b}\bar{c}.$$

3.4 MINTERMS AND MAXTERMS†

Definition 3.4.1 A product term of n variables is called a *minterm* of n variables.

Thus $\bar{a}\bar{b}\bar{c}$ and $a\bar{b}\bar{c}$ are minterms of 3 variables.

Since, each variable can appear in a minterm in one of the two forms, true or complemented, the number of all possible minterms of n variables is 2^n.

If a 0 is written for a complemented variable and a 1 for the uncomplemented variable, then each minterm can be expressed as a binary number. Each minterm is then designated as m_i, where the subscript i is the decimal value of the binary number. These are shown in Table 3-1 for the three variable case.

Definition 3.4.2 A sum term of n variables is called a *maxterm* of n variables.

† Minterms and maxterms are also called *fundamental products* and *fundamental sums* respectively.

TABLE 3-1

Minterms (n = 3)	Corresponding binary number representations	Corresponding symbolic representations
$\bar{a}\bar{b}\bar{c}$	000	m_0
$\bar{a}\bar{b}c$	001	m_1
$\bar{a}b\bar{c}$	010	m_2
$\bar{a}bc$	011	m_3
$a\bar{b}\bar{c}$	100	m_4
$a\bar{b}c$	101	m_5
$ab\bar{c}$	110	m_6
abc	111	m_7

Thus $\bar{a} \vee b \vee c \vee \bar{d}$ and $a \vee b \vee c \vee d$ are maxterms of 4 variables.

Like the minterms, the number of maxterms is also 2^n for n variables. It is customary to designate each maxterm by M_i, where i is the decimal equivalent of the binary number representation of a maxterm as shown in Table 3-2 for the 3 variable case. It should be noted that in the maxterms we have written a 0 for a variable in the true form, and a 1 for a variable in the complemented form. This is the opposite of what we had followed in case of minterms.

It can be easily verified that the complement of m_i is M_i and *vice-versa*, that is,

$$\bar{m}_i = M_i \tag{3.1}$$

TABLE 3-2

Maxterms (n = 3)	Corresponding binary number representations	Corresponding symbolic representations
$a \vee b \vee c$	000	M_0
$a \vee b \vee \bar{c}$	001	M_1
$a \vee \bar{b} \vee c$	010	M_2
$a \vee \bar{b} \vee \bar{c}$	011	M_3
$\bar{a} \vee b \vee c$	100	M_4
$\bar{a} \vee b \vee \bar{c}$	101	M_5
$\bar{a} \vee \bar{b} \vee c$	110	M_6
$\bar{a} \vee \bar{b} \vee \bar{c}$	111	M_7

and

$$\overline{M}_i = m_i. \tag{3.2}$$

For $n \leqslant 3$, a minterm or a maxterm can be represented on a Venn diagram. It is interesting to see that *a minterm occupies a minimum area on the Venn diagram, whereas a maxterm occupies the maximum area.* (Fig.3-1.)

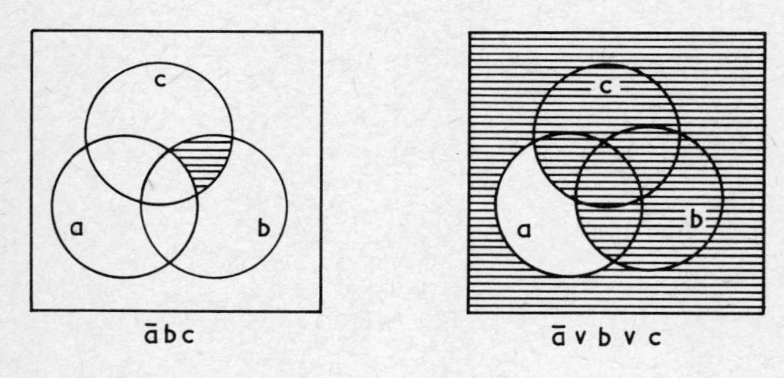

Figure 3-1 m_3 and M_4 on the Venn diagram.

From the definitions of the minterm and maxterm it is evident that *the disjunctive canonical form of a Boolean function is a sum of minterms, and the conjunctive canonical form is a product of maxterms.* Thus

$$f_1 = \bar{a}\bar{b}\bar{c} \vee \bar{a}\bar{b}c \vee abc$$

$$= m_0 \vee m_1 \vee m_7$$

and

$$F_2 = (a \vee b \vee c)\,(\bar{a} \vee \bar{b} \vee c)\,(\bar{a} \vee \bar{b} \vee \bar{c})$$

$$= M_0 M_6 M_7.$$

This has led to another convenient way of writing the canonical forms, wherein the m's or M's are not written at all. Thus

$$f_1 = \vee(0, 1, 7),$$

and

$$F_2 = \Pi(0, 6, 7).$$

Here the decimal numbers (which are the subscripts of the m's or M's) identify the particular terms, and $\vee$ indicates that the terms are minterms and the function is a summation, whereas Π indicates that the terms are maxterms and the function is a product.

The minterms and maxterms have certain distinctive properties which are very useful in the manipulation of Boolean functions. These are now discussed in the following theorems.

Theorem 3.4.1 The sum of all 2^n minterms of n variables is 1. That is,

$$\bigvee_{i=0}^{2^n-1} m_i = 1.$$

Proof: By Cor. 3.3.1A, 1 is the sum of 2^n product terms of n variables. Again by Def. 3.4.1 each of these 2^n product terms is a minterm. Hence the theorem is proved. ▲

Theorem 3.4.2 The product of any two n-variable minterms, which are different, is 0. That is,

$$m_i m_j = 0 \text{ when } i \neq j.$$

Proof: Since $i \neq j$, there must be at least one variable which appears in the true form in one of the minterms and in the complemented form in the other. Hence

$$\begin{aligned} m_i m_j &= (x_1 x_2 \ldots x_r \ldots x_n)(x_1 x_2 \ldots \bar{x}_r \ldots x_n) \\ &= (x_1 x_2 \ldots x_{r-1} x_{r+1} \ldots x_n)(x_r \bar{x}_r) = 0. \quad ▲ \end{aligned}$$

Theorem 3.4.3 Let the set of all the 2^n minterms of n-variables be partitioned into two subsets, and let f_1 be the sum of members of one partition, so that $f_i = \bigvee m_i$ and f_j be the sum of members of the other partition so that $f_j = \bigvee m_j$, then

$$f_i = \bar{f}_j.$$

Proof: By the complementarity theorem

$$f_i = \bar{f}_j$$

if

$$f_i \vee f_j = 1 \text{ and } f_i f_j = 0.$$

Now,

$$f_i \vee f_j = \bigvee m_i \vee \bigvee m_j = \bigvee_{p=0}^{2^n-1} m_p = 1 \qquad \text{by Th.3.4.1}$$

and

$$f_i f_j = \bigvee_{i \neq j} m_i m_j = 0 \qquad \text{by Th.3.4.2}$$

therefore,

$$f_i = \overline{f_j}. \quad \blacktriangle$$

Theorem 3.4.4 $m_i^D = M_{2^n-1-i}$.

Proof: When the dual of a minterm is taken, its .'s are changed to v's. Consequently, the product form of a minterm, becomes a sum form, which is the form of a maxterm. Again, due to dualization the variable which is in true form in the minterm remains in the true form in the maxterm. Now, while expressing a literal in 0 or 1, since the convention followed in the maxterm is opposite to that followed in the minterm, the 0's and 1's in the binary representation of the minterm are changed to 1's and 0's respectively in the binary representation of the maxterm. Thus if the minterm m_i is changed into the maxterm M_p, then the arithmetic sum of i and p will be the binary number with all 1's. The decimal equivalent of this number is $2^n - 1$. Hence

$$i + p = 2^n - 1$$

or,

$$p = 2^n - 1 - i.$$

Hence

$$m_i^D = M_{2^n-1-i}. \quad \blacktriangle$$

The above theorems show the important properties of minterms. Maxterms also exhibit similar properties. These are stated in the following four theorems.

Theorem 3.4.5 The product of all 2^n maxterms of n variables is 0. That is,

$$\prod_{i=0}^{2^n-1} M_i = 0.$$

Theorem 3.4.6 Sum of any two n-variable maxterms, which are different, is 1. That is,

$$M_i \vee M_j = 1, i \neq j.$$

Theorem 3.4.7 Let the set of all the n-variable maxterms be partitioned into two subsets, and let F_i be the product of all members of one partition, so that $F_i = \Pi M_i$, and F_j be the product of all members of the other partition, so that $F_j = \Pi M_j$, then

$$F_i = \overline{F_j}.$$

Theorem 3.4.8 $M_i^D = m_{2^n-1-i}$.

Although all these theorems can be proved independently, this is not really necessary. It can be easily seen that these theorems regarding maxterms are the duals of the theorems regarding minterms.

3.5 INTERCONVERSION BETWEEN THE DISJUNCTIVE AND CONJUNCTIVE CANONICAL FORMS

It has been mentioned earlier that the disjunctive canonical form is the same as the sum of minterms and the conjunctive canonical form is the same as the product of maxterms. It is interesting to note that if a function is expressed in the DCF, it can also be expressed in the CCF, and *vice-versa*. This is illustrated by working out a few examples.

Example 3.5.1 Express the following functions in the other type of canonical form.

a) $f_1 = \bar{a}bc \vee a\bar{b}\bar{c} \vee abc$

b) $f_2 = \bar{a}\bar{b}\bar{c}\bar{d} \vee \bar{a}b\bar{c}d \vee \bar{a}bc\bar{d}$

c) $F_3(abc) = M_3M_4M_5.$

Solution:

a) $f_1(abc) = m_3 \vee m_4 \vee m_7$

The complementary function, $f_j(abc) = m_0 \vee m_1 \vee m_2 \vee m_5 \vee m_6$ by Th.3.4.3. Again

$$\bar{f_j} = \bar{m}_0\bar{m}_1\bar{m}_2\bar{m}_5\bar{m}_6$$

$$= M_0M_1M_2M_5M_6.$$

Now, by Th. 3.4.3, $f_i = \bar{f_j}$, hence, $f_1 = M_0M_1M_2M_5M_6$.

b) $f_2(abcd) = m_0 \vee m_5 \vee m_6 = \vee\,(0, 5, 6)$

hence the complementary function,

$$f_j(abcd) = \vee\,(1, 2, 3, 4, 7, 8, 9, 10, 11, 12, 13, 14, 15).$$

Now,

$$\bar{f_j} = \bar{m}_1\bar{m}_2\bar{m}_3\bar{m}_4\bar{m}_7\bar{m}_8\bar{m}_9\bar{m}_{10}\bar{m}_{11}\bar{m}_{12}\bar{m}_{13}\bar{m}_{14}\bar{m}_{15}$$

$$\therefore\ f_2 = \bar{f_j} = \Pi(1, 2, 3, 4, 7, 8, 9, 10, 11, 12, 13, 14, 15).$$

c) $F_3(abc) = M_3M_4M_5 = \Pi(3, 4, 5)$

hence the complementary function,

$$F_j(abc) = \Pi(0, 1, 2, 6, 7).$$

Now

$$\bar{F_j} = \vee(0, 1, 2, 6, 7)$$

$$\therefore\ F_3 = \bar{F_j} = \vee(0, 1, 2, 6, 7).$$

The result of these examples can be stated in the form of a theorem as follows.

Theorem 3.5.1 A Boolean function expressed as a sum of minterms or as a product of maxterms can be converted into the other form as given by

$$\vee m_i = \Pi M_j$$

$$\Pi M_i = \vee m_j.$$

where the subset $\{i\}$ and the subset $\{j\}$ are two partitions of the entire set of 2^n subscripts of either m's or M's. The procedure to be followed in proving this theorem will be obvious from the worked out examples above.

We are now in a position to state a very important theorem.

Theorem 3.5.2 Any Boolean function can be expressed in both the disjunctive and the conjunctive canonical forms.

Proof: The validity of the theorem follows as a direct consequence of Theorems 3.3.3 and 3.5.1. ▲

3.6 TRUTH TABLE

Let $f(abc) = m_0 \vee m_3 \vee m_5 \vee m_7$ be a Boolean function expressed as a sum of minterms. Now, f can also be written as follows:

$$f = 1 \cdot m_0 \vee 0 \cdot m_1 \vee 0 \cdot m_2 \vee 1 \cdot m_3 \vee 0 \cdot m_4 \vee 1 \cdot m_5 \vee 0 \cdot m_6 \vee 1 \cdot m_7.$$

This shows that the function will be 1, only when any one of the minterms m_0, m_3, m_5, or m_7 is 1. If any other minterm becomes 1 the function is 0. The variables of the function, a, b, and c, can form 2^3 or 8 different combinations, since each of them can be either in the true or in the complemented form.

These eight different combinations can be written in the form of a table as shown in Table 3-3. The particular function then takes the form of a column of

TABLE 3-3

	abc	f
m_0	000	1
m_1	001	0
m_2	010	0
m_3	011	1
m_4	100	0
m_5	101	1
m_6	110	0
m_7	111	1

1's and 0's. 1 is written against those rows representing the minterms with coefficient 1, and 0 for those rows representing minterms with coefficient 0. This is known as the *truth table* representation of the function. If a function is given as a product of maxterms, it can first be expressed as a sum of minterms, and then the truth table for the function can be written.

3.7 DOUBLE COMPLEMENTATION

So far, we have expressed Boolean functions as functions of Boolean variables in the true form. But nothing prevents the function from being expressed with some or all of the variables in the complemented form. For example, let $f(abc) = \vee(0, 1, 3, 6)$. Now, let $a = x$, $b = \bar{y}$, and $c = z$ then the function can be written as

$$f(abc) = f(x\bar{y}z) = \vee(0, 1, 3, 6).$$

The truth table representation of this function is as shown in Table 3-4(A). Note that the variable heading the second column of the truth table is $\bar{y}$ instead of y. If we now wish to change $\bar{y}$ to y at the head of the column, then the 0's of the column are to be changed into 1's and *vice-versa*. In this way the Boolean form of a row remains unchanged, and therefore, the function remains the same. For example the fifth row is 100, and its Boolean form is $x\bar{\bar{y}}\,\bar{z} = xy\bar{z}$. When $\bar{y}$ is replaced by y the binary form of the 5th row changes to 110, and the Boolean form becomes $xy\bar{z}$, which is the same as the form obtained previously. This process of complementing the variable at the head of a truth table column and then, also complementing the 0's and 1's of that column is called *double complementation.* As has been shown, the process of double complementation does not alter the intrinsic Boolean function although it may look like a new function in terms of its constituent minterms. Thus the function of Table 3-4(A) is

TABLE 3-4

(A)

$x\bar{y}z$	f
000	1
001	1
010	0
011	1
100	0
101	0
110	1
111	0

(B)

xyz	f
010	1
011	1
000	0
001	1
110	0
111	0
100	1
101	0

$f_1 = f(x\bar{y}z) = \vee(0, 1, 3, 6)$. After double complementation, the same function becomes,

$$f_2 = f(xyz) = \vee(2, 3, 1, 4) = \vee(1, 2, 3, 4).$$

but $f_1 = f_2$.

3.8 CHARACTERISTIC NUMBER: OCTAL DESIGNATION

When the truth table for a Boolean function is written, the particular function appears as an array of 1's and 0's in a single column. If this column is now written as a row with the topmost element of the column at the extreme right of the row, then we get a binary number, which is called the characteristic number of the function. Thus the function of Table 3.4(A) has the characteristic number

01001011.

Again, if the above function is expressed as a sum of minterms (in terms of *abc* or $x\bar{y}z$), it will be

$$f = m_0 \vee m_1 \vee m_3 \vee m_6.$$

This can be re-written as follows:

$$f = 0 \cdot m_7 \vee 1 \cdot m_6 \vee 0 \cdot m_5 \vee 0 \cdot m_4 \vee 1 \cdot m_3 \vee 0 \cdot m_2 \vee 1 \cdot m_1 \vee 1 \cdot m_0.$$

The characteristic number is then the coefficients of the minterms arranged as above, that is, in a descending order from left to right. The above approaches indicate as to how the characteristic number of a function can easily be written when the function either has been defined by a truth table, or has been expressed in its DCF.

This also shows a mathematical way of defining the characteristic number. Let the array of 1's and 0's under f in the truth table (where the decimal designations of the rows must be in order, 0, 1, 2 etc.) be considered as a column matrix. Let it be denoted by Y. Then, by the familiar transpose operation on Y, we can change this column matrix into a row matrix. Let us now define a new operation, which we shall call the reverse or reflection operation.

Definition 3.8.1 The *reverse* or *reflection operation,* denoted by the superscript R, reverses the elements of a column or a row matrix, from top to bottom, or from right to left.

Thus,

$$\begin{bmatrix} x_1 \\ x_2 \\ \cdot \\ \cdot \\ \cdot \\ \cdot \\ x_m \end{bmatrix}^R = \begin{bmatrix} x_m \\ \cdot \\ \cdot \\ \cdot \\ \cdot \\ x_2 \\ x_1 \end{bmatrix}$$

and,

$$[x_1 x_2 \ldots . x_n]^R = [x_n \ldots . x_2 x_1].$$

It can easily be verified that if

$$(P)^R = Q, \text{ then } (Q)^R = P \tag{3.3}$$

also

$$(P^R)^T = (P^T)^R. \tag{3.4}$$

Definition 3.8.2 If Y is the column matrix denoting a Boolean function in its truth table, then its *characteristic number*, N is given by the row matrix,

$$N = (Y^T)^R = (Y^R)^T.$$

It is interesting to see now that once the characteristic number of a function is known, it is very easy to write its CCF, i.e., the product-of-maxterms form.

Theorem 3.5.1 gives the relation between the DCF and CCF of a Boolean function, it is

$$\vee m_i = \Pi M_j.$$

Now, in the characteristic number, there are 1's for the i terms. Hence the 0's are j terms. Therefore those maxterms which appear in the CCF can be found by knowing the values of j's. Thus the CCF can be written by noting the locations of 0's either in the Y matrix of a truth table or in the characteristic number.

Example 3.8.1 A three variable Boolean function has a characteristic number 10101100. Find its DCF and CCF.

Solution: The eight positions will be occupied by the minterms and maxterms as follows:

m_7	m_6	m_5	m_4	m_3	m_2	m_1	m_0
M_7	M_6	M_5	M_4	M_3	M_2	M_1	M_0
1	0	1	0	1	1	0	0

Those minterms where the characteristic number has 1's will appear in the DCF of the function, and those maxterms where the characteristic number has 0's will appear in the CCF of the function. Hence,

$$f = m_2 \vee m_3 \vee m_5 \vee m_7 \quad \text{(DCF)}$$

$$= M_0 M_1 M_4 M_6 \quad \text{(CCF).}$$

This shows that the characteristic number is a very convenient way of designating a function as it has inherent in it the DCF, the CCF, and also the truth table representation of the function.

A compact way of remembering or writing the characteristic number is to express the binary number as an octal number. From our knowledge of interconversion between the binary and octal forms (Sec.0.7, Chapter 0), this is easily carried out by bunching the bits of binary number in groups of 3. In the three variable case, since there is no ninth position, it is filled by a 0, while converting from binary to octal, and the 0 at the ninth position is omitted after converting from octal to binary.

Example 3.8.2 Express the following function in its CCF and find its octal designation.

$f = \vee(2, 3, 6)$.

Solution: It is a three variable function. Therefore, its characteristic number will be determined by writing 1's below m_2, m_3, and m_6, and 0's below others

m_7	m_6	m_5	m_4	m_3	m_2	m_1	m_0
0	1	0	0	1	1	0	0

Noting the 0's in the characteristic number, the CCF of the function is $M_0 M_1 M_4 M_5 M_7$. Since, $(001, 001, 100)_2 = (114)_8$, its octal designation is 114.

Example 3.8.3 Express the function, whose octal designation is 256, in both types of canonical forms.

Solution: $(256)_8 = (010, 101, 110)_2$. Therefore, it is a three variable function and its characteristic number (omitting the 0 at the extreme left) is 10101110. Therefore, its DCF $= m_1 \vee m_2 \vee m_3 \vee m_5 \vee m_7$, and CCF $= M_0 M_4 M_6$.

3.9 COMPLEMENTARY AND DUAL FUNCTIONS

Theorem 3.4.3 points out that to get the complement of a function, we have simply to change the 0's into 1's and 1's into 0's in the characteristic number. Let this operation be indicated by the superscript, C, and be defined as follows:

Definition 3.9.1 Let $A = [a_{ij}]$ be an $m \times n$ matrix, then $A^C = B = [b_{ij}]$ where B is also an $m \times n$ matrix and $b_{ij} = \overline{a_{ij}}$. It can easily be verified that if

$$A^C = B, \text{ then } B^C = A. \tag{3.5}$$

Again, if P is a column or a row matrix, then

$$(P^C)^R = (P^R)^C. \tag{3.6}$$

Also

$$A^C \vee A = [1] \tag{3.7}$$

and

$$A^C \cdot A = [0], \tag{3.8}$$

where $[1]$ and $[0]$ are matrices having the same order as A and whose elements are all 1's and 0's respectively.

Theorem 3.9.1 If N is the characteristic number of a function, then the characteristic number of the complementary function is N^C.

Proof: The validity of this theorem follows from the definition of the characteristic number, and Th. 3.4.3. ▲

The dual of a function can also be found by the following theorem.

Theorem 3.9.2 If N is the characteristic number of a function, then the characteristic number of its dual is $(N^R)^C$ or $(N^C)^R$.

Proof: Let a function f have a characteristic number N, then the function $\overline{f}$ has the characteristic number N^C and let the function whose characteristic number is $(N^C)^R$ be $\overline{f}_R$. Also, let $f = \vee m_i$ then $\overline{f} = \vee m_j$ by Theorem 3.4.3 and $\overline{f}_R = \vee m_{2^n-1-j}$. Now,

$$\begin{aligned} f^D &= \Pi m_i^D \\ &= \Pi M_{2^n-1-i} = \vee m_{2^n-1-j} \\ &= \overline{f}_R. \end{aligned}$$

Hence,

$$f^D = (N^C)^R = (N^R)^C. \quad ▲$$

Example 3.9.1 Find the complement and dual of the octal function 2563.

Solution: $(2563)_8 = (010, 101, 110, 011)_2$. Hence, this is a four variable function, and

$N = (0, 000, 010, 101, 110, 011)$.

Therefore, characteristic number of complementary function $\bar{f} = N^C = (1, 111, 101, 010, 001, 100)$, therefore, octal of $\bar{f}$ is 175214, characteristic number of the dual function $f^D = (N^C)^R = (0, 011, 000, 101, 011, 111$ Therefore, octal of f^D is 30537.

3.10 SELF-DUAL (SD) FUNCTIONS

Definition 3.10.1 If a Boolean function is equal to its dual, then the function is called a self-dual (SD) function.

If N be the characteristic number of an SD function, then by Th. 3.9.2

$$N = (N^C)^R = (N^R)^C. \tag{3.9}$$

Example 3.10.1 Verify if the octal function 115 is an SD function.

Solution: $(115)_8 = (001, 001, 101)_2$

$\therefore N = (01001101)$

$$\begin{aligned} \therefore (N^C)^R &= (10110010)^R \\ &= (01001101) \\ &= N \end{aligned}$$

$\therefore$ 115 is an SD function.

Alternatively:

$f = m_0 \vee m_2 \vee m_3 \vee m_6$

$$\begin{aligned} \therefore f^D &= m_0^D m_2^D m_3^D m_6^D \\ &= M_{7-0} M_{7-2} M_{7-3} M_{7-6} \\ &= M_7 M_5 M_4 M_1 \\ &= m_0 \vee m_2 \vee m_3 \vee m_6 \end{aligned}$$

$\therefore f = f^D$ and f is an SD function.

Theorem 3.10.1 There are exactly $2^{2^{n-1}}$ n-variable SD functions.

Proof: For an SD function,

$N = (N^R)^C$

r,

$$N^C = N^R. \tag{3.10}$$

et N be the binary number

$$b_{2^n-1} \cdots\cdots b_1 b_0.$$

hen, from Eqn.(3.10)

$$(b_{2^n-1} \cdots b_1 b_0)^C = (b_{2^n-1} \cdots b_1 b_0)^R$$

r,

$$\bar{b}_{2^n-1} \cdots\cdots \bar{b}_1 \bar{b}_0 = b_0 b_1 \cdots b_{2^n-1}. \tag{3.11}$$

qn.(3.11) can be satisfied, if

$$\bar{b}_{2^n-1} = b_0$$

$$\bar{b}_{2^n-1-1} = b_1.$$

hat is, in general, if

$$\bar{b}_{2^n-1-i} = b_i \quad 0 \leqslant i \leqslant 2^{n-1} - 1. \tag{3.12}$$

hus, in an SD function the values of $b_0, b_1, \ldots, b_{2^{n-1}-1}$ determines the values f $b_{2^n-1}, \ldots, b_{2^{n-1}}$. Since, the number of combinations of $b_0 \ldots b_{2^{n-1}-1}$ is 2^{n-1}, the number of n-variable SD functions is also $2^{2^{n-1}}$. ▲

OROLLARY **3.10.1A** If the characteristic number N of an SD function is vided into two halves, the left half N_l and the right half N_r, so that

$$N = N_l N_r$$

en

$$N_l = (N_r^C)^R = (N_r^R)^C$$

nd,

$$N_r = (N_l^C)^R = (N_l^R)^C.$$

Proof: The validity of this corollary follows from Eqn.(3.12). ▲

11 ALGEBRA OF BOOLEAN MATRICES

Chapter 2, we have discussed several operations concerning Boolean matrices. this chapter we have encountered three more operations, namely, transposition, versal, and complementation of a matrix. The transposition operation is the me as is found in ordinary matrix algebra.

It must be mentioned here that whereas the transpose and complement oper: tions are defined for a Boolean matrix with m rows and n columns, the reverse operation is defined only for a column or a row matrix. It should also be notice that the matrix A^C as defined in this section is different from the matrix $\overline{A}$ as defined in Chapter 2. Thus, as we have two ways of defining the product of tw matrices, so also we have two definitions for the complement operation.

The rules for these three operations can be summarized in the following equations:

Let A be an $m \times n$ Boolean matrix, [1] and [0] be matrices whose every element is 1 or 0 respectively, and X is either a column or a row matrix, then

$$(A^T)^T = A \qquad (3.1$$

$$(A^C)^C = A \qquad (3.1$$

$$(X^R)^R = X \qquad (3.1$$

$$(A^T)^C = (A^C)^T \qquad (3.1$$

$$(X^T)^R = (X^R)^T \qquad (3.1$$

$$(X^C)^R = (X^R)^C \qquad (3.1$$

$$A^C \vee A = [1] \qquad (3.1$$

$$A^C \cdot A = [0]. \qquad (3.2$$

REFERENCES

Chu, Y.: *Digital Computer Design Fundamentals,* McGraw-Hill Book Co., New York, 1962.

Phister, M. Jr.: *Logical Design of Digital Computers,* John Wiley and Sons, Inc., New York, 1959.

PROBLEMS

3.1 Express the following functions in the canonical forms:

a) $ab \vee c$

b) $\overline{a}bc \vee ab$

c) $\overline{a} \vee abcd$

d) $(\overline{a} \vee b)(\overline{b} \vee \overline{c})$

e) $(b \vee c)(a \vee b \vee c)$

f) $\overline{a \vee b} \vee \overline{c(a \vee b)}$

g) $a(b \vee c) \vee (c \vee a)(\overline{a} \vee b)$.

3.2 Write the following functions in the conjunctive canonical form:

a) $f_1 = \vee(1, 2, 4, 7)$

b) $f_2 = \vee(0, 2, 4, 6)$

c) $f_3 = \vee(1, 4, 7, 8, 11)$

d) $f_4 = \vee(5, 8, 12, 13, 15)$.

3.3 Express the following functions as sums of minterms:

a) $F_1 = M_1M_2M_3M_7$

b) $F_2 = M_0M_5M_9$

c) $F_3 = M_3M_4M_{10}M_{13}M_{15}$

d) $F_4 = M_1M_{12}$.

3.4 Write the DCF's of the following functions:

a	b	c	.	f_1	f_2	f_3	f_4
0	0	0	.	0	1	1	0
0	0	1	.	0	1	1	0
0	1	0	.	1	0	1	1
0	1	1	.	1	1	0	1
1	0	0	.	0	0	0	1
1	0	1	.	0	1	0	0
1	1	0	.	1	0	0	1
1	1	1	.	1	1	1	0

3.5 Find the complements of the following functions first by De Morgan's theorem, and then by Th. 3.4.3. Finally show that they turn out to be the same.

a) $f_1 = \vee(0, 2, 3)$

b) $f_2 = \vee(2, 3, 5, 7)$

c) $f_3 = \vee(2, 4, 6, 9, 12)$

d) $f_4 = \vee(3, 7, 8, 10, 13, 14, 15)$.

3.6 Write the expressions in problem 3.1 as a) sum-of-minterms, and b) product-of-maxterms.

3.7 Write the following functions in both types of canonical forms:

a	b	c	d	.	f_1	f_2	f_3	f_4
0	0	0	0	.	0	1	0	0
0	0	0	1	.	1	0	1	0
0	0	1	0	.	0	0	1	0
0	0	1	1	.	1	0	0	0
0	1	0	0	.	0	1	0	1
0	1	0	1	.	1	1	0	1
0	1	1	0	.	1	0	1	1
0	1	1	1	.	1	0	1	0
1	0	0	0	.	0	0	1	0
1	0	0	1	.	1	1	0	1
1	0	1	0	.	0	1	0	1
1	0	1	1	.	1	0	1	0
1	1	0	0	.	1	1	0	0
1	1	0	1	.	1	0	1	1
1	1	1	0	.	0	1	0	1
1	1	1	1	.	1	0	1	1

3.8 Express the following functions as sum of minterms, when the variables of the functions are all in the true form:

a) $f_1(ab\bar{c}) = \vee(1, 2, 5, 7)$

b) $f_2(\bar{a}\bar{b}c) = \vee(0, 1, 6)$

c) $f_3(\bar{a}\bar{b}cd) = \vee(1, 2, 3, 6, 8, 9, 11)$

d) $f_4(\bar{a}\bar{b}\bar{c}\bar{d}) = \vee(2, 5, 8, 10, 15)$.

3.9 Find if any two of the following functions are equal:

a) $f_1(abcd) = \vee(1, 2, 5, 9, 13, 15)$

b) $f_2(\bar{a}\bar{b}cd) = \vee(1, 3, 5, 8, 13, 14)$

c) $f_3(ab\bar{c}d) = \vee(0, 1, 4, 5, 9, 11)$.

3.10 Find the characteristic numbers of functions f_1 through f_4 of problem 3.2, and express them in octal numbers. With the help of the characteristic numbers express them as products of maxterms.

3.11 Solve problem 3.3 with the concept of characteristic number.

3.12 Express the following octal functions in both types of canonical forms:

a) 231; b) 5; c) 47; d) 126.

3.13 Find the octal designations of the following functions:

a) $bc \vee \bar{b}\bar{c}$

b) $\bar{a}\bar{b}c$

c) $\bar{a}b \vee a\bar{b} \vee c$

d) $abc \vee a\bar{b}\bar{c} \vee \bar{a}b\bar{c} \vee \bar{a}\bar{b}c$.

3.14 Find the complements and the duals of the following functions:

a) 23; b) 104; c) 251; d) 374; e) 666; f) 12534.

3.15 Show that if a 3-variable function is given in the octal form, its complement can be found by subtracting the octal number from 377. What will be the corresponding number in a 4-variable function?

3.16 Apply the result of problem 3.15 to find the complements of the functions of problem 3.14.

3.17 Show that if $f_1 \geqslant f_2$, then

$$f_1^D \leqslant f_2^D.$$

3.18 If either $f \leqslant g$, or $g \leqslant f$, then f and g are said to be *comparable*. Show that f_1 and f_2 are comparable, where

$$f_1 = x_1 \vee x_2$$

and

$$f_2 = x_1 \vee x_2 x_3.$$

3.19 If either $f \geqslant f^D$ or $f \leqslant f^D$, then f is said to be *dual comparable*. Show that $f = x_1 \vee x_2$ is dual comparable.

Chapter 4

MINIMIZATION METHODS

4.1 INTRODUCTION

One of the main objects of a logic designer is to realize a Boolean function with minimum cost. Let us study this aspect of the problem with an example. Consider the realization of the following Boolean function:

$$f = a\bar{b} \vee ab \vee \bar{a}bc. \tag{4.1}$$

Figure 4-1(A) gives the realization of this function with series–parallel contact network, and Fig.4-1(B) shows the realization with two level AND and OR gates.

The first circuit needs 7 contacts. The second circuit needs 10 diodes if the gates are diode gates. But the function f can be simplified or minimized as follows:

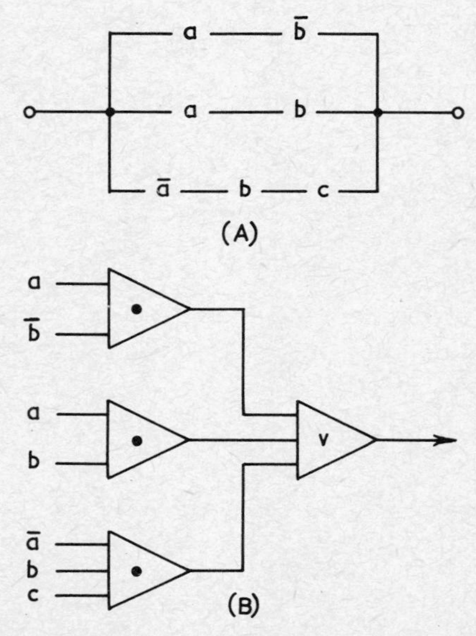

Figure 4-1 Realization of $f = a\bar{b} \vee ab \vee \bar{a}bc$ (A) by contact network, and (B) by gate network.

$$f = a\bar{b} \vee ab \vee \bar{a}bc$$
$$= a(\bar{b} \vee b) \vee \bar{a}bc$$
$$= a \vee \bar{a}bc$$
$$= a \vee bc. \qquad (4.2)$$

So the same function can be realized by the circuits as shown in Fig.4-2. These two circuits require less number of contacts and diodes. These are, therefore, cheaper. However, to obtain these more economical circuits, it is necessary to reduce the function f as given by Eqn.(4.1) to the form as given in Eqn.(4.2). It can be seen that in both these forms, the function is expressed as a sum of a number of product terms. But the number of literals in Eqn.(4.2) is less than that in Eqn.(4.1). One of the aims, therefore, must be to obtain a form for function f where the number of literals is the minimum. For every function, there exists such a form, and it is called the *minimal sum-of-products (MSP) form* of the function. Similarly, every function can also be expressed in the *minimal product-of-sums (MPS) form.* The various methods of achieving these minimal forms are discussed in this chapter.

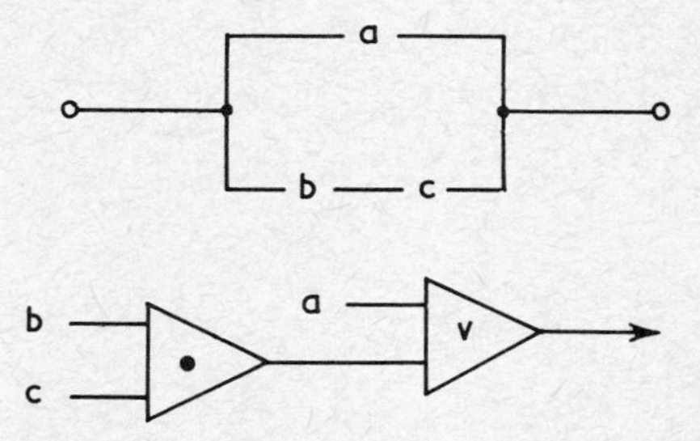

Figure 4-2 Realization of the function of Fig.4-1 by more economical circuits.

The method by which Eqn.(4.1) has been reduced to Eqn.(4.2) is the algebraic method. In this method the function is simplified algebraically by utilizing the theorems, notably the following:

$$ab \vee a\bar{b} = a \qquad (4.3)$$

$$a \vee ab = a \qquad (4.4)$$

$$a \vee \bar{a}b = a \vee b. \qquad (4.5)$$

This method is however not very efficient as there is no definite rule to follow, but only experience and intelligent guess can lead to a minimal form. Consequently, this method has only an academic value, and is not used in practice, where systematic procedures leading to the MSP or MPS forms are sought for. However, before proceeding to discuss various methods of obtaining the minimal forms, it is essential to know how a given Boolean function can be represented.

4.2 GEOMETRICAL REPRESENTATION OF BOOLEAN FUNCTIONS: THE *n*-CUBE

Consider the 8 minterms of three variables, the variables being x, y, and z. Then each minterm may be considered to be a point in the three dimensional space as shown in Fig.4-3(A). The binary bits designating a minterm give the x, y, and z co-ordinates of the point. Thus the eight minterms make up the 3 dimensional cubical structure known as a 3-cube. Similarly, the 4 minterms of

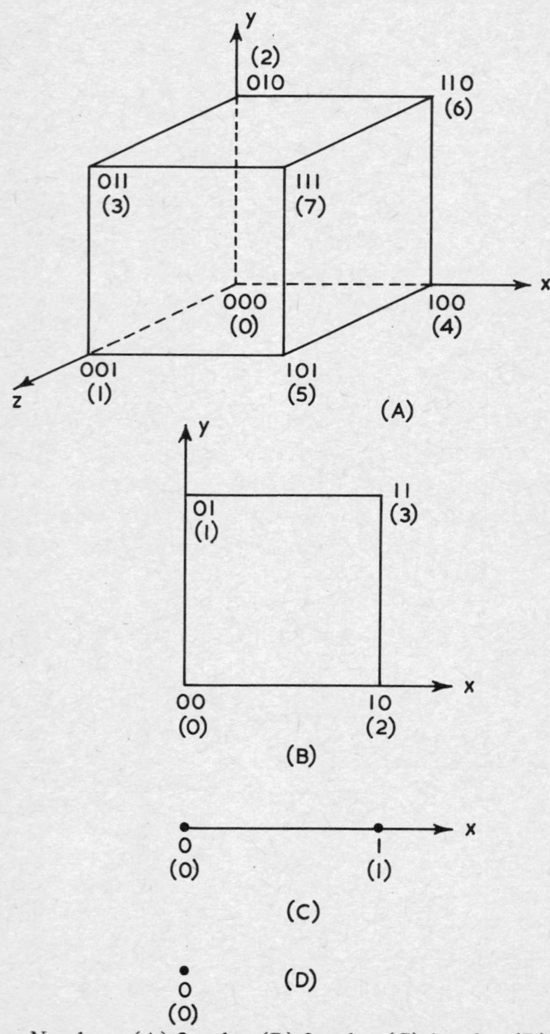

Figure 4-3 *N*-cubes. (A) 3-cube, (B) 2-cube, (C) 1-cube, (D) 0-cube.

two variables, and the two minterms of one variable define a 2-cube and a 1-cube respectively (Figures 4-3(B) and (C)). Generalizing this concept, the 2^n minterms of n-variables define an n-cube, each minterm being a vertex of the n-cube. The vertex occupied by a minterm for which the function is 1 is known as a *true vertex,* whereas a vertex where the function is 0 is called a *false vertex.* While it is possible to draw on a paper the 0, 1, 2, and 3-cubes, an n-cube for which n is greater than 3, cannot be drawn.

Each face of a 3-cube forms a 2-subcube, and each edge of a 3-cube forms a 1-subcube. Thus each n-cube has a number of p-subcubes $(0 \leqslant p < n)$. A three variable Boolean function can be minimized very easily by recognizing the subcubes formed by the true vertices of the 3-cube.

Example 4.2.1 Minimize $f(xyz) = \vee(2, 3, 4, 6, 7)$.

Solution: For this example, the vertices 2, 3, 4, 6, and 7 of the 3-cube are the true vertices and have been shown by the bold dots in Fig.4-4. It can be easily seen that the vertices, 2, 3, 6, and 7 form a 2-subcube, which can be represented by the single literal y. (In the geometrical figure the 2-subcube represents a plane whose equation is $y = 1$.) Also, the vertices 6 and 4 form a 1-subcube, $x\bar{z}$ (the equation of the straight line joining the vertices 4 and 6 is $x = 1, z = 0$). Hence,

$$f(xyz) = \vee(2, 3, 4, 6, 7) = y \vee x\bar{z}.$$

Although the 3-cube comes very handy in minimizing a three-variable function, it is obvious that the recognition of subcubes for n-cubes with $n > 3$ becomes difficult. For this reason, this method is not very much used. Nevertheless, the concept of an n-cube is very helpful in understanding the various techniques for handling Boolean functions. Some of the properties of an n-cube may be summarized as follows:

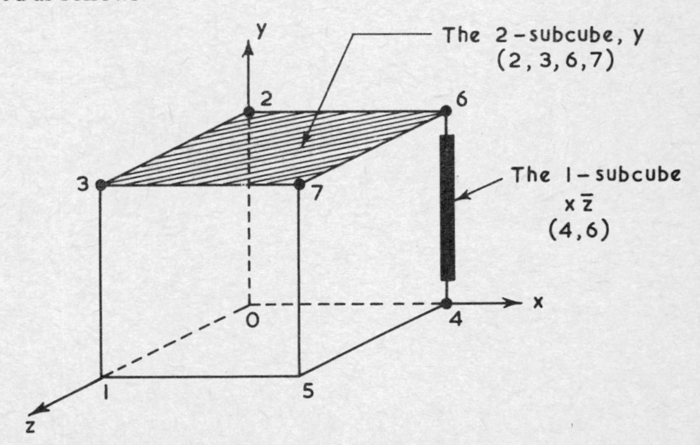

Figure 4-4 Minimization of $f(xyz) = \vee(2, 3, 4, 6, 7)$.

1) An n-cube has 2^n vertices. Each vertex represents a minterm, that is, a product term of n variables.

2) Each vertex of an n-cube is adjacent to n other vertices of the n-cube.

3) 2^p $(0 < p < n)$ mutually adjacent vertices form a p-subcube of the n-cube. A p-subcube represents a product term of $(n - p)$ variables.

4.3 VEITCH–KARNAUGH MAP (VEITCH, 1952; KARNAUGH, 1953)

In this section we shall see how a function can be represented on a map. Such a map was first suggested by Veitch (1952) and then was reorganized by Karnaugh (1953). We know from Th.3.5.2 that any Boolean function can be expressed in the disjunctive canonical form or as a sum of minterms. We also know that the number of minterms of n variables is 2^n. The Veitch–Karnaugh map for n variables has 2^n cells, so that each cell represents one minterm. Theoretically it is possible to draw a V–K map for any number of variables. In practice a map for more than 6 variables is seldom used, whereas 2, 3, and 4 variable maps are extensively used.

A two-variable map is shown in Fig.4-5. Figure 4-5(A) shows that the variable a appears as $\bar{a}$ in the 1st column and as a in the second column. Similarly the variable b appears as $\bar{b}$ in the 1st row and as b in the second row. Usually columns and rows for only the true form of the variables are so marked; the rest are meant by implication. The four minterms $\bar{a}\bar{b}$ through ab are located as shown. The columns and rows can also be designated by the binary forms of the variables as shown in Fig.4-5(B). Here, the variable a appears as 0 in the 1st column, and as 1 in the 2nd column. Similarly the variable b appears as 0 in the first row, and as 1 in the second row. As a result, cells for the four minterms in the binary form 00 through 11 get fixed up as shown in Fig.4-5(B). The two maps as shown in (A) and (B) differ only in nomenclature, but they depict and represent the same thing. Each cell of the map can also be recognized by the decimal designation of the minterm it represents. This is shown in Fig.4-5(C). Remembering the decimal designations of the cells can be very helpful in plotting a Boolean function on the map, when the function is given as a sum of minterms, and the minterms themselves are given in decimal designations. The corresponding forms of maps for three and four variables are shown in Fig.4-6 and 4-7.

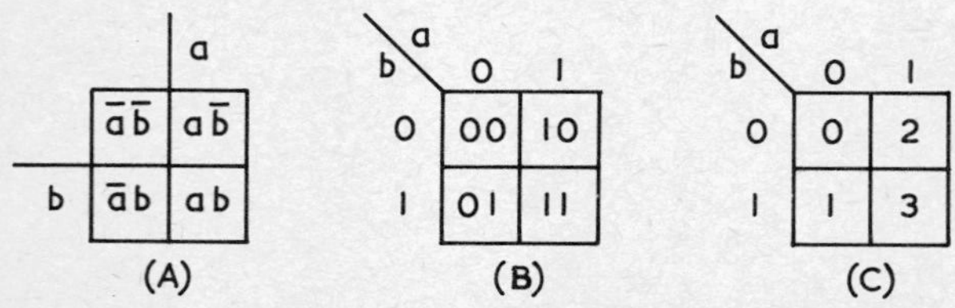

Figure 4-5 Two-variable Veitch–Karnaugh map.

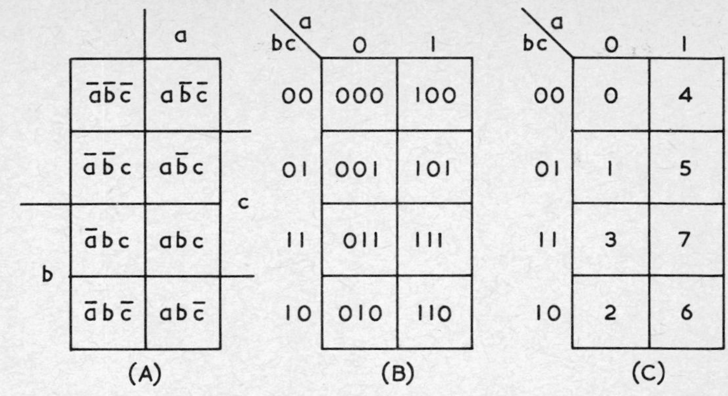

Figure 4-6 Three-variable Veitch–Karnaugh map.

cd \ ab	00	01	11	10
00	0	4	12	8
01	1	5	13	9
11	3	7	15	11
10	2	6	14	10

Figure 4-7 Four-variable Veitch–Karnaugh map.

One important point that must be noticed in the various V–K maps is the following. Looking at Fig.4-6(B), the variables *bc*, takes up the values 00, 01, 11, 10, in the 1st, 2nd, 3rd and 4th rows respectively. In this arrangement the designation of only one variable changes while going from one row to its adjacent row. Similar procedure has been followed in case of the 4 rows and 4 columns of the 4-variable map too. We can now define a very important concept on these maps.

Definition 4.3.1 Two rows/columns of a Veitch–Karnaugh map are said to be *adjacent* to each other if their variable coordinates differ in one variable only.

It should be noted that all rows/columns in the 2, 3 and 4 variable maps which are physically adjacent are adjacent by this definition also. In addition, according to the above definition, the top and bottom rows of the 3 and 4 variable maps, the extreme left and extreme right columns of the 4-variable map are also adjacent, although they do not appear so physically on these two-dimensional diagrams.

If a Boolean function is given in its DCF, it can be very easily represented on the map by writing 1's on those cells for which the function is 1. No 0's are usually written on those cells for which the function is 0, but this is meant by implication.

Example 4.3.1 Represent the following functions on the Veitch–Karnaugh map.

a) $f_1(abc) = \vee(0, 1, 5, 7)$

b) $f_2(abcd) = \vee(2, 4, 6, 9, 11, 12, 15)$.

Solution: The functions f_1 and f_2 are as shown in Fig.4-8(A) and (B).

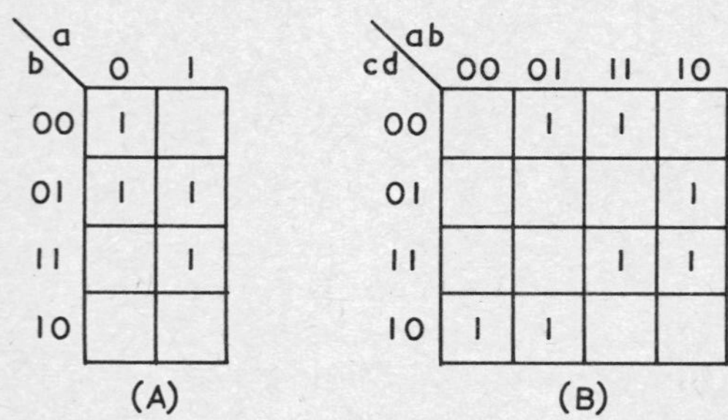

Figure 4-8 (A) $f_1(abc) = \vee(0, 1, 5, 7)$ plotted on a three-variable map. (B) $f_2(abcd) = \vee(2, 4, 6, 9, 11, 12, 15)$ plotted on a four-variable map.

If a function is not given in its DCF, it can be first expressed in its DCF and then plotted directly. For this purpose let us identify some typical functions as plotted on 4-variable maps of Fig.4-9. The function of Fig.4-9(A) is

$$f_1(abcd) = \bar{a}b\bar{c}\bar{d} \vee \bar{a}b\bar{c}d.$$

As could be seen these two minterms differ only in one variable. This is because they are in the same column, which means their designations for the variables a and b will be the same viz, $\bar{a}b$. As they are on adjacent rows, they will differ only in one of the variables c or d. Here they differ in variable d.

$$\therefore f_1 = \bar{a}b\bar{c}(\bar{d} \vee d)$$
$$= \bar{a}b\bar{c}.$$

Thus f_1 becomes a product term of 3 variables. This will be true for any other pair of 1's plotted on two adjacent rows/columns. The variable which differs among the two rows/columns gets itself eliminated from the term. This leads to the conclusion that a *three variable product term can be plotted on a 4-variable map by writing two 1's on two adjacent rows or columns as need be.*

The function of Fig.4-9(B), can be considered to be two three-variable product terms plotted on two columns but in the same two rows as indicated by

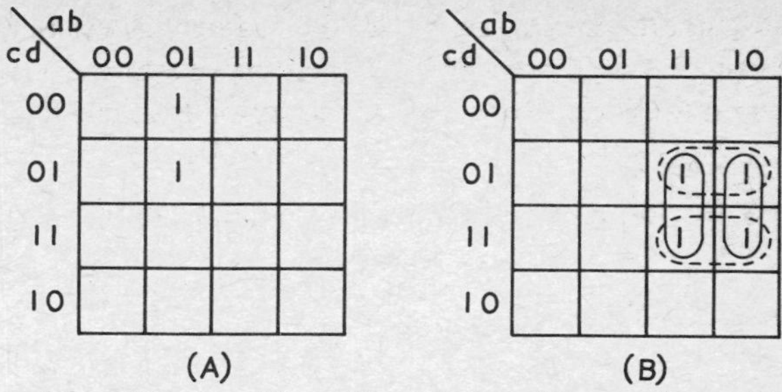

Figure 4-9 Cluster of two 1's and four 1's on a 4-variable map.

the solid circles. Then $f_2 = abd \vee a\bar{b}d = ad(b \vee \bar{b}) = ad$. The four 1's can also be considered as two three-variable product terms indicated by the two dotted circles. These are then $a\bar{c}d$ and acd. But f_2 again turns out to be the same, for

$$a\bar{c}d \vee acd = ad.$$

Thus this cluster of four 1's represents a product term of two variables. Here again, the variables absent are those which differ among the rows and columns. Thus *a two-variable product term can be plotted on a four-variable map by a cluster of four 1's shared by two adjacent rows and two adjacent columns.*

Similarly, it can be shown that *a cluster of eight 1's on two adjacent rows or columns of a 4-variable map represents a single variable which remains invariant*

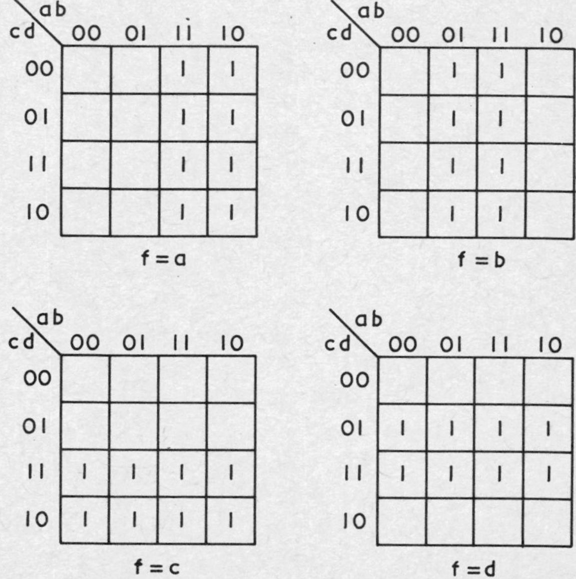

Figure 4-10 Variables a, b, c, and d plotted on a 4-variable map.

on all these cells. Figure 4-10 shows the clusters of eight representing the variables, a, b, c and d on a four-variable V–K map. These show that the Veitch–Karnaugh map is an ingenious Venn diagram where the rectangles with eight cells as shown represent a variable. The entire area containing the 16 cells represents the universal set. A product term can also be plotted by considering it as intersection of these rectangular areas. Thus the term $\bar{a}bc$ is the intersection of $\bar{a}$, b and c rectangles. Therefore, it can be plotted by two 1's defining the intersection as shown in Fig.4-11.

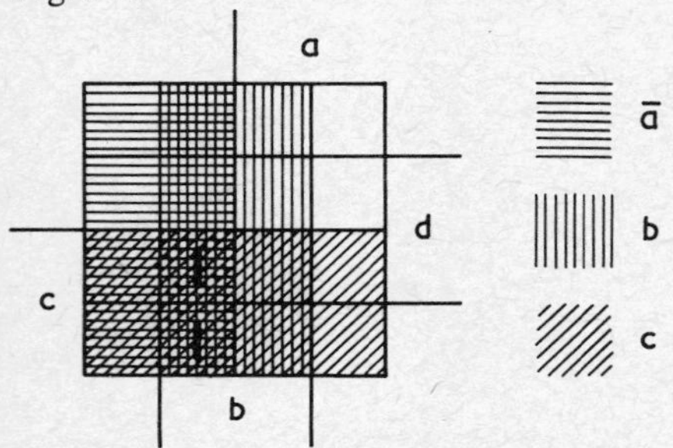

Figure 4-11 $\bar{a}bc$ as intersection of the three rectangles representing $\bar{a}$, b, and c.

The general rule for plotting a product term of m literals on an n-variable map ($1 \leqslant m \leqslant n$) can be stated as follows:

Rule 4.3.1: Recognize the 2^{n-1} cells on which the first literal has the value 0 or 1, (i.e., complemented or true) in the given term. Out of these 2^{n-1} cells recognize the 2^{n-2} cells on which the second literal has the given value (0 or 1). Repeat this procedure until the 2^{n-m} cells for which the mth literal with the given value are recognized. Plotting 1's on these 2^{n-m} cells completes the plotting of the product term of m literals.

Many problems regarding Boolean functions can be solved by simply plotting them on the map. A given function can be expressed in the canonical form by plotting it on the map.

Example 4.3.2 Express the following functions in their DCF's and determine if there exists any relation between them.

$f_1 = b(\bar{a} \vee \bar{c} \vee d) \vee \bar{a}cd$

$f_2 = \bar{a}b\bar{c}\bar{d} \vee d(ab \vee bc)$

$f_3 = \bar{a}b\bar{d} \vee bd \vee \bar{a}\bar{b}cd \vee ab\bar{c}\bar{d}$

Solution: Expressing in the SP form,

$f_1 = \bar{a}b \vee b\bar{c} \vee bd \vee \bar{a}cd$

$f_2 = \bar{a}b\bar{c}\bar{d} \vee abd \vee bcd.$

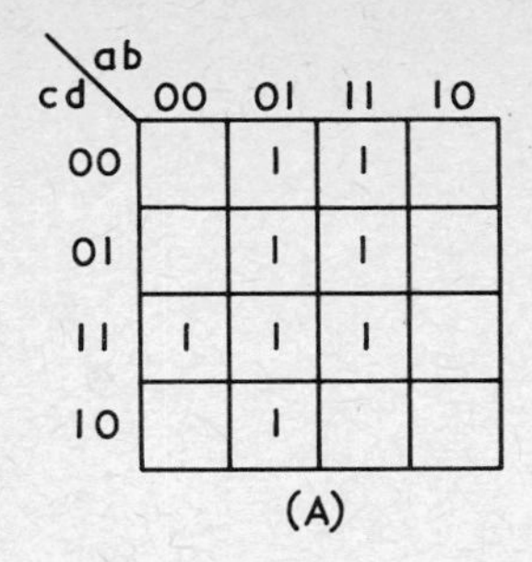

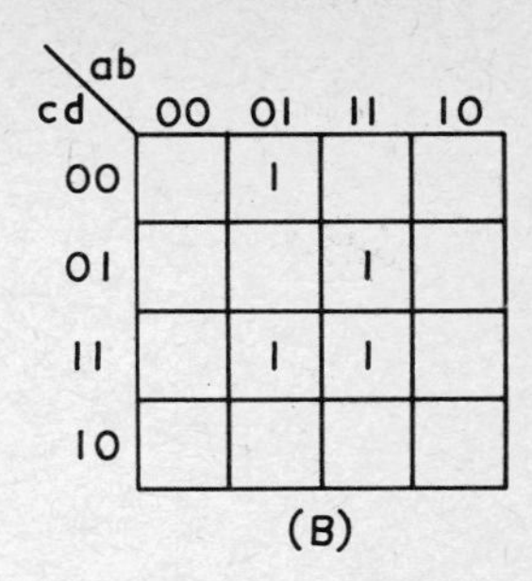

cd \ ab	00	01	11	10
00		1	1	
01		1	1	
11	1	1	1	
10		1		

(C)

Figure 4-12 (A) $f_1 = \bar{a}b \vee b\bar{c} \vee bd \vee \bar{a}cd$, (B) $f_2 = \bar{a}b\bar{c}\bar{d} \vee abd \vee bcd$, and (C) $f_3 = \bar{a}b\bar{d} \vee bd \vee \bar{a}\bar{b}cd \vee ab\bar{c}\bar{d}$ plotted on maps.

f_1, f_2 and f_3 are now plotted on maps as shown in Fig.4-12. Hence the DCF's of the functions are

$$f_1 = \vee(3, 4, 5, 6, 7, 12, 13, 15)$$

$$f_2 = \vee(4, 7, 13, 15), \text{ and}$$

$$f_3 = \vee(3, 4, 5, 6, 7, 12, 13, 15).$$

If the sets of minterms constituting f_1, f_2 and f_3 are $\{m(f_1)\}$, $\{m(f_2)\}$, and $\{m(f_3)\}$ respectively, then it can be seen that

$$\{m(f_1)\} = \{m(f_3)\}$$

$$\therefore f_1 = f_3.$$

Again $\{m(f_2)\} \subset \{m(f_1)\}$

$$\therefore f_2 \leqslant f_1 = f_3$$

or $\quad f_2 \leqslant f_1$

and $\; f_2 \leqslant f_3.$

4.4 PRIME IMPLICANTS

Consider the Boolean function $f = a\bar{b}\bar{c} \vee ac \vee bc$. The function is a sum of three product terms. Let t be any of these terms, then

$$t \vee f = f.$$

Hence by Def.1.6.1, $t \leqslant f$, and t implies f. For this reason t is called an *implicant* of the function.

Definition 4.4.1 A product term which implies a Boolean function is called an *implicant* of the given function.

In the above mentioned function, each of the terms $a\bar{b}\bar{c}$, ac and bc is an implicant of the function, since each one of them is included in f. From the implicant bc, let the literal c be dropped. Then the remainder of the term b is no longer included in the function. This can be verified from the fact that

$$b = \vee(2, 3, 6, 7) \text{ and}$$

$$f = \vee(3, 4, 5, 7)$$

$$\therefore b \nleqslant f.$$

In such a case the term bc is called a prime implicant.

Definition 4.4.2 An implicant from which no literal can be removed without altering its implicant status is known as a *prime implicant.*

Theorem 4.4.1 Every implicant of a Boolean function implies a prime implicant of the function.

Proof: The theorem holds good for every implicant which is itself a prime implicant. If an implicant is not a prime implicant by itself, then one or more literals can be removed from it until it becomes a prime implicant. Therefore the prime implicant will be a product term T_p whereas the implicant was a product term $T_p\,T_q$, the term T_q being the product of the literals removed. Since $T_p \vee T_p\,T_q = T_p$

$$\therefore T_p \geqslant T_p\,T_q. \quad \blacktriangle$$

Theorem 4.4.2 Every Boolean function can be expressed as a sum of prime implicants only.

Proof: Let the given Boolean function f be expressed as a sum of products, and let a term T_i be one which is not a prime implicant. Let the sum of the rest of the terms be F. Then, $f \geqslant F$, and $f = F \vee T_i$. Again let T_i imply the prime implicant P_i so that $T_i \leqslant P_i$. Now, let $f_1 = F \vee P_i$. Since, $T_i \leqslant P_i$ $\therefore f \leqslant f_1$. Again, since P_i is a prime implicant of the given function f, $P_i \leqslant f$. Also $F \leqslant f$, $\therefore F \vee P_i \leqslant f$ or $f_1 \leqslant f$. But $f \leqslant f_1$, $\therefore f_1 = f$. Hence the value of the given function remains unaltered if the term T_i is replaced by its prime implicant P_i. Similarly all other terms which are not prime implicants can be replaced by the corresponding prime implicants without altering the value of the given Boolean function. ▲

It is interesting to observe as to how the implicants and prime implicants look on a Veitch–Karnaugh map. Consider the Boolean function $f = a\bar{b}\bar{c} \vee ac \vee bc$. It can be checked by algebraic verification that the terms ac and bc are prime implicants, whereas the term $a\bar{b}\bar{c}$ is an implicant included in the prime implicant $a\bar{b}$. Now, let the function be plotted on the map, as has been shown in Fig.4-13. The prime implicants ac and bc are the clusters of two. But the implicant $a\bar{b}\bar{c}$ is a cluster of one. It can be easily seen that the 1 representing $a\bar{b}\bar{c}$ can combine with the 1 immediately below it to form a cluster of two. This is shown by the dotted curve and it represents the term $a\bar{b}$ which is the prime implicant. Thus *the prime implicants are always the biggest clusters that can be formed by combining 2, 4, 8 or 2^m 1's. An implicant is a sub-cluster of a bigger cluster.*

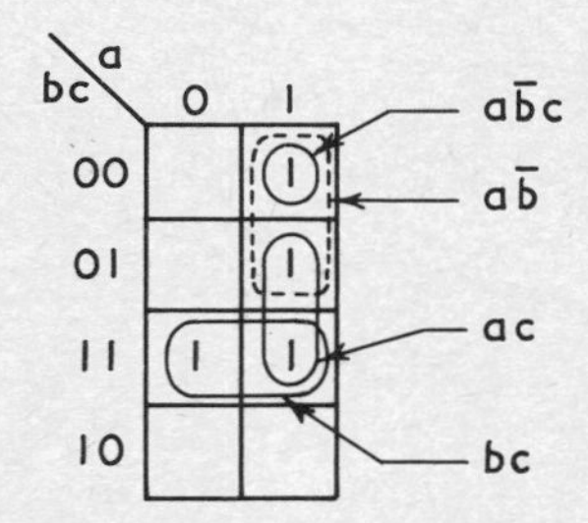

Figure 4-13 Implicants and prime implicants on the map.

Obviously the cost function of a circuit realizing an arbitrary function can be reduced if it can be reduced to the DNF which is a sum of prime implicants only. The next question that must be investigated is whether *all* the prime implicants are necessary to express a function; or if *some* of them can be completely removed. Let us consider the function in our present example. The function can be written as follows, as a sum of prime implicants only:

$$f = a\bar{b} \vee ac \vee bc.$$

If now the term ac is removed from this function, it will still have the same plot as the given function. This clearly demonstrates that not all prime implicants are essential. Hence the problem of minimization does not end by finding the prime implicants only. After the prime implicants are found, it is necessary to determine which ones are essential and which ones are redundant. For this purpose let the following terms be defined.

Definition 4.4.3 When a cluster of 1's indicates a prime implicant on the map, each of the minterms represented by a 1 within the cluster is said to *subsume*

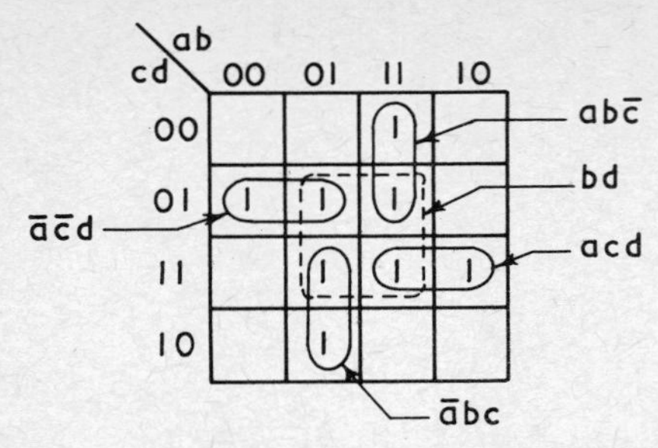

Figure 4-14 Essential and redundant prime-implicants.

the prime implicant, and the prime implicant is said to *cover* each of these minterms.

Definition 4.4.4 If among the minterms subsuming a prime implicant, there is at least one which is covered by this and only this prime implicant then the prime implicant is an *essential prime implicant (EPI).*

In Fig.4-14 the terms $\bar{a}\bar{c}d$, $ab\bar{c}$, acd, and $\bar{a}bc$ are all essential prime implicants.

Definition 4.4.5 If each of the minterms subsuming a prime implicant is covered by other essential prime implicants, then the prime implicant is a *redundant prime implicant (RPI).*

In Fig.4-14 the prime implicant bd is an RPI.

Definition 4.4.6 A prime implicant, which is neither an essential nor a redundant prime implicant, is a *selective prime implicant (SPI).* Among the minterms subsuming such a prime implicant, there is at least one which is covered neither by any EPI nor by this and only this prime implicant.

It is obvious from the definition that the particular minterm is also covered by another selective prime implicant. Thus the existence of one implies the existence of a second, and two SPI's appear as two inter-connecting links of a chain. There may be cases, where such a chain may be constituted by a number of SPI's. The set of SPI's constituting a chain can always be divided into two subsets, each of which covers the minterms need to be covered by the SPI's. As

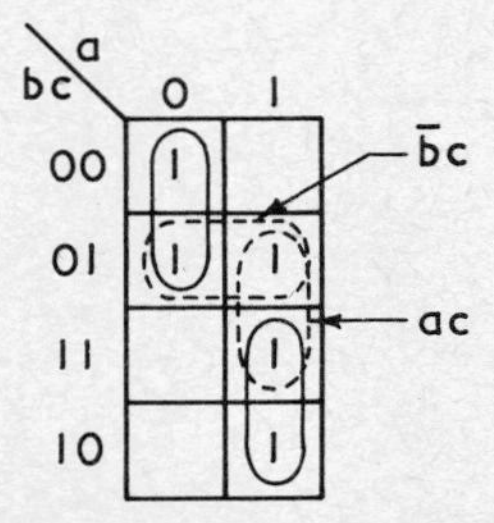

Figure 4-15 Chain of selective prime-implicants.

a result, the MSP form of the given Boolean function will consist of the EPI's and the subset of the SPI's having the smallest number of literals. If both the subsets have the same number of literals, then any one can be included, and the function has two valid MSP forms as defined below. There may be cases, especially when the number of variables is large, where the number of SPI chains may be more than one, giving rise to a number of valid MSP forms for the same function.

In Fig.4-15 the prime implicants $\bar{b}c$ and ac constitute the two interconnecting links of an SPI chain. In this pair the minterm m_5 is covered by no other prime implicant but this pair, and none of which is an essential prime implicant by Def.4.4.4.

Definition 4.4.7 When a Boolean function is expressed as a sum-of-products so that the total number of literals in the whole expression is minimum, the function is said to be expressed in its *minimal sum-of-products (MSP)* form.

Obviously a circuit which realizes a function in its MSP form will be very economical. Thus reduction to the MSP form is a very important goal to the circuit designer. The techniques or algorithms by which this goal is achieved are known as minimization methods.

Theorem 4.4.3 The MSP form of a Boolean function a) must contain all essential prime implicants, b) must not contain the redundant prime implicants, and c) must contain the subset with the smallest number of literals of the chain or chains of selective prime implicants.

Proof: The validity of this theorem is obvious from the definitions of the essential, redundant, and selective prime implicants. ▲

4.5 THE MAP METHOD (VEITCH, 1952; KARNAUGH, 1953)

Following the discussions in the preceding sections, it must be evident that it will be very easy to recognize the EPI's and the cheaper subset of the SPI's on a map. Hence the map method is very extensively used to minimize a Boolean function of up to 4 variables. Referring back to Fig.4-14, it can be seen that the temptation to form a bigger cluster must be avoided, as it may turn out to be an RPI. Hence, the following procedure is suggested.

Step 1: Plot the function on the map.

Step 2: Circle the 1's which do not combine with any other 1. These are EPI's.

Step 3: Locate the 1's, if any, which combine with only one 1. Circle these two 1's. It is also an EPI.

Step 4: Locate the uncovered 1's, if any, which form only one cluster of four 1's. Circle these four 1's. This is also an EPI.

Step 5: Repeat this procedure to determine the EPI's of eight 1's and so on.

Step 6: If after steps 4 and 5 there is still some term/terms uncovered, then look for SPI's, and choose the subset with smallest number of literals.

Example 4.5.1 Minimize $f(abcd) = \vee(2, 4, 5, 8, 10, 12, 13)$. The function has been plotted on the four variable map of Fig.4-16. To cover the term 2,

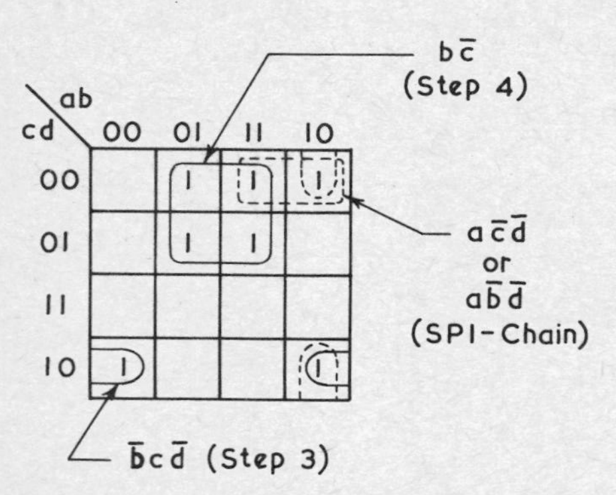

Figure 4-16 Minimization of $f(abcd) = \vee(2, 4, 5, 8, 10, 12, 13)$.

the prime implicant $\bar{b}c\bar{d}$ is chosen by following step 3. Step 4 yields the EPI $b\bar{c}$. The term 8 is covered by the pair of SPI's $a\bar{c}\bar{d}$ and $a\bar{b}\bar{d}$. Since both have the same number of literals, any one of them may be chosen. Thus this function has two valid MSP forms:

$f(abcd) = \bar{b}c\bar{d} \vee b\bar{c} \vee a\bar{c}\bar{d}$ or,

$f(abcd) = \bar{b}c\bar{d} \vee b\bar{c} \vee a\bar{b}\bar{d}$.

Example 4.5.2 Minimize $f(abcd) = \vee(1, 3, 5, 7, 9, 10, 13, 14, 15)$. The function has been plotted on the four variable map of Fig.4-17. *Step 3* yields

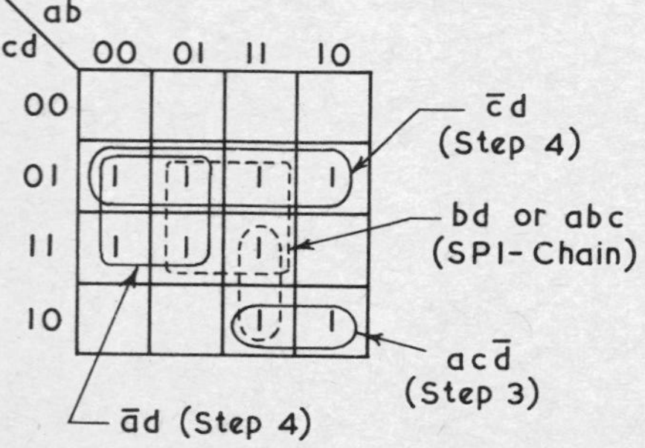

Figure 4-17 Minimization of $f(abcd) = \vee(1, 3, 5, 7, 9, 10, 13, 14, 15)$.

the EPI $ac\bar{d}$, and *Step 4* yields the EPI's $\bar{c}d$ and $\bar{a}d$. *Step 6* gives the pair of SPI's bd and abc. Since, bd has the smaller number of literals, it must be included in the MSP form. Hence here the function has only one valid MSP form, which is as follows:

$$f(abcd) = ac\bar{d} \vee \bar{c}d \vee \bar{a}d \vee bd.$$

Example 4.5.3 Minimize $f(wxyz) = \vee(2, 3, 5, 7, 8, 10, 12, 13)$. In this example, the function has no EPI, but only SPI's. Both the subsets have equal number of literals. Hence it has two valid MSP forms. (Figure 4-18.)

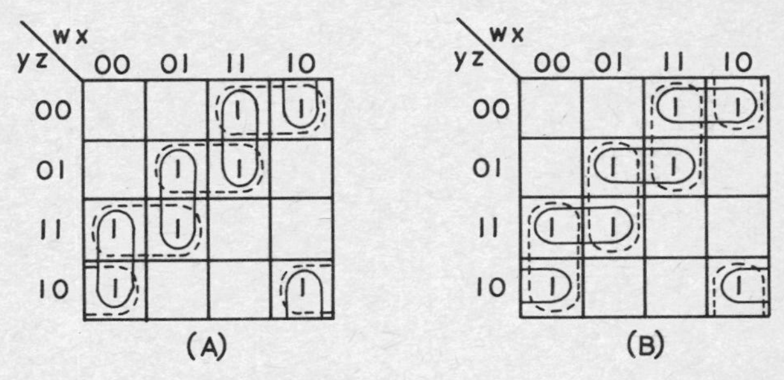

Figure 4-18 Minimization of $f(wxyz) = \vee(2, 3, 5, 7, 8, 10, 12, 13)$.

4.6 MINIMAL PRODUCT-OF-SUMS (MPS) FORM

Once the function has been plotted on a V–K map, it is also possible to determine the MPS form of the function from the techniques already acquired. Take for example the function

$$f(abcd) = \vee(0, 2, 6, 7, 8, 10, 12, 14, 15).$$

The MSP form of the function (Fig.4-19(A)) is

$$f(abcd) = a\bar{d} \vee \bar{b}\bar{d} \vee bc.$$

By Th.3.4.3 the complement of f, $\bar{f}$ is the function which will occupy the 0 cells. So minimizing the 0's of Fig.4-19(A) the MSP form of $\bar{f}$ is (Fig.4-19(B)).

$$\bar{f}(abcd) = \bar{a}b\bar{c} \vee \bar{c}d \vee \bar{b}d.$$

Taking the complement (by De Morgan's theorem)

$$\bar{\bar{f}} = f = (a \vee \bar{b} \vee c)\,(c \vee \bar{d})\,(b \vee \bar{d})$$

This is the MPS form of the original function f. In this case, the MSP and the MPS forms contain different number of literals, namely six and seven. But, there may be cases where the numbers of literals in both forms are the same.

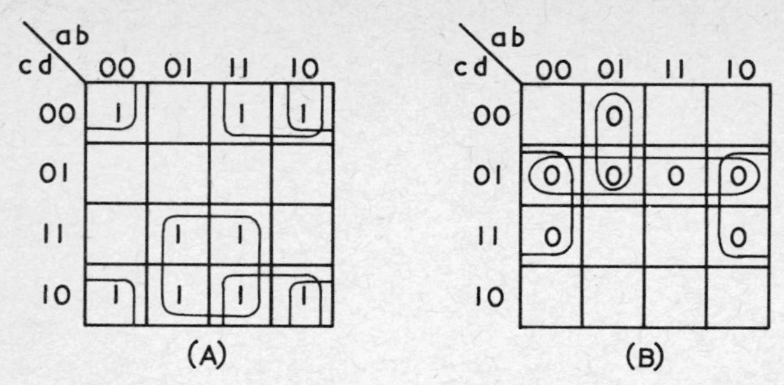

Figure 4-19 MSP and MPS forms of $f(abcd) = \vee(0, 2, 6, 7, 8, 10, 12, 14, 15)$.

The reader may verify this by working out both the MSP and MPS forms of the function, $\vee(0, 1, 12, 15)$.

It is worthwhile mentioning here that neither the MSP nor the MPS form can always claim to be the form containing the least number of literals. In fact, a mixed form may have a smaller number of literals. In our above example, the MSP form can be further factored to produce a mixed form as given below with 5 literals only.

$$f(abcd) = (a \vee \bar{b})\bar{d} \vee bc.$$

4.7 DON'T-CARE TERMS

To understand the significance of don't-care terms let us consider the design of a circuit. It has four input and one output lines. On the four input lines, say *a, b, c,* and *d* lines are incident the 0's and 1's corresponding to the BCD code of the ten decimal numbers 0 through 9. Thus when the values of the a, b, c, and d lines are say 0, 1, 1, and 1 respectively, then the number 0111, that is, 7, is said to be incident at the input of the circuit. The circuit is to produce an output as soon as any of the numbers, 0, 2, 4, 6, and 8 appears at its input. Thus we can write the truth table of this circuit (Fig.4-20) as shown in Table 4-1.

Consequently, f is given by

$$f(abcd) = \vee(0, 2, 4, 6, 8).$$

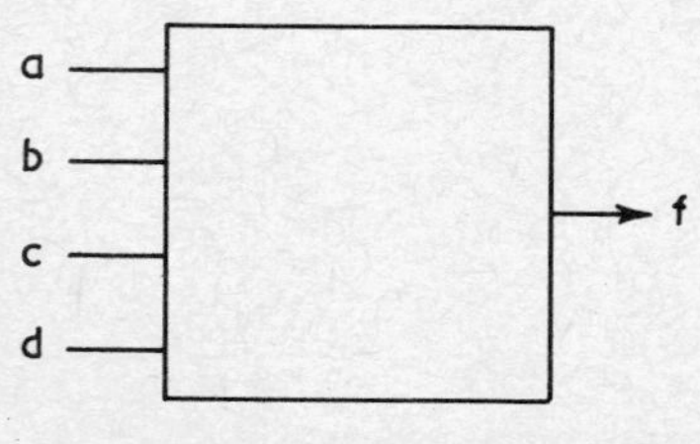

Figure 4-20

TABLE 4-1

	a	b	c	d	f
0	0	0	0	0	1
1	0	0	0	1	0
2	0	0	1	0	1
3	0	0	1	1	0
4	0	1	0	0	1
5	0	1	0	1	0
6	0	1	1	0	1
7	0	1	1	1	0
8	1	0	0	0	1
9	1	0	0	1	0

The function minimizes to (Fig.4-21)

$$f = \bar{a}\bar{d} \vee \bar{b}\bar{c}\bar{d} \tag{4.6}$$

and the resultant design is shown in Fig.4-22. It can be seen that the circuit needs four inverters, two AND, and one OR gate. However, in our design we have not taken into consideration one fact, which is that the six combinations of 0's and 1's corresponding to the decimal numbers 10 through 15 do not appear at the input lines at any time. These combinations are called the don't-care terms, and are denoted by ϕ's in the truth table and the map. Thus the truth table must be re-written as shown in Table 4-2. The function is now written as

$$f(abcd) = \vee(0, 2, 4, 6, 8)$$
$$\vee\phi(10, 11, 12, 13, 14, 15).$$

The plotting of the function is as shown in Fig.4-23. It can now be seen that no malfunction of the circuit will occur if the ϕ's on cells 10, 12, and 14 are taken to be 1's, since these numbers will never appear at the input. On the other hand, by considering these ϕ's as 1's the function can be minimized to an MSP form containing fewer literals than that given in Eqn.(4-6). Here

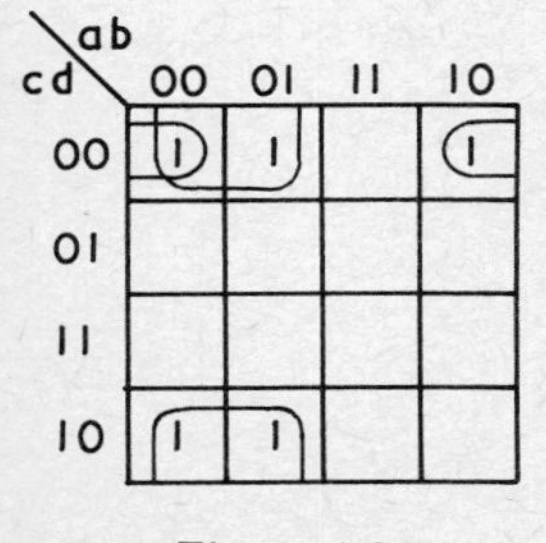

Figure 4-21

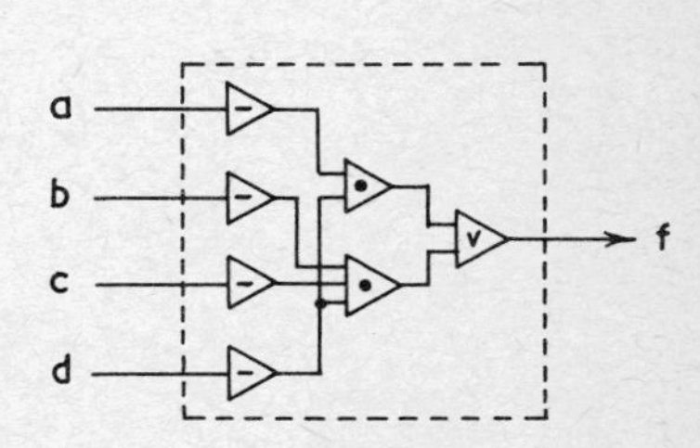

Figure 4-22

TABLE 4-2

	a	b	c	d	f
0	0	0	0	0	1
1	0	0	0	1	0
2	0	0	1	0	1
3	0	0	1	1	0
4	0	1	0	0	1
5	0	1	0	1	0
6	0	1	1	0	1
7	0	1	1	1	0
8	1	0	0	0	1
9	1	0	0	1	0
10	1	0	1	0	ϕ
11	1	0	1	1	ϕ
12	1	1	0	0	ϕ
13	1	1	0	1	ϕ
14	1	1	1	0	ϕ
15	1	1	1	1	ϕ

$$f = \bar{d}. \tag{4.7}$$

The resultant circuit is shown in Fig.4-24. The saving in the cost of the design is obvious. Thus the don't-care terms are very useful to the logic designer inasmuch as they help in reducing the cost of the design. Consequently one must be very careful as not to overlook them.

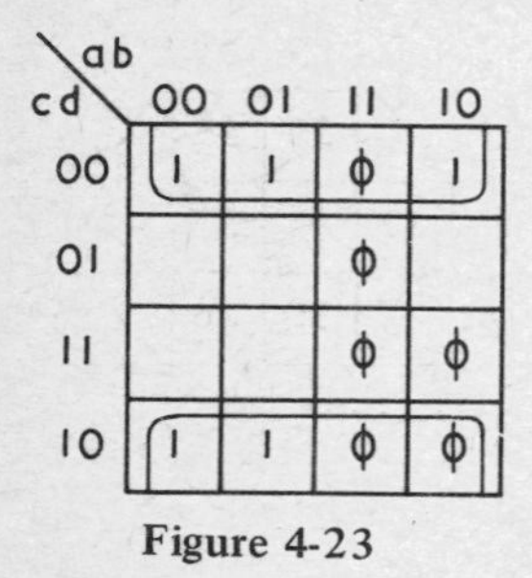

Figure 4-23

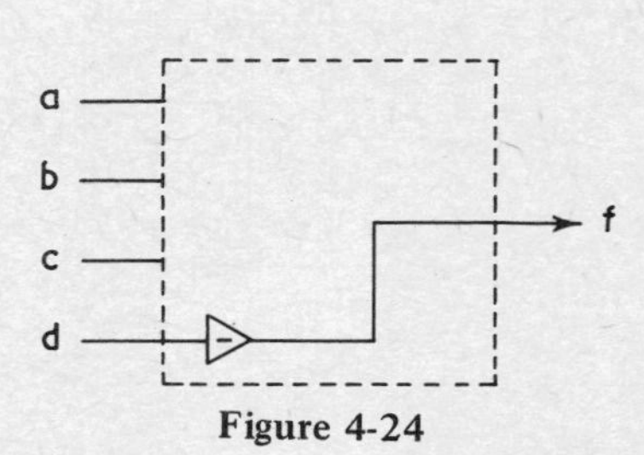

Figure 4-24

4.8 FIVE AND SIX VARIABLE MAPS

The five-variable map is shown in Fig.4-25. It essentially consists of two four variable maps. In one of the four-variable ($wxyz$) map v is 0, whereas in the other four variable ($wxyz$) map v is 1. The decimal designations of the cells can be easily computed and they are as shown in the figure. After a function has been plotted on the map, the 1's in any of the four variable map can be combined in

V = 0

yz \ wx	00	01	11	10
00	0	4	12	8
01	1	5	13	9
11	3	7	15	11
10	2	6	14	10

V = 1

yz \ wx	00	01	11	10
00	16	20	28	24
01	17	21	29	25
11	19	23	31	27
10	18	22	30	26

Figure 4-25 Five-variable Veitch–Karnaugh map.

the same way as is done on a four variable map. In addition, any two 1's which occupy relatively the same position on the two maps can be combined. Thus two 1's on cells 5 and 21, or on cells 14 and 30 can be combined. Extending this argument, eight 1's on cells 0, 1, 3, 2, and 16, 17, 19, 18 will also combine. Thus a good rule to effect combination between the 1's belonging to the two four variable maps is to look for similar patterns occupying identical positions on the two maps.

The six variable map is composed of 4 four-variable maps. Let the variables be u, v, w, x, y, and z. Then, in each of the four variable maps the variables w, x, y, and z have the same co-ordinates. They differ in the co-ordinates of the

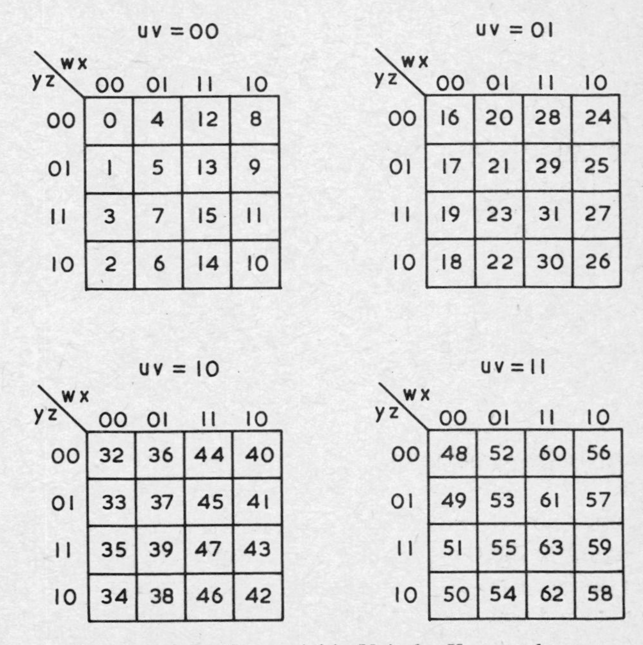

Figure 4-26 Six-variable Veitch–Karnaugh map.

variables, u and v. These as well as the decimal designations of the 64 cells have been shown in Fig.4-26. Here also similar patterns occupying identical positions on any two maps which are either horizontally or vertically (but not diagonally) adjacent will combine. Also a similar pattern occupying the same position on *all four* maps will also combine.

Example 4.8.1 Minimize $f(uvwxyz) = \vee(0, 1, 4, 5, 6, 11, 14, 15, 16, 17, 20, 21, 22, 30, 32, 33, 36, 37, 48, 49, 52, 53, 59, 63)$. The solution is depicted in Fig.4-27 and is as follows:

$$f(uvwxyz) = \bar{w}\bar{y} \vee \bar{u}\bar{v}wyz \vee \bar{u}xy\bar{z} \vee uvwyz.$$

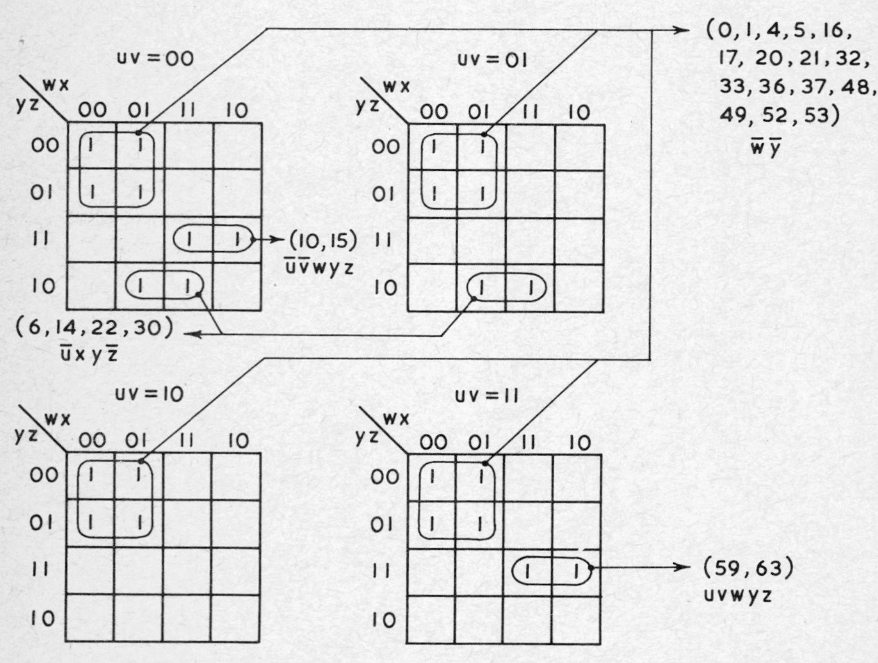

Figure 4-27 Minimization of $f(uvwxyz) = \vee(0, 1, 4, 5, 6, 11, 14, 15, 16, 17, 20, 21, 22, 30, 32, 33, 36, 37, 48, 49, 52, 53, 59, 63)$.

A few exercises on the five and six variable maps will convince the reader that the determination of prime implicants becomes very inconvenient as the complexity of the given function is increased. For this reason, the five and six variable maps are very seldom used. To handle functions of five or more variables, the tabular methods described in the subsequent sections are recommended.

4.9 THE TABULAR METHOD: QUINE (1952), McCLUSKEY (1956)

In the method suggested by Quine, the function is first expressed as a sum of minterms which are arranged in a table. Each minterm is then compared with each of the rest to determine if any literal can be eliminated by the application of the formula,

$$aB \vee \bar{a}B = B, \tag{4.8}$$

where a is a single variable, and B is a product term of one or more variables. Later McCluskey improved the method by pointing out that it is not necessary to compare each term with all the remaining terms of the table, but need be compared with only a limited number of terms. This considerably reduces the labor involved in carrying out the comparisons which must be exhaustive. According to the modification suggested by McCluskey, the minterms are written by their binary number designations where a 0 is written for a literal in the complemented form, and a 1 for a literal in the true form. Once, this is done, we may define what may be called the weight of a term.

Definition 4.9.1 The *weight* of a term is defined as the number of 1's in the binary designation of the term.

Thus the weight of the minterm m_{14} is 3, since m_{14}, when written in the binary form, appears as 1110.

It will now be evident that whenever a *single* literal is removed by the combination of two terms as per Eqn.(4.8), the weights of the two terms must differ by only *one*. Hence in the Quine–McCluskey method, the table of the minterms is so partitioned that each partition contains the minterms of a particular weight only. The partition having the minterms with least weight comes at the top. Each minterm of this partition is then compared with each of the next partition only, and so on. In this process each minterm of weight w is compared with each minterm of weight $w - 1$, and $w + 1$. This makes the job of comparison less tedious, retaining at the same time the nature of comparison quite exhaustive. Let the starting table, where all minterms are written in their binary forms and are partitioned according to their weights be called Table T_0. Each term of the topmost partition, say, of weight w is now compared with each term of the next partition of weight $w + 1$, to see if two terms (which are of n variables) can generate a term of $n - 1$ variables. Thus in a five variable case, the minterms 1, and 3 (written as 00001 and 00011 in Table T_0) combine to generate the term (1, 3) to be written as 000-1 in Table T_1. A check mark (X) is placed after the terms which combine. Many times a term may combine with more than one term; but one check mark is adequate. After the comparison between the terms of weight w, and $w + 1$ is over, each term of the partition of weight $w + 1$ is compared with the terms of the next partition of weight $w + 2$. This procedure is repeated until all comparisons are completed. The terms generated by this comparison form the next table which is called Table T_1. All terms of Table T_1 are of $n - 1$ variables, and also automatically get partitioned according to their weights. Following the same procedure Table T_2 is formed from Table T_1. However, two pairs of terms of Table T_1 may generate the same term on Table T_2. But there is no need to wri them twice, although all the four terms on Table T_1 must be checked off. Let this and other features of the method be now illustrated by working out a few examples.

Example 4.9.1 Minimize $f(vwxyz) = \vee(1, 3, 5, 6, 7, 13, 15, 16, 17, 18, 19, 22, 25, 27, 28, 29, 30, 31)$.

Solution: Tables T_0, T_1 and T_2 for this example are shown in Fig.4-28. It ma be seen that the terms (1, 3) and (5, 7) in Table T_1 combine to generate the ter (1, 3, 5, 7) in Table T_2. Also the same term is generated by the terms (1, 5) and (3, 7) in Table T_1. However, due to the idempotence law the term need appear only once in Table T_2. To avoid multiple appearances of the same tern in Tables T_2, T_3 etc., all terms are written in numerical order. In this example Table T_3 cannot be formed. The terms which have not combined with any other term in these tables are the prime implicants of this function. These are now marked as A, B, C, etc. in all tables.

The next task is to select the essential terms from among these prime implicants. For this purpose other tables called the *prime implicant (PI) tables* are formed. First PI table to be formed will be called Table P_0. Subsequent PI tables will be called P_1, P_2 etc. For this example the PI table P_0 is shown in Fig.4-29. In this table each minterm of Table T_0 heads a column and each of the prime implicants heads a row. Since the PI A(6, 7) covers the minterms 6 and 7 X's are placed in this row under the columns of 6 and 7. This is done for all rows.

In the PI table P_0, essential prime implicants are now selected as given by the following definition. EPI's are then marked with asterisks in table P_0.

Definition 4.9.2 In PI table P_0, if there exists a column under which there is one and only one X, then the row of this X is headed by an *essential prime implicant (EPI)*.

After an EPI is determined, all minterms covered by it are circled. It may be mentioned that the minterms appear at the head of the columns, and it is covered by an EPI, when a X appears at the intersection of the column of the minterm and the row of the EPI. After this step, if there still remain some uncircled columns, then the PI table P_1 is formed by removing the circled column and starred rows. For this example table P_1 is shown in Fig.4-30. In this table dominated rows and redundant prime implicants are struck out as given by the following definitions.

Definition 4.9.3 In PI table P_1, if there are two rows G and H, such that G

T_0

	v	w	x	y	z	
1	0	0	0	0	1	X
16	1	0	0	0	0	X
3	0	0	0	1	1	X
5	0	0	1	0	1	X
6	0	0	1	1	0	X
17	1	0	0	0	1	X
18	1	0	0	1	0	X
7	0	0	1	1	1	X
13	0	1	1	0	1	X
19	1	0	0	1	1	X
22	1	0	1	1	0	X
25	1	1	0	0	1	X
28	1	1	1	0	0	X
15	0	1	1	1	1	X
27	1	1	0	1	1	X
29	1	1	1	0	1	X
30	1	1	1	1	0	X
31	1	1	1	1	1	X

T_1

	v	w	x	y	z	
1 3	0	0	0	-	1	X
1 5	0	0	-	0	1	X
1 17	-	0	0	0	1	X
16 17	1	0	0	0	-	X
16 18	1	0	0	-	0	X
3 7	0	0	-	1	1	X
3 19	-	0	0	1	1	X
5 7	0	0	1	-	1	X
5 13	0	-	1	0	1	X
6 7	0	0	1	1	-	A
6 22	-	0	1	1	0	B
17 19	1	0	0	-	1	X
17 25	1	-	0	0	1	X
18 19	1	0	0	1	-	X
18 22	1	0	-	1	0	C
7 15	0	-	1	1	1	X
13 15	0	1	1	-	1	X
13 29	-	1	1	0	1	X
19 27	1	-	0	1	1	X
22 30	1	-	1	1	0	D
25 27	1	1	0	-	1	X
25 29	1	1	-	0	1	X
28 29	1	1	1	0	-	X
28 30	1	1	1	-	0	X
15 31	-	1	1	1	1	X
27 31	1	1	-	1	1	X
29 31	1	1	1	-	1	X
30 31	1	1	1	1	-	X

T_2

	v	w	x	y	z
1 3 5 7	0	0	-	-	1
1 3 17 19	-	0	0	-	1
16 17 18 19	1	0	0	-	-
5 7 13 15	0	-	1	-	1
17 19 25 27	1	-	0	-	1
13 15 29 31	-	1	1	-	1
25 27 29 31	1	1	-	-	1
28 29 30 31	1	1	1	-	-

Figure 4-28 Tables T_0, T_1 and T_2 for Example 4.9.1.

has X's under all those columns where H has X's, and in addition G has X under at least one column where H does not have any X then the row H is said to be *dominated* by the row G. This is expressed as $H \subset G$ or $G \supset H$.

There may be two rows with equal number of X's appearing under the same columns. In such a case any one of them may be retained in PI table P_1.

					P0																	
					1	(16)	3	5	6	(17)	(18)	7	13	(19)	22	25	(28)	15	27	(29)	(30)	(3
A			6	7					X			X										
B			6	22					X						X							
C			18	22							X				X							
D			22	30											X						X	
E	1	3	5	7	X		X	X				X										
F	1	3	17	19	X		X			X				X								
* G	16	17	18	19		(X)				X	X			X								
H	5	7	13	15				X				X	X					X				
I	17	19	25	27						X				X		X			X			
J	13	15	29	31									X					X		X		X
K	25	27	29	31												X			X	X		X
* L	28	29	30	31													(X)			X	X	X

Figure 4-29 PI table P_0 for Example 4.9.1.

					P1									
					1	3	5	6	7	13	22	25	15	27
A			6	7				X	X					
B			6	22				X			X			
~~C~~			~~18~~	~~22~~							X			
~~D~~			~~22~~	~~30~~							X			
E	1	3	5	7	X	X	X		X					
~~F~~	~~1~~	~~3~~	~~17~~	~~19~~	X	X								
H	5	7	13	15			X		X	X			X	
I	17	19	25	27								X		X
~~J~~	~~13~~	~~15~~	~~29~~	~~31~~						X			X	
~~K~~	~~25~~	~~27~~	~~29~~	~~31~~								X		X

$C = D \subset B$ $F \subset E$ $I = K$ $J \subset H$

Figure 4-30 PI table P_1 for Example 4.9.1.

P_2

					1	3	5	6	7	13	22	25	15	27
A			6	7				X	X					
* B			6	22				X			(X)			
* E	1	3	5	7	(X)	(X)	X		X					
* H	5	7	13	15			X		X	(X)			(X)	
* I	17	19	25	27								(X)		(X)

Figure 4-31 PI table P_2 for Example 4.9.1.

Definition 4.9.4 In PI Table P_1, if there is any row which has no X under any column, then the prime implicant heading the row is a *redundant prime implicant (RPI)*.

After the dominated and redundant rows are eliminated from table P_1, we get PI table P_2, as shown in Fig.4.31. Essential terms are determined in this table following the procedure of table P_0. However here the essential terms are not EPIs, but SPIs. After the minterms have been circled in the table, if there still remain some uncovered columns, then the procedure of tables P_1 and P_2 are to be repeated in tables P_3, P_4, and so on. In this example the operation gets terminated in table P_2. The MSP form turns out to be

$$f(vwxyz) = (16, 17, 18, 19) \vee (28, 29, 30, 31)$$

$$\vee\, (6, 22) \vee (1, 3, 5, 7) \vee (5, 7, 13, 15) \vee (17, 19, 25, 27)$$

$$= v\bar{w}\bar{x} \vee vwz \vee vwx \vee \bar{w}xy\bar{z} \vee \bar{v}\bar{w}z \vee \bar{v}xz \vee v\bar{x}z$$

There exist many Boolean functions where essential terms cannot be determined in table P_0, or P_2 etc., as there may not be any 'single-cross' column in the table. In such cases, a new technique called the *branching method* is to be employed. Let this be illustrated by the following example.

Example 4.9.2 Minimize $f(wxyz) = \vee\,(1, 2, 3, 5, 6, 7, 8, 9, 10, 12, 13, 14)$.

Solution: After tables T_0, T_1 and T_2 (not shown here) are formed, the PI table P_0 is as shown in Fig.4-32. It may be seen that there is no column in table P_0 with only one X. This means that the function does not have any EPI. All columns have two X's. To start with we may choose any prime implicant to be an essential one. Selecting the PI A, we mark it with an asterisk, and circle the columns which are covered by it. We now derive PI table P_1 from P_0 after removing the circled columns and starred prime impli-

P_0

					(1)	2	8	(3)	(5)	6	9	10	12	(7)	13	14
* A	1	3	5	7	X			X	X					X		
B	1	5	9	13	X				X		X				X	
C	2	3	6	7		X		X		X				X		
D	2	6	10	14		X				X		X				X
E	8	9	12	13			X				X		X		X	
F	8	10	12	14			X					X	X			X

Figure 4-32 PI table P_0 for Example 4.9.2.

cants (Fig.4-33). The table P_2 is derived from P_1 after removing the dominated rows (Fig.4-34). In table P_2 the columns 2, 6, 9, and 13 have only one X each. Hence the prime implicants D and E are starred, and the columns 2, 6, 10, 14 and 8, 9, 12, 13 are circled. With this all columns are circled, and the operation terminates. The MSP form of the function is then given by

$$f(wxyz) = (1, 3, 5, 7) \vee (2, 6, 10, 14) \vee (8, 9, 12, 13)$$
$$= \bar{w}z \vee y\bar{z} \vee w\bar{y}.$$

If in table P_0, we would have elected to choose the PI B, then we would have followed in an identical manner another branch of the solution. Therefore, the function has another valid MSP form. This is

$$f(wxyz) = (1, 5, 9, 13) \vee (2, 3, 6, 7) \vee (8, 10, 12, 14)$$
$$= \bar{y}z \vee \bar{w}y \vee w\bar{z}.$$

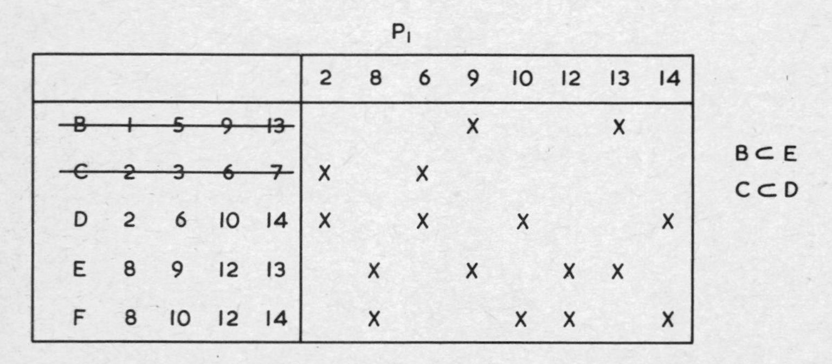

P_1

					2	8	6	9	10	12	13	14
~~B~~	~~1~~	~~5~~	~~9~~	~~13~~				X			X	
~~C~~	~~2~~	~~3~~	~~6~~	~~7~~	X		X					
D	2	6	10	14	X		X		X			X
E	8	9	12	13		X		X		X	X	
F	8	10	12	14		X			X	X		X

B ⊂ E
C ⊂ D

Figure 4-33 PI table P_1 for Example 4.9.2.

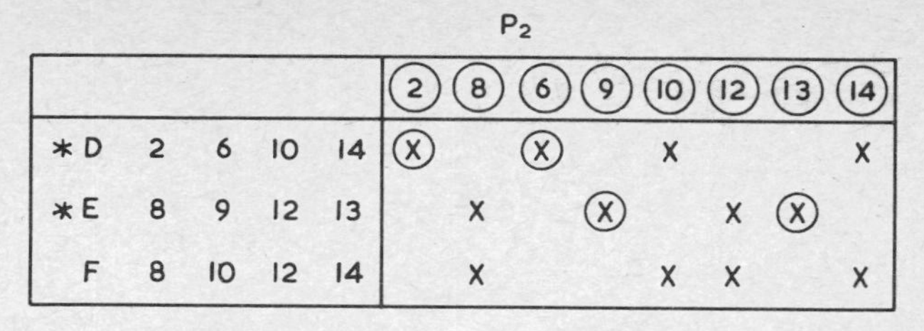

P_2

	(2)	(8)	(6)	(9)	(10)	(12)	(13)	(14)
*D 2 6 10 14	(X)		(X)		X			X
*E 8 9 12 13		X		(X)		X	(X)	
F 8 10 12 14		X			X	X		X

Figure 4-34 PI table P_2 for Example 4.9.2.

4.10 THE ADJACENCY METHOD: BISWAS (1971)

In the Quine–McCluskey (Q–M) method as described above, after all the prime implicants are established the essential ones are chosen by one or more successive prime implicant tables. In a method expounded by Biswas (1971) the need for PI tables is entirely eliminated. We shall call this method the adjacency method.

Like the Quine–McCluskey method, in this method also the function to be minimized is first expressed as a sum of minterms, and then the minterms are arranged as rows of a combination table in ascending order of their weights. The resulting table is designated as *Table* T_0. Successive tables are called *Table* T_1, *Table* T_2, etc. Table T_1 is formed from Table T_0, and Table T_2 from Table T_1. In general Table T_a is formed from Table T_{a-1}.

In the adjacency method, Table T_1 is formed in the same way as in the Q–M method. However, in this method an additional quantity which has been called the *degree of adjacency* is determined for each minterm during the process of forming Table T_1 from Table T_0.

Definition 4.10.1 Each time a term in Table T_0 combines with another term in Table T_0 to generate a term in Table T_1, an X is placed after each one of the two combining terms. After the formation of Table T_1 has been completed, the number of X's after a term in Table T_0 is the *degree of adjacency* of the term. It can be easily verified that the degree of adjacency of a minterm of n variables lies between 0 and n, both values inclusive.

A theorem which is the heart of this method can be stated as follows:

Theorem 4.10.1 If a minterm with a degree of adjacency a can generate a term in Table T_a, then that term in Table T_a is an essential prime implicant.

In the Q–M method all EPI's are detected in a PI table by the application of the Def.4.9.2. In the method presented here, all EPI's are detected in a Table T_a by the application of Th.4.10.1.

The method can now be illustrated by a few typical examples. The various steps are as follows:

Step 1: Form Table T_0 as in the Q–M method.

Step 2: Form Table T_1 as in the Q–M method. Also determine the degree of adjacency of each minterm in Table T_0 during the process of forming Table T_1 from Table T_0.

Step 3: Detect any EPI in Table T_0 which satisfies Th.4.10.1. In this Table, minterms exhibiting zero degree of adjacency satisfy Th.4.10.1. Circle and tick this term.

Step 4: Detect any EPI in Table T_1 which satisfies Th.4.10.1. Circle and tick this term. This term is to be considered absent when forming Table T_2 from Table T_1.

Step 5: Note the minterms subsuming the EPI. Circle these minterms in Table T_0. If in so doing, all terms in Table T_0 are circled, then the next table need not be formed.

Step 6: If after step 5 is completed there is/are still some uncircled term/terms in Table T_0, then form Table T_2 from Table T_1 as in the Q–M method, but with this modification. Detected EPI's (circled and ticked) are considered absent. Thus these neither initiate any comparison with other terms, nor are available to combine with any other term.

Step 7: Repeat Step 4 and Step 5 on Table T_2 and Table T_0 respectively.

Step 8: Repeat Step 6 to form Table T_3 from Table T_2. These steps are now repeated until no further table need/can be formed.

Step 9: a) Note the prime implicants, if any, in all tables, which have neither been checked nor ticked. Mark them as A, B, C etc. These constitute the redundant and selective prime implicants.

b) In each of these, cancel the terms which have already been covered by the EPI's (that is, those terms which have been circled in Table T_0). If in so doing, all terms subsuming a prime implicant are cancelled, then the particular prime implicant is an RPI. The rest are SPI's.

c) Note the degree of adjacency of the uncovered terms on Table T_0. Group them in different sections of a Table called the SPI Table according to their degrees of adjacency. Then select the SPI's to be included in the MSP form/forms following the procedure as will be evident while minimizing the function of the following examples.

Example 4.10.1 Minimize the function of Example 4.9.1 by the adjacency method.

Solution: Tables T_0 and T_1 are formed (Fig.4-35). Since no term has exhibited an adjacency of either 0 or 1, there is no EPI in Table T_0 or Table T_1. Form Table T_2 from Table T_1 (Fig.4-35). As the terms 16, 6, and 28 have exhibited an adjacency of 2nd degree, they are likely to generate EPI's in Table T_2. Consequently the terms (16, 17, 18, 19) and (28, 29, 30, 31) in Table T_2 are

T_0

	v	w	x	y	z	1	2	3	4	5
1	0	0	0	0	1	X	X	X		
(16)	1	0	0	0	0	X	X			
3	0	0	0	1	1	X	X	X		
5	0	0	1	0	1	X	X	X		
6	0	0	1	1	0	X	X			
(17)	1	0	0	0	1	X	X	X	X	
(18)	1	0	0	1	0	X	X	X		
7	0	0	1	1	1	X	X	X	X	
13	0	1	1	0	1	X	X	X		
(19)	1	0	0	1	1	X	X	X	X	
22	1	0	1	1	0	X	X	X		
25	1	1	0	0	1	X	X	X		
(28)	1	1	1	0	0	X	X			
15	0	1	1	1	1	X	X	X		
27	1	1	0	1	1	X	X	X		
(29)	1	1	1	0	1	X	X	X	X	
(30)	1	1	1	1	0	X	X	X		
(31)	1	1	1	1	1	X	X	X	X	

T_1

	v	w	x	y	z	
1 3	0	0	0	-	1	X
1 5	0	0	-	0	1	X
1 17	-	0	0	0	1	X
16 17	1	0	0	0	-	X
16 18	1	0	0	-	0	X
3 7	0	0	-	1	1	X
3 19	-	0	0	1	1	X
5 7	0	0	1	-	1	X
5 13	0	-	1	0	1	X
6 7	0	0	1	1	-	A
6 22	-	0	1	1	0	B
17 19	1	0	0	-	1	X
17 25	1	-	0	0	1	X
18 19	1	0	0	1	-	X
~~18~~ 22	1	0	-	1	0	C
7 15	0	-	1	1	1	X
13 15	0	1	1	-	1	X
13 29	-	1	1	0	1	X
19 27	1	-	0	1	1	X
22 ~~30~~	1	-	1	1	0	D
25 27	1	1	0	-	1	X
25 29	1	1	-	0	1	X
28 29	1	1	1	0	-	X
28 30	1	1	1	-	0	X
15 31	-	1	1	1	1	X
27 31	1	1	-	1	1	X
29 31	1	1	1	-	1	X
30 31	1	1	1	1	-	X

T_2

	v	w	x	y	z	
1 3 5 7	0	0	-	-	1	E
1 3 ~~17~~ ~~19~~	-	0	0	-	1	F
(16 17 18 19)	1	0	0	-	-	✓
5 7 13 15	0	-	1	-	1	G
~~17~~ ~~19~~ 25 27	1	-	0	-	1	H
13 15 ~~29~~ ~~31~~	-	1	1	-	1	I
25 27 ~~29~~ ~~31~~	1	1	-	-	1	J
(28 29 30 31)	1	1	1	-	-	✓

Figure 4-35 Tables T_0, T_1 and T_2 for Example 4.10.1.

EPI's. These are circled and ticked. The terms 16, 17, 18, 19, 28, 29, 30 and 31 are also circled in Table T_0. As there are some terms in Table T_0 which are still uncovered, attempt is made to form Table T_3; but this cannot be formed. This takes us to step 9.

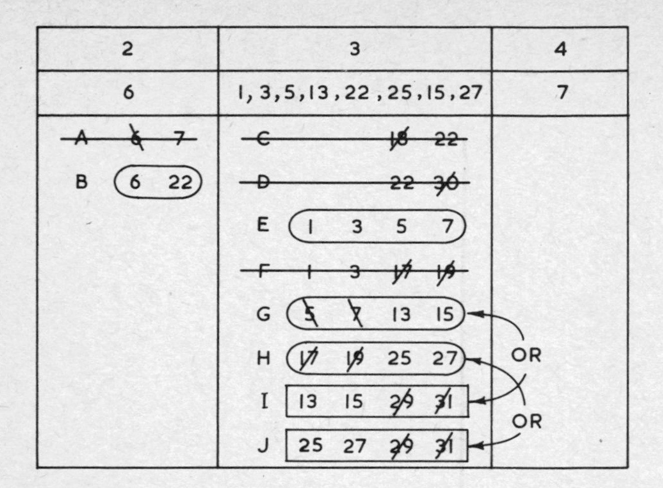

B ⊃ C = D → C̸, D̸, Ⓑ → A ⊂ E AND G → A̸
E ⊃ F → F̸, Ⓔ → G = I, Ⓖ OR Ⓘ
H = J, Ⓗ OR Ⓙ

Figure 4-36 SPI table for Example 4.10.1.

The SPI Table for this example is shown in Fig.4-36. The term 6 has an adjacency of 2nd degree. The SPI's A and B include 6. Hence these are written under 2. The terms 1, 3, 5, 13, 22, 25 and 27 have adjacency of 3rd degree, and the SPI's C, D, E, F, G, H, I and J are required to cover these terms. Hence these are written under 3. Similarly the term 7 and the SPI's (none) are written under 4. There is no uncovered term exhibiting an adjacency of 5th degree. It may be mentioned that there cannot be any SPI with an adjacency of 1st degree.

Comparing SPI's B, C and D we see that both C and D have only 22 as the uncovered minterm, whereas the uncovered minterms of B are 6 and 22. Hence B dominates C and D. C and D are therefore struck out and B is circled, since it alone covers the term 22. After B is chosen as an essential term, the appearance of 6 in other SPI's is cancelled. This removes A as it becomes dominated by both E and G. Again comparing E and F, we find that E dominates F. E also turns out to be an essential SPI. The terms 5 and 7 in G now get cancelled. This makes both G and I equal, and any one can be selected to cover 13 and 15. Similarly, any one of H and J can be chosen to cover 25 and 27. Thus all the SPI's are accounted for by applying the concept of *relative dominance* among the SPI's. With this the procedure terminates, and the MSP form turns out to be

$$f(vwxyz) = (16, 17, 18, 19) \vee (28, 29, 30, 31)$$
$$\vee (6, 22) \vee (1, 3, 5, 7) \vee \ (5, 7, 13, 15) \text{ or } (13, 15, 29, 31)$$

$\vee$ (17, 19, 25, 27) or (25, 27, 29, 31)

$= v\bar{w}\bar{x} \vee vwx \vee \bar{w}xyz \vee \bar{v}\bar{w}z \vee (\bar{v}xz \text{ or } wxz) \vee (v\bar{x}z \text{ or } vwz)$.

Example 4.10.2 Minimize $f(abcde) = \vee(1, 5, 7, 8, 9, 10, 11, 17, 21, 23, 24, 25, 26, 29, 30, 31)$.

Solution: Form Tables T_0, T_1, and T_2 (Fig.4-37); terms exhibiting adjacency of 2nd degree are 7, 11, and 30. Hence in Table T_2 the terms (8, 9, 10, 11),

T_0

	a	b	c	d	e	1	2	3	4	5
1	0	0	0	0	1	X	X	X		
(8)	0	1	0	0	0	X	X	X		
(5)	0	0	1	0	1	X	X	X		
(9)	0	1	0	0	1	X	X	X	X	
(10)	0	1	0	1	0	X	X	X		
17	1	0	0	0	1	X	X	X		
24	1	1	0	0	0	X	X	X		
(7)	0	0	1	1	1	X	X			
(11)	0	1	0	1	1	X	X			
(21)	1	0	1	0	1	X	X	X	X	
25	1	1	0	0	1	X	X	X	X	
26	1	1	0	1	0	X	X	X		
(23)	1	0	1	1	1	X	X	X		
29	1	1	1	0	1	X	X	X		
30	1	1	1	1	0	X	X			
31	1	1	1	1	1	X	X	X		

T_1

	a	b	c	d	e	
1 5	0	0	-	0	1	X
1 9	0	-	0	0	1	X
1 17	-	0	0	0	1	X
8 9	0	1	0	0	-	X
8 10	-	1	0	0	0	X
8 24	-	1	0	0	0	X
5 7	0	0	1	-	1	X
5 21	-	0	1	0	1	X
9 11	0	1	0	-	1	X
9 25	-	1	0	0	1	X
10 11	0	1	0	1	-	X
10 26	-	1	0	1	0	X
17 21	1	0	-	0	1	X
17 25	1	-	0	0	1	X
24 25	1	1	0	0	-	X
24 26	1	1	0	-	0	X
7 23	-	0	1	1	1	X
21 23	1	0	1	-	1	X
21 29	1	-	1	0	1	X
25 29	1	1	-	0	1	X
26 30	1	1	-	1	0	A
23 31	1	-	1	1	1	X
29 31	1	1	1	-	1	X
30 31	1	1	1	1	-	B

T_2

	a	b	c	d	e	
1 ~~5~~ 17 ~~21~~	-	0	-	0	1	C
1 ~~9~~ 17 25	-	-	0	0	1	D
(8 9 10 11)	0	1	0	-	-	✓
~~8~~ ~~9~~ 24 25	-	1	0	0	-	E
~~8~~ ~~10~~ 24 26	-	1	0	-	1	F
(5 7 21 23)	-	0	1	-	1	✓
17 ~~21~~ 25 29	1	-	-	0	1	G
~~21~~ ~~23~~ 29 31	1	-	1	-	1	H

Figure 4-37 Tables T_0, T_1 and T_2 for Example 4.10.2.

and (5, 7, 21, 23) are circled and ticked as EPI's. The subsuming minterms are circled in Table T_0. As Table T_3 cannot be formed, the SPI Table as shown in Fig.4-38 is formed. Applying the principle of relative dominance o SPI's in a sequence as shown below the table, the MSP form is

$$f(abcde) = (1, 5, 17, 21) \vee (8, 9, 10, 11) \vee (5, 7, 21, 23)$$

$$\vee (26, 30) \vee (8, 9, 24, 25) \vee (21, 23, 29, 31)$$

$$= \bar{b}\bar{d}e \vee \bar{a}b\bar{c} \vee \bar{b}ce \vee abd\bar{e} \vee b\bar{c}\bar{d} \vee ace.$$

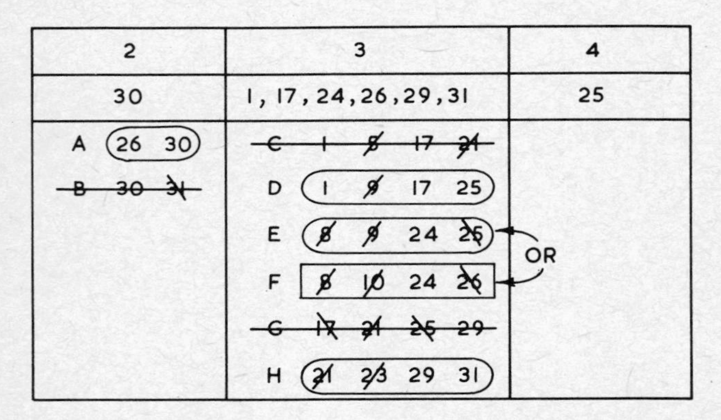

Figure 4-38 SPI table for Example 4.10.2.

Following another sequence (Fig.4-39) the MSP form is given by

$$f(abcde) = (8, 9, 10, 11) \vee (5, 7, 21, 23)$$

$$\vee (30, 31) \vee (1, 9, 17, 25) \vee (8, 10, 24, 26) \vee (17, 21, 25, 29) \text{ or}$$

$$(21, 23, 29, 31)$$

$$= \bar{a}b\bar{c} \vee \bar{b}ce \vee abcd \vee \bar{c}\bar{d}e \vee b\bar{c}\bar{e} \vee \quad (a\bar{d}e \text{ or } ace).$$

The example below illustrates the procedure to be followed when the applic tion of the principle of relative dominance of SPI's may not be adequate.

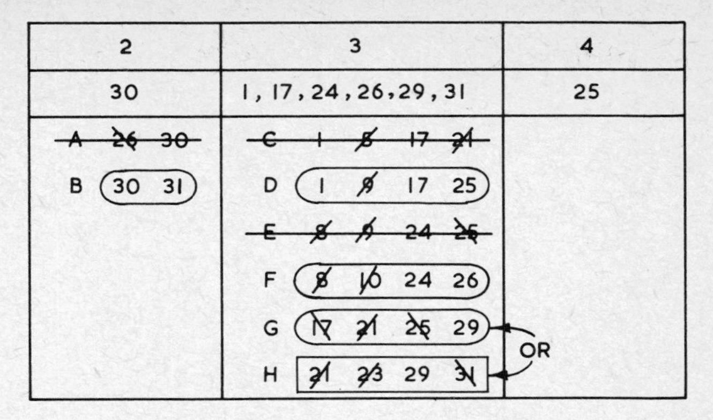

D⊃C → C̸, (D) → F⊃E → E̸, (F) → B⊃A ---

---→ A̸, (B) → G = H, (G) OR (H)

Figure 4-39 SPI table for Example 4.10.2.

Example 4.10.3 Minimize the function $f(abcde) = \vee(0, 4, 12, 16, 19, 24, 27, 28, 29, 31)$.

Solution: Form Tables T_0 and T_1 (Fig.4-40). Only the term 19 has an adjacency of the first degree. So the term (19, 27) in Table T_1 is an EPI. This term is circled and ticked. The terms 19 and 27 are circled on Table T_0. Table T_2 cannot be formed. The SPI Table is now formed with two sections headed by adjacency values of 2 and 3 only (Fig.4-41). In SPI H, the term 27 is cancelled. Consequently, it is dominated by the SPI I, and is, there-

T_0

	a	b	c	d	e	1	2	3	4	5
0	0	0	0	0	0	X	X			
4	0	0	1	0	0	X	X			
16	1	0	0	0	0	X	X			
12	0	1	1	0	0	X	X			
24	1	1	0	0	0	X	X			
(19)	1	0	0	1	1	X				
28	1	1	1	0	0	X	X	X		
(27)	1	1	0	1	1	X	X			
29	1	1	1	0	1	X	X			
31	1	1	1	1	1	X	X			

T_1

	a	b	c	d	e	
0 4	0	0	–	0	0	A
0 16	–	0	0	0	0	B
4 12	0	–	1	0	0	C
16 24	1	–	0	0	0	D
12 28	–	1	1	0	0	E
24 28	1	1	–	0	0	F
(19 27)	1	–	0	1	1	✓
28 29	1	1	1	0	–	G
2̸7̸ 31	1	1	–	1	1	H
29 31	1	1	1	–	1	I

Figure 4-40 Tables T_0 and T_1 for Example 4.10.3.

2	3
0, 4, 16, 12, 24, 29, 31	28
A : 0 4	
B : 0 16	
C : 4 12	
D : 16 24	
E : 12 28	
F : 24 28	
~~G : 28 29~~	
~~H : 29 31~~	
I : (29 31)	

$I \supset H \rightarrow \not{H}$, (I) $\rightarrow G \subset E$ AND F ---→ $\not{\emptyset}$, CLOSED SPI CHAIN

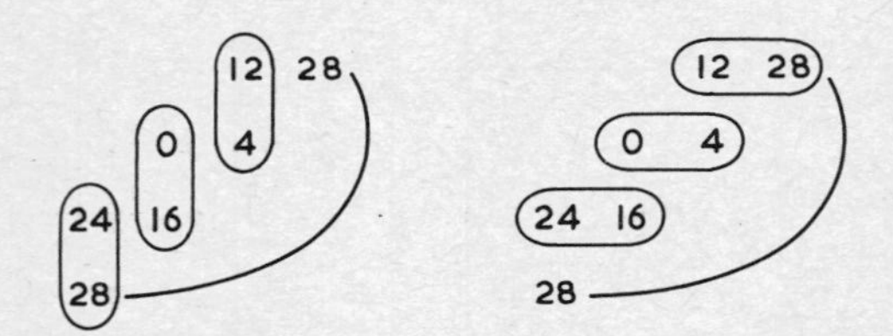

Figure 4-41 SPI table and chain for Example 4.10.3.

fore, struck out. The SPI I now becomes an essential SPI, since it alone cover 31. Consequently 29 of G is cancelled, and it becomes dominated by E and F. The principle of relative dominance of SPI's cannot be applied any more. We must proceed now to form the SPI chain. An SPI chain is formed by writing one of the two interconnecting links horizontally and the other vertically. In this example the SPI chain formed by the SPI's A, B, C, D, E and F is also shown in Fig.4-41. An interesting feature of this chain is that it is closed. *The formation of a closed SPI chain is the indication of the existence of cyclic prime implicants.* In such cases the set of SPIs constituting the close chain can be divided into two subsets, each covering all the subsuming minterms. We may call one the horizontal and the other the vertical subset. The two subsets really represent the two solutions of the familiar branching methc

T_0

	a	b	c	d	1	2	3	4
1	0	0	0	1	X	X	X	
2	0	0	1	0	X	X	X	
8	1	0	0	0	X	X	X	
3	0	0	1	1	X	X	X	
5	0	1	0	1	X	X	X	
6	0	1	1	0	X	X	X	
9	1	0	0	1	X	X	X	
10	1	0	1	0	X	X	X	
12	1	1	0	0	X	X	X	
7	0	1	1	1	X	X	X	
13	1	1	0	1	X	X	X	
14	1	1	1	0	X	X	X	

T_1

		a	b	c	d	
1	3	0	0	-	1	X
1	5	0	-	0	1	X
1	9	-	0	0	1	X
2	3	0	0	1	-	X
2	6	0	-	1	0	X
2	10	-	0	1	0	X
8	9	1	0	0	-	X
8	10	1	0	-	0	X
8	12	1	-	0	0	X
3	7	0	-	1	1	X
5	7	0	1	-	1	X
5	13	-	1	0	1	X
6	7	0	1	1	-	X
6	14	-	1	1	0	X
9	13	1	-	0	1	X
10	14	1	-	1	0	X
12	13	1	1	0	-	X
12	14	1	1	-	0	X

T_2

				a	b	c	d	
1	3	5	7	0	-	-	1	
1	5	9	13	-	-	0	1	
2	3	6	7	0	-	1	-	
2	6	10	14	-	-	1	0	
8	9	12	13	1	-	0	-	
8	10	12	14	1	-	-	0	

Figure 4-42 Tables T_0, T_1 and T_2 for Example 4.10.3.

The function has two valid MSP forms, which are as follows:

$$f(abcde) = (19, 27) \vee (29, 31) \vee (4, 12) \vee (0, 16) \vee (24, 28)$$

$$= a\bar{c}de \vee abce \vee \bar{a}c\bar{d}\bar{e} \vee \bar{b}\bar{c}\bar{d}\bar{e} \vee ab\bar{d}\bar{e}$$

and

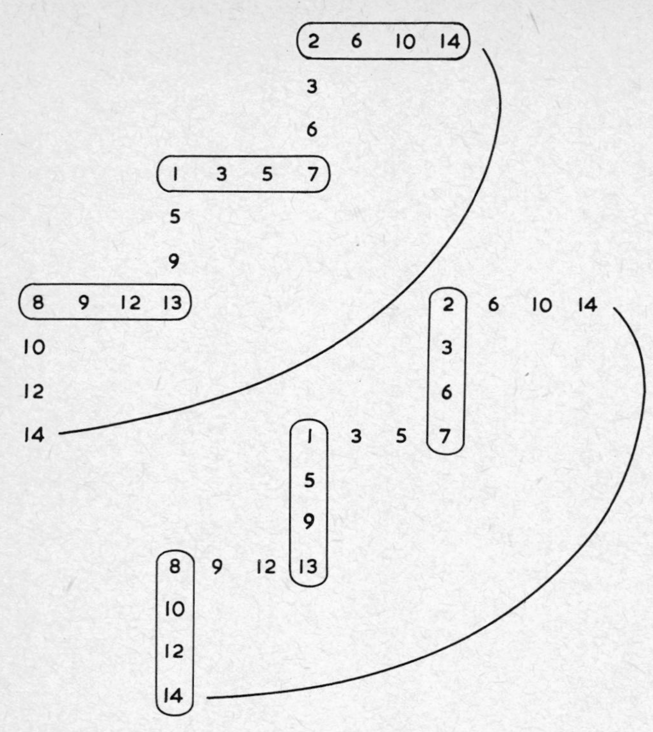

Figure 4-43 SPI chain for Example 4.10.3.

$$f(abcde) = (19, 27) \vee (29, 31) \vee (12, 28) \vee (0, 4) \vee (16, 24)$$

$$= a\bar{c}de \vee abce \vee bc\bar{d}\bar{e} \vee \bar{a}\bar{b}\bar{d}\bar{e} \vee a\bar{c}\bar{d}\bar{e}.$$

Example 4.10.4 Minimize the function of Example 4.9.2 by the adjacency method.

Form Tables T_0, T_1, and T_2 (Fig.4-42). In this example all terms in Table T_0 have adjacencies of 3rd degree. But Table T_3 cannot be formed. Hence th function does not have any EPI. Table T_2 may serve as the SPI Table. After the SPI chain is formed, it can be seen to be a closed one (Fig.4-43). Hence any of the two subsets, horizontal or vertical, can be chosen. Consequently t function has two valid MSP forms. These are:

$$f(abcd) = (2, 6, 10, 14) \vee (1, 3, 5, 7) \vee (8, 9, 12, 13)$$

$$= c\bar{d} \vee \bar{a}d \vee a\bar{c}$$

and

$$f(abcd) = (2, 3, 6, 7) \vee (1, 5, 9, 13) \vee (8, 10, 12, 14)$$
$$= \bar{a}c \vee \bar{c}d \vee a\bar{d}.$$

Don't-care terms can be taken care of by observing the following rules.

1) Don't-care terms are treated like other minterms while forming Table T_0. Once Table T_0 is formed, these terms are cancelled on Table T_0 itself.

2) Although cancelled, these terms initiate comparison and combine with other terms to generate a term in the next table.

3) While moving from one table to the next, the terms remain cancelled. Thus, the terms $\not{3}$ and 7 of Table T_0 generate the term ($\not{3}$, 7) in Table T_1.

4) A term in any table, whose *all* constituent minterms are cancelled does not qualify to be reckoned either as an essential or as a selective prime implicant.

The minimizing technique as presented in this section has the merit of enabling one to choose the essential prime implicants from a Table T_a as soon as it is formed. In the Quine–McCluskey method only prime implicants can be chosen from a Table T_a, after forming Table T_{a+1}. Thus it is very likely that more tables are to be formed in the Q–M method than the method presented here. Also this method does not need the successive prime implicant tables from which rows and columns are eliminated to select essential terms. The introduction of the concept of selective prime implicants and their chains may prove very useful.

REFERENCES

Biswas, N. N.: 'Minimization of Boolean functions', *IEEE Trans. Computers,* **C-20**, pp.925–929, August 1971.

Karnaugh, M.: 'The map method for the synthesis of combinational logic circuits', *AIEE Trans (Commun. Electron),* **72**, pp.593–598, Nov. 1953.

Krieger, M.: *Basic Switching Circuit Theory,* Macmillan, New York, 1967.

McCluskey, E. J.: 'Minimization of Boolean functions', *Bell System Tech. J.,* **35**, No.6, pp.1417–1444, Nov. 1956.

——: *Introduction to the Theory of Switching Circuits,* McGraw Hill Book Co. New York, 1965.

Quine, W. V.: 'The problem of simplifying truth functions' *Am. Math. Monthly,* **59**, pp.521–531, Oct. 1952.

Scheinman, A. H.: 'A method for simplifying Boolean functions' *Bell System Tech. J.* **41**, pp.1337–1346, July 1962.

Veitch, E. W.: 'A chart method for simplifying truth functions', *Proc. Assoc. of Compt. Mach.,* pp.127–133, 1952.

PROBLEMS

4.1 Minimize the following functions from their n-cubical structures:

a) $f(xy) = \vee(0, 2, 3)$

b) $f(xyz) = \vee(0, 2, 4, 6)$

c) $f(xyz) = \vee(1, 2, 5, 6, 7)$

d) $f(xyz) = \vee(1, 3, 5, 7)$.

4.2 Represent the following functions on n-cubes, and thereby write them as sums of minterms.

a) $f_1 = x \vee \bar{y}$

b) $f_2 = x_1 \vee \bar{x}_2 x_3$

c) $f_3 = \bar{x}_1 \vee x_1 \bar{x}_2 x_3$

d) $f_4 = \bar{x} \vee y \vee z$.

4.3 Given the functions $f_1 = \bar{a}b \vee \bar{a}d \vee bc$, $f_2 = \bar{a}b \vee \bar{a}\bar{b}d \vee abc$, and $f_3 = \bar{a}\bar{c}d \vee bcd \vee \bar{a}b\bar{d}$, show that $f_1 = f_2$ and $f_3 < f_1$.

4.4 Find the DCFs of the following functions

a) $f = \bar{a} \vee bc$

b) $f = \bar{a}b \vee a\bar{b} \vee c$

c) $f = a \vee \bar{c}d$

d) $f = \bar{a}\bar{b}\bar{c} \vee bd$.

4.5 Find the CCFs of the functions of problem 4.4.

4.6 Plot the following functions on maps and determine all the prime implicants and then classify them as essential, redundant, and selective prime implicants.

a) $\vee(0, 1, 4, 6, 7)$

b) $\vee(1, 3, 4, 5)$

c) $\vee(0, 2, 4, 6, 8, 9, 10, 11)$

d) $\vee(2, 4, 5, 10, 11, 12, 13, 14, 15)$.

4.7 Minimize the following functions on the map.

a) $\vee(2, 3, 6, 7)$

b) $\vee(0, 2, 5, 6, 7)$

c) $\vee(1, 2, 3, 4, 7, 8, 10)$

d) $\vee(5, 7, 12, 13, 14, 15)$

e) $\vee(0, 2, 6, 8, 10, 14)$

f) $\vee(1, 3, 5, 7, 9, 11, 13, 15)$

g) $\vee(0, 2, 4, 6, 8, 10, 12, 14)$

h) $\vee(0, 2, 5, 7, 8, 10, 13, 15)$

i) $\vee(0, 1, 2, 3, 4, 6, 8, 9, 10, 11, 12, 14)$.

4.8 Determine the MSP forms of the following functions

a) $\vee(8, 9, 11, 12, 14) \vee \phi(1, 5, 7, 13, 15)$

b) $\vee(4, 8, 14, 15) \vee \phi(0, 1, 6, 9, 10, 12, 13)$

c) $\vee(1, 2, 6, 10, 12) \vee \phi(3, 5, 7, 13)$.

4.9 Determine the MPS forms of the functions of problem 4.8.

4.10 Redesign the networks of Fig.2-18 so that the cost will be minimum.

4.11 Implement the following functions with diode AND and OR gates. Assume that each variable is available in both true and complemented forms. The circuit must have the minimum number of diodes.

a) $f_1 = \vee(0, 1, 3, 4, 7, 10) \vee \phi(2, 6, 12, 13, 14, 15)$

b) $f_2 = \vee(2, 4, 6, 9, 12, 13, 14) \vee \phi(1, 5, 7, 10)$

c) $f_3 = \vee(0, 2, 9, 10, 11, 12) \vee \phi(3, 4, 6, 13, 14, 15)$

d) $f_4 = \vee(2, 8, 10, 12, 14) \vee \phi(3, 6, 9, 13, 15)$.

4.12 Minimize the following functions on map

a) $f(vwxyz) = \vee(2, 3, 4, 6, 8, 10, 12, 14, 16, 17, 18, 19, 20, 23, 25, 26, 27, 28, 30, 31)$

b) $f(vwxyz) = \vee(1, 5, 7, 10, 15, 21, 22, 23, 25, 28, 30) \vee \phi(14, 16, 17, 18, 29, 31)$

c) $f(uvwxyz) = \vee(0, 2, 4, 5, 8, 10, 11, 12, 14, 15, 18, 19, 20, 21, 22, 23, 26, 27, 29, 30, 31, 32, 33, 35, 36, 38, 39, 42, 44, 50, 51, 52, 54, 57, 58, 59, 60, 62, 63)$

d) $f(uvwxyz) = \vee(1, 2, 3, 4, 6, 8, 9, 10, 12, 13, 14, 16, 17, 18, 19, 21, 24, 25, 27, 28, 29, 30, 31, 33, 34, 35, 36, 37, 38, 39, 40, 42, 43, 44, 50, 51) \vee \phi(56–63)$.

4.13 Minimize the functions of problem 4.7 by the Quine-McCluskey method.

4.14 Minimize the following functions by the Quine–McCluskey method.

a) $\vee(0, 2, 6, 8, 10, 12, 14, 15, 18, 20, 21, 26, 30, 31)$

b) $\vee(1, 3, 9, 10, 12, 13, 15, 18, 19, 20, 26, 27, 29, 30)$

c) $\vee(2, 3, 8–11, 18–25, 28, 29)$

d) $\vee(2, 3, 8–11, 14, 18–25, 28, 29)$

e) $\vee(6–10, 15–19, 21, 30, 31)$

f) $\vee(0, 3, 7–12, 19, 20, 21, 23, 24, 25, 26, 30)$

g) $\vee(0–7, 12–15, 16–19, 24–27, 40–47, 56–59)$.

4.15 Minimize the functions of problem 4.7 by the adjacency method.

4.16 Minimize the functions of problem 4.12 by the adjacency method.

4.17 Minimize the functions of problem 4.14 by the adjacency method. Which ones of these exhibit cyclic prime implicants?

4.18 Minimize the following function by the adjacency method:

$$f = \vee(2, 3, 5, 7, 8, 10, 12, 13, 17, 19, 20, 21, 26, 28, 30).$$

4.19 Can a self-dual function be recognized as such by plotting it on the map? [Hint: see Section 3.10.]

Chapter 5

SYMMETRIC FUNCTIONS

5.1 INTRODUCTION

The symmetric functions constitute an important subclass of the switching functions. By virtue of the symmetry possessed by these functions, they exhibit many interesting properties. The study of these properties is not only helpful in synthesizing such functions by special methods, but also throws important light in the structure of Boolean functions in general.

Definition 5.1.1 A Boolean function is a *totally symmetric function (TSF)*, if and only if it remains invariant (unchanged) by any permutation of its variables.

It can be easily verified that the function $f(x_1x_2) = \bar{x}_1\bar{x}_2 \vee x_1x_2$ does not change if x_1 and x_2 are permuted. Hence this is a totally symmetric function. Similarly the function $f(x_1x_2x_3) = \bar{x}_1\bar{x}_2\bar{x}_3 \vee x_1x_2x_3$ can be recognized as a TSF.

Definition 5.1.2 When a Boolean function remains invariant by any permutation between some (and not all) of its variables, then the function is known as a *partially symmetric function (PSF)*.

Thus, the function $f = x_1 \vee x_2x_3$, although not a TSF, is symmetric in x_2 and x_3, and is, therefore, a PSF.

In some cases, it may be possible to recognize a partially or a totally symmetric function simply by looking at it. But this may not be the case always. For example, let us consider the function

$$f(x_1x_2x_3) = \bar{x}_1x_2\bar{x}_3 \vee x_1\bar{x}_2x_3.$$

This function does not appear to be symmetric at the first sight. Now put $y_1 = x_1$, $y_2 = \bar{x}_2$ and $y_3 = x_3$ in the above function, so that it can be rewritten as

$$f(y_1y_2y_3) = \bar{y}_1\bar{y}_2\bar{y}_3 \vee y_1y_2y_3.$$

Obviously the function now appears to be symmetric, but the variables of symmetry are y_1, y_2 and y_3, that is, x_1, $\bar{x}_2$ and x_3. This example shows that the detection of symmetry is not always very simple. Systematic procedures for detection of symmetric functions have been discussed in later sections.

5.2 ELEMENTARY SYMMETRIC FUNCTION

Before defining an elementary symmetric function, we proceed to prove the following theorem which states a condition which is sufficient to make a Boolean function totally symmetric.

Theorem 5.2.1 If the disjunctive canonical form of a Boolean function contains *all* minterms of only one particular weight, then the function is totally symmetric.

Proof: When a pair of variables is permuted in a minterm m_i of weight w say, the minterm may still remain the same minterm m_i, or become another minterm, m_j; but in both cases its weight remains w. Since the function contains *all* minterms of weight w, it remains invariant for any permutation between its variables. ▲

As an example, in a three variable case, the minterms with weight 2, are m_3, m_5 and m_6. Therefore, the function

$$f(x_1x_2x_3) = m_3 \vee m_5 \vee m_6$$
$$= \bar{x}_1x_2x_3 \vee x_1\bar{x}_2x_3 \vee x_1x_2\bar{x}_3$$

is totally symmetric.

It should be mentioned that while Th.5.2.1 states a condition which is sufficient to make a function a TSF, it is not a necessary condition. In other words, not all TSFs are necessarily the sum of minterms of only *one* particular weight. In fact such TSFs which have minterms of only one weight are a subset of the set of all TSFs, and these are known as elementary symmetric functions.

Definition 5.2.1 A totally symmetric function whose DCF is the sum of all minterms of only one weight, is known as an *elementary symmetric function* (ESF).

An ESF of minterms of weight a, and with $x_1, x_2, \ldots$ and x_n as variables of symmetry is written as

$$S_a(x_1x_2 \ldots x_n).$$

The number a is often referred to as the a-number of the symmetric function. Thus the function $f(x_1x_2x_3) = m_3 \vee m_5 \vee m_6$ is written as

$$S_2(x_1x_2x_3)$$

showing that 2 is the a-number, and x_1, x_2, and x_3 are the variables of symmetry. It is now evident that in an n-cube there are $n + 1$ ESFs as the number of possible weights (and hence of possible a-numbers) is $n + 1$, namely $0, 1, 2, \ldots n$.

Theorem 5.2.2 An elementary symmetric function $S_a(x_1x_2 \ldots x_n)$ becomes 1, if and only if exactly a number of variables become 1.

Proof: The proof of this theorem follows from the fact that the ESF $S_a(x_1x_2 \ldots x_n)$ contains *all* minterms of weight a, and no minterm of weight other than a. ▲

Theorem 5.2.3 The number of minterms in (and consequently the number of rows in the tabular form of) an ESF $S_a(x_1x_2 \ldots x_n)$ is given by the binomial coefficient,

$$\binom{n}{a} = \frac{n!}{(n-a)!\,a!}.$$

Proof: The number of minterms in $S_a(x_1x_2 \ldots x_n)$
= number of minterms of weight a
= number of combinations of a 1's and $(n - a)$ 0's.

$$= {}^nC_a = \frac{n!}{(n-a)!\,a!} = \binom{n}{a}. \quad ▲$$

Theorem 5.2.4 In the tabular form of an ESF all columns have equal number of 1's and 0's.

Proof: Without any loss of generality let us prove it for the columns headed by the variables x_1 and x_2. Under these columns, the various combinations of 0's and 1's that may appear in any row, are 00, 11, 01 or 10. The number of 1's and 0's in these two columns will be equal, if the number of rows with 01 under x_1x_2 is equal to the number of rows with 10 under x_1x_2. Now, the ESF can be expressed as

$$F = \bar{x}_1\bar{x}_2F_a \vee x_1x_2F_b \vee \bar{x}_1x_2F_c \vee x_1\bar{x}_2F_d,$$

where F_a, F_b, F_c and F_d are all functions of $x_3x_4 \ldots x_n$ and are the sums of all rows with 00, 11, 01 and 10 respectively under x_1x_2. As the function F is totally symmetric, F_c must equal F_d. Hence the number of rows of F_c must equal the number of rows of F_d. This proves the theorem. ▲

Again, it must be mentioned here that Th.5.2.4 gives only a necessary, but not a sufficient condition for a function to be an ESF. For example, the function $f(x_1x_2x_3) = (3, 6, 9, 12)$ has equal number of 1's and 0's in each column, but is not an ESF.

Theorem 5.2.5 The ratio of number of 1's to that of 0's of each column of the ESF $S_a(x_1x_2 \ldots x_n)$ is given by

$$N(1)/N(0) = \binom{n-1}{a-1} \Big/ \binom{n-1}{a}$$

Proof: Number of rows in the tabular form of the ESF = $\binom{n}{a}$.
Number of 1's in each row = a, ∴ total number of 1's in the table = $a\binom{n}{a}$.

$\because$ By Th.5.2.4, the 1's are equally shared by all the n columns, the number of 1's per column

$$N(1) = \frac{a\binom{n}{a}}{n} = \binom{n-1}{a-1}.$$

Then the number of 0's in each column

$$N(0) = \text{number of rows} - N(1)$$

$$= \binom{n}{a} - \binom{n-1}{a-1} = \binom{n-1}{a}$$

since, by the recursion relation for binomial coefficients,

$$\binom{p}{q} + \binom{p}{q+1} = \binom{p+1}{q+1}.$$

Hence,

$$N(1)/N(0) = \binom{n-1}{a-1} \Big/ \binom{n-1}{a}. \qquad \blacktriangle$$

Theorem 5.2.6 The elementary symmetric functions $S_0(x_1x_2 \ldots x_n)$ and $S_n(x_1x_2 \ldots x_n)$ are single term symmetric functions and can be expanded as follows.

$$S_0(x_1x_2 \ldots x_n) = \bar{x}_i S_0(x_1x_2 \ldots x_{i-1}, x_{i+1} \ldots x_n)$$

$$S_n(x_1x_2 \ldots x_n) = x_i S_{n-1}(x_1x_2 \ldots x_{i-1}, x_{i+1} \ldots x_n).$$

Proof: Number of minterms in $S_0(x_1x_2 \ldots x_n)$

$$= \binom{n}{0} = 1.$$

$\therefore$ It has only one term and its weight is 0. Hence,

$$S_0(x_1x_2 \ldots x_n) = \bar{x}_1\bar{x}_2 \ldots \bar{x}_n$$

$$= \bar{x}_i(\bar{x}_1 \ldots \bar{x}_{i-1}\bar{x}_{1+1} \ldots \bar{x}_n)$$

$$= \bar{x}_i S_0(x_1x_2 \ldots x_{i-1}x_{i+1} \ldots x_n).$$

The other part can also be proved in a similar manner. $\blacktriangle$

Theorem 5.2.7 An elementary symmetric function $S_a(x_1x_2 \ldots x_n)$, where $0 < a < n$ can be expanded as

$$\bar{x}_i S_a(x_1 \ldots x_{i-1}, x_{i+1} \ldots x_n) \vee$$

$$x_i S_{a-1}(x_1 \ldots x_{i-1}, x_{i+1} \ldots x_n).$$

Proof: Expanding $S_a(x_1 \ldots x_n)$ about x_i

$$S_a(x_1 \ldots x_n) = \bar{x}_i f(x_1 \ldots x_{i-1}, x_{i+1} \ldots x_n)$$
$$\vee\, x_i g(x_1 \ldots x_{i-1}, x_{i+1} \ldots x_n),$$

where $f(x_1 \ldots x_{i-1}, x_{i+1} \ldots x_n)$ is the sum of all rows of the tabular form of $S_a(x_1 \ldots x_n)$, where there are 0's under the column of x_i. Hence, f is a function of $(n-1)$ variables, and has $\binom{n-1}{a}$ number of rows each of weight a.

$\therefore f$ is the symmetric function

$$S_a(x_1 \ldots x_{i-1}, x_{i+1} \ldots x_n).$$

Similarly it can be shown that g is the symmetric function,

$$S_{a-1}(x_1 \ldots x_{i-1}, x_{i+1} \ldots x_n). \quad \blacktriangle$$

Theorem 5.2.8 An elementary symmetric function $S_a(x_1 \ldots x_n)$ can also be expressed as $S_{n-a}(\bar{x}_1 \ldots \bar{x}_n)$.

Proof: The tabular form of the ESF $S_a(x_1 \ldots x_n)$ has $\binom{n}{a}$ rows of weight a. Double complementing all columns, the tabular form will have $\binom{n}{a}$ rows each of weight $(n-a)$.

$$\because \binom{n}{a} = \binom{n}{n-a},$$

the new tabular form also represents an ESF, which can be written as

$$S_{n-a}(\bar{x}_1 \ldots \bar{x}_n).$$

Hence

$$S_a(x_1 \ldots x_n) = S_{n-a}(\bar{x}_1 \ldots \bar{x}_n). \quad \blacktriangle$$

Theorem 5.2.8 shows that any ESF can have two designations. Thus,

$$S_2(x_1 x_2 x_3) = S_1(\bar{x}_1 \bar{x}_2 \bar{x}_3)$$

and

$$S_3(x_1 x_2 \bar{x}_3 x_4 \bar{x}_5) = S_2(\bar{x}_1 \bar{x}_2 x_3 \bar{x}_4 x_5).$$

5.3 SYMMETRIC FUNCTIONS

Very often the term symmetric function is used to mean a totally symmetric function, which has been defined in Def.5.1.1. In this section, some of the important properties of TSFs will be developed. It may be noticed that elementary symmetric functions are also TSFs.

Theorem 5.3.1 A Boolean function is totally symmetric if and only if it can be expressed either as an ESF or as a sum of ESFs with the same variables of symmetry.

Proof: i) *Sufficiency:* If the function can be expressed as a sum of ESFs, then the function remains invariant for all permutations of variables, as each of the constituent ESF remains invariant.

ii) *Necessity:* Let the Boolean function be written in its partitioned tabular form. Now, if the function cannot be expressed as a sum of ESFs, then there exists at least one partition of row weight say w, where the number of rows is less than $\binom{n}{w}$. Hence the partition does not have *all* minterms of weight w, and therefore cannot remain invariant for all permutations among the variables. This proves the necessity of the function being expressed as a sum of ESFs. ▲

Let a TSF $f(x_1 \ldots x_n)$ be expressed as follows

$$f(x_1 \ldots x_n) = S_{a_1}(x_1 \ldots x_n) \vee S_{a_2}(x_1 \ldots x_n) \vee \ldots S_{a_k}(x_1 \ldots x_n),$$

then $f(x_1 \ldots x_n)$ is written as

$$S_{a_1, a_2, \ldots a_k}(x_1 \ldots x_n).$$

Thus, we may define the a-numbers in the following way.

Definition 5.3.1 The *a-numbers* of a totally symmetric function are the row weights in the tabular form of the function, where the variables heading the columns of the table are the variables of symmetry.

COROLLARY **5.3.1A** The ratio of number of 1's to that of 0's of all columns of a totally symmetric function is the same.

Proof: Follows from Theorems 5.3.1 and 5.2.4. ▲

This theorem and its corollary tell us how a given Boolean function can be identified to be a symmetric function.

Example 5.3.1 Determine if the following function is a TSF.

$$f(x_1x_2x_3x_4) = \vee\,(1, 2, 4, 7, 8, 13, 14).$$

Solution: The tabular form of the function is shown in Table I of Fig.5-1. The function exhibits two reciprocal ratios. Doubly complementing column x_2, we get Table II, wherein the row-weights are 2, and 0. Since,

$$N(2) = 6 = \binom{4}{2}$$

$$N(0) = 1 = \binom{4}{0}$$

the function is a TSF, and can be written as follows

I				II				
x_1	x_2	x_3	x_4	x_1	$\bar{x}_2$	x_3	x_4	r_w
0	0	0	1	0	1	0	1	2
0	0	1	0	0	1	1	0	2
0	1	0	0	0	0	0	0	0
0	1	1	1	0	0	1	1	2
1	0	0	0	1	1	0	0	2
1	1	0	1	1	0	0	1	2
1	1	1	0	1	0	1	0	2
$\frac{3}{4}$	$\frac{4}{3}$	$\frac{3}{4}$	$\frac{3}{4}$	$\frac{3}{4}$	$\frac{3}{4}$	$\frac{3}{4}$	$\frac{3}{4}$	

Figure 5-1

$$f = S_{02}(x_1\bar{x}_2x_3x_4). \tag{5.1}$$

By Th.5.3.6 the function $f = S_{02}(x_1\bar{x}_2x_3x_4)$ can also be written as

$$f = S_{24}(\bar{x}_1x_2\bar{x}_3\bar{x}_4). \tag{5.2}$$

A consequence of Th.5.3.6 (which states that all TSFs can be designated in two alternative forms) is that a flexible approach is available in the method of detection. For example, if in the above example we would have taken the ratio 4/3 as correct, then we would have obtained the function as given by Eqn.(5.2) first and as given by Eqn.(5.1) next. This shows that we arrive at the same result, no matter which way we start.

For functions whose columns have equal number of 1's and 0's, it is not possible to recognize the columns which need double complementation. Hence this type of TSFs (which have been called the 'unity-ratio' symmetric functions) has been dealt with in a subsequent section.

Theorem 5.3.2 A totally symmetric function $S_{a_1,a_2...a_k}(x_1 \ldots x_n)$ becomes 1, if and only if exactly $a_m(m = 1, 2, \ldots, k)$ of the variables of symmetry become 1 simultaneously.

Proof: The validity of this theorem follows from the Theorems 5.3.1 and 5.2.2. ▲

Following this theorem one can provide an alternative definition of a-numbers than that given in Def. 5.3.1 above.

Definition 5.3.2 For a totally symmetric function there exists a set of integers $a_k(0 \leqslant k \leqslant n)$ such that the function becomes 1 if and only if exactly $a_m(m = 1, 2, \ldots, k)$ of the variables of symmetry becomes 1 simultaneously. The integers a_k are known as the *a-numbers* of the given TSF.

It may be observed that a symmetric function is uniquely defined, and its canonical structure is completely known, as soon as its a-numbers and the variables of symmetry are specified.

Theorem 5.3.3 The Boolean sum of two symmetric functions with the same variables of symmetry is also a symmetric function with the same variables of symmetry and with a-numbers as given by

$$S_{\{a_i\}}(x_1 \ldots x_n) \vee S_{\{a_j\}}(x_1 \ldots x_n)$$
$$= S_{\{a_i\} \cup \{a_j\}}(x_1 \ldots x_n).$$

Proof: The proof of this theorem follows from Th.5.3.1. ▲

Thus,

$$S_{125}(x_1 x_2 x_3 \bar{x}_1 x_5) \vee S_{24}(x_1 x_2 x_3 \bar{x}_4 x_5)$$
$$= S_{1245}(x_1 x_2 x_3 \bar{x}_4 x_5).$$

Theorem 5.3.4 The Boolean product of two symmetric functions with the same variables of symmetry is also a symmetric function with the same variables of symmetry and with a-numbers as given by

$$S_{\{a_i\}}(x_1 \ldots x_n)\, S_{\{a_j\}}(x_1 \ldots x_n)$$
$$= S_{\{a_i\} \cap \{a_j\}}(x_1 \ldots x_n).$$

Proof: The validity of this theorem also follows from Th.5.3.1. ▲

Thus,

$$S_{1235}(x_1 \bar{x}_2 x_3 \bar{x}_4 x_5)\, S_{245}(x_1 \bar{x}_2 x_3 \bar{x}_4 x_5)$$
$$= S_{25}(x_1 \bar{x}_2 x_3 \bar{x}_4 x_5).$$

Theorem 5.3.5 The complement of a symmetric function is also a symmetric function with a-numbers and variables of symmetry as given by

$$\bar{S}_{\{a_i\}}(x_1 \ldots x_n)$$
$$= S_{\{a_j\}}(x_1 \ldots x_n),$$

where

$$a_i \neq a_j$$

and

$$\{a_i\} \cup \{a_j\} = \{0, 1, \ldots, n\}.$$

Proof: Let $S_{\{a_i\}}(x_1 \ldots x_n) = \vee\, m_i$

$$\therefore \bar{S}_{\{a_i\}}(x_1 \ldots x_n) = \vee\, m_j,$$

where

$$m_i \neq m_j$$

and

$$\{m_i\} \cup \{m_j\} = \{m_0, m_1, \ldots, m_{2n-1}\}.$$

Now, $\{m_i\}$ has all minterms of weight $\{a_i\}$

$\therefore$ $\{m_j\}$ must have all minterms of weight $\{a_j\}$

$\therefore \vee m_j = S_{\{a_j\}}(x_1 \ldots x_n)$. ▲

Theorem 5.3.6 A symmetric function $S_{a_1,a_2 \ldots a_k}(x_1 \ldots x_n)$ can also be expressed as

$$S_{n-a_1, \ldots, n-a_k}(\bar{x}_1 \ldots \bar{x}_n).$$

Proof: The validity of this theorem follows from Th.5.3.1 and Th.5.2.8. ▲

Theorem 5.3.7 A symmetric function $S_{a_1,a_2,\ldots a_k}(x_1 \ldots x_n)$ can be expanded as

$$\bar{x}_i S_{a_1,a_2 \ldots a_k}(x_1 \ldots x_{i-1}, x_{i+1} \ldots x_n)$$
$$\vee x_1 S_{a_1-1,a_2-1,\ldots a_k-1}(x_1 \ldots x_{i-1}, x_{i+1} \ldots x_n),$$

where $a_1 - 1$ and a_k are dropped in the expansion if $a_1 = 0$ and $a_k = n$ respectively.

Proof: Follows from Theorems 5.2.6 and 5.2.7. ▲

5.4 THE UNITY RATIO SYMMETRIC FUNCTIONS

Definition 5.4.1 When each column of the tabular form of a TSF has equal number of 1's and 0's then the TSF will be called a *unity ratio totally symmetric function.*

The a-numbers and the number of 1's and 0's of such functions (up to 6 variables) are given in the Table of Fig.5-2. It may be noticed that some of the functions in the Table have been circled. These form an interesting subclass among the unity-ratio symmetric functions, inasmuch as they remain invariant for all combinations of true and complemented variables of symmetry. To demonstrate this point, let us take the function

$$F = S_{024}(x_1x_2x_3x_4) = \vee\,(0, 3, 5, 6, 9, 10, 12, 15).$$

In Table (I) of Fig.5-3, the function is written in its tabular form. In Table (II), the x_1 column has been doubly complemented. Consequently there are four rows with weight 1, and four rows with weight 3. Since,

$\frac{N(1)}{N(0)}$	α-numbers when n is				
	2	3	4	5	6
1:1	(1 / 0 2)	0 3	0 4	0 5	0 6
2:2		(1 3 / 0 2)			
3:3		1 2	2		
4:4			(1 3 / 0 2 4)		
5:5			0 1 3 4	1 4	
6:6				0 1 4 5	1 5
7:7			1 2 3		0 1 5 6
8:8				(1 3 5 / 0 2 4)	
10:10				2 3	3
11:11				0 2 3 5	2 5 6 0 3 6 0 1 4
15:15				1 2 3 4	2 4
16:16					(1 3 5 / 0 2 4 6)
17:17					0 1 3 5 6
21:21					2 3 5 6 1 2 4 5 0 1 3 5
22:22					0 1 2 4 5 6
25:25					2 3 4
26:26					0 2 3 4 6
31:31					1 2 3 4 5

Figure 5-2 Unity-ratio symmetric functions of up to 6 variables.

(I)

x_1	x_2	x_3	x_4	r_w
0	0	0	0	0
0	0	1	1	2
0	1	0	1	2
0	1	1	0	2
1	0	0	1	2
1	0	1	0	2
1	1	0	0	2
1	1	1	1	4
$\frac{4}{4}$	$\frac{4}{4}$	$\frac{4}{4}$	$\frac{4}{4}$	

(II)

$\bar{x}_1$	x_2	x_3	x_4	r_w
1	0	0	0	1
1	0	1	1	3
1	1	0	1	3
1	1	1	0	3
0	0	0	1	1
0	0	1	0	1
0	1	0	0	1
0	1	1	1	3
$\frac{4}{4}$	$\frac{4}{4}$	$\frac{4}{4}$	$\frac{4}{4}$	

(III)

$\bar{x}_1$	$\bar{x}_2$	x_3	x_4	r_w
1	1	0	0	2
1	1	1	1	4
1	0	0	1	2
1	0	1	0	2
0	1	0	1	2
0	1	1	0	2
0	0	0	0	0
0	0	1	1	2
$\frac{4}{4}$	$\frac{4}{4}$	$\frac{4}{4}$	$\frac{4}{4}$	

Figure 5-3

$$\binom{4}{1} = 4 \text{ and } \binom{4}{3} = 4,$$

the function as depicted in Table (II) is a TSF, and $F = S_{13}(\bar{x}_1 x_2 x_3 x_4)$. By doubly complementing any other column of Table (II), the function gets changed into as shown in Table (III). This can easily be checked to be a TSF of the designation,

$$F = S_{024}(\bar{x}_1 \bar{x}_2 x_3 x_4).$$

Thus the function oscillates between two alternative a-number designations, namely S_{13} and S_{024}, by double complementation of any one column. There are two such functions in each n-cube. Their ratio of number of 1's to that of 0's is given by

$$N(1) : N(0) = 2^{(n-2)} : 2^{(n-2)}.$$

One of the functions has all the odd a-numbers, and the other all the even a-numbers [Biswas, 1969 (a,b)]. The two functions can be formally defined as follows.

Definition 5.4.1 Let the entire set of $(n + 1)$ a-numbers $0, 1, 2, \ldots, n$ be partitioned into two subsets $\{a_{\text{odd}}\}$ and $\{a_{\text{even}}\}$ so that $\{a_{\text{odd}}\} = \{a | a$ is odd, and $1 \leqslant a \leqslant n\}$ and $\{a_{\text{even}}\} = \{a | a$ is even, and $0 \leqslant a \leqslant n\}$ and let two TSFs F_o and F_e be defined as follows

$$F_o = S_{\{a_{\text{odd}}\}}(x_1 x_2 \ldots x_n)$$

$$F_e = S_{\{a_{\text{even}}\}}(x_1 x_2 \ldots x_n).$$

It can be shown that F_o and F_e are complements of each other. Further it can be verified that

$$F_o = S_{\{a_{\text{odd}}\}}(x_1 x_2 \ldots x_p \ldots x_q \ldots x_n) \tag{5.3}$$

$$= S_{\{a_{\text{even}}\}}(x_1 x_2 \ldots \bar{x}_p \ldots x_q \ldots x_n) \tag{5.4}$$

$$= S_{\{a_{\text{odd}}\}}(x_1 x_2 \ldots \bar{x}_p \ldots \bar{x}_q \ldots x_n) \tag{5.5}$$

and

$$F_e = S_{\{a_{\text{even}}\}}(x_1 x_2 \ldots x_p \ldots x_q \ldots x_n) \tag{5.6}$$

$$= S_{\{a_{\text{odd}}\}}(x_1 x_2 \ldots \bar{x}_p \ldots x_q \ldots x_n) \tag{5.7}$$

$$= S_{\{a_{\text{even}}\}}(x_1 x_2 \ldots \bar{x}_p \ldots \bar{x}_q \ldots x_n). \tag{5.8}$$

Equations (5.3) through (5.8) reveal the criteria by which a given function in this subclass can be identified to be either F_o or F_e. Let k be the number of variables of symmetry in the complemented form; then

a) A given function in this class is F_o if both k and row-weights are not odd/even simultaneously.

b) A given function is F_e if both k and row-weights are odd/even simultaneously.

The a-numbers and variables of symmetry of a given Boolean function belonging to this special class of unity-ratio TSFs can easily be determined as soon as the function is written in its tabular form. However, other unity-ratio symmetric functions cannot be so easily detected. As has been pointed out earlier they also defy the procedure as applied in detecting the function of Example 5.3.1. The special techniques for the detection and identification of such functions are described in the following two sections.

(I)

x_1	x_2	x_3	x_4	x_5	r_w
0	0	0	1	1	2
0	0	1	0	1	2
0	0	1	1	0	2
0	1	0	0	0	1
0	1	1	1	1	4
1	0	0	0	0	1
1	0	1	1	1	4
1	1	0	0	1	3
1	1	0	1	0	3
1	1	1	0	0	3
$\frac{5}{5}$	$\frac{5}{5}$	$\frac{5}{5}$	$\frac{5}{5}$	$\frac{5}{5}$	

(II)

x_2	x_3	x_4	x_5	r_w	r'_w with x_2	x_3	x_4	x_5
0	0	1	1	2	3			
0	1	0	1	2	3 $N(3) = 4$			
0	1	1	0	2	3 $N(0) = 1$			
1	0	0	0	1	0			
1	1	1	1	4	3			
$\frac{2}{3}$	$\frac{3}{2}$	$\frac{3}{2}$	$\frac{3}{2}$					

(III)

x_2	x_3	x_4	x_5	r_w	r'_w with x_2	x_3	x_4	x_5
0	0	0	0	0	1 $N(1) = 4$			
0	1	1	1	3	4 $N(4) = 1$			
1	0	0	1	2	1			
1	0	1	0	2	1			
1	1	0	0	2	1			
$\frac{3}{2}$	$\frac{2}{3}$	$\frac{2}{3}$	$\frac{2}{3}$					

Figure 5-4

5.5 THE DECOMPOSITION METHOD: McCLUSKEY (1956)

Let the method be illustrated by working out an example.

Example 5.5.1 Is the following function a TSF? If yes, determine its *a*-numbers and variables of symmetry.

$f(x_1x_2x_3x_4x_5) = \vee\,(3, 5, 6, 8, 15, 16, 23, 25, 26, 28)$

Solution: Write the function in its tabular form (Fig.5-4). The ratio of the number of 1's to that of 0's of its columns is 5:5. It has 3 rows of weight 2, 3 rows of weight 3, and 2 rows each of weight 1 and 4. But $\binom{5}{2} = 10$. Hence the function cannot be a TSF with x_1, x_2, x_3, x_4 and x_5 as variables of symmetry. Since all columns have the ratio 5:5, no indication can be obtained as to which columns need double complementation, if at all the function is a TSF. So, we decompose the function around the variable x_1, and study the two functions. The tabular forms of these two functions are obtained from Table (I), and are shown in Tables (II) and (III). The column ratios of Table (II) indicate that the column of x_2 should be double complemented. On so doing the new row weights r'_w will be as shown. Now,

$$\binom{4}{3} = 4, \text{ and } \binom{4}{0} = 1.$$

Hence, the function is a TSF with $\bar{x}_2$, x_3, x_4 and x_5 as variables of symmetry, and it can be written as

$S_{03}(\bar{x}_2x_3x_4x_5)$.

A similar procedure in Table (III) detects the function to be a TSF with the designation:

$S_{14}(\bar{x}_2x_3x_4x_5)$.

$\therefore$ The original function, f, can be written as

$$f = \bar{x}_1\, S_{03}(\bar{x}_2x_3x_4x_5) \vee x_1\, S_{14}(\bar{x}_2x_3x_4x_5)$$

$$= \bar{\bar{x}}_1\, S_{14}(\bar{x}_2x_3x_4x_5) \vee \bar{x}_1\, S_{03}(\bar{x}_2x_3x_4x_5)$$

$$= S_{14}(\bar{x}_1\bar{x}_2x_3x_4x_5) \qquad \text{by Th.5.3.7}$$

$$= S_{14}(x_1x_2\bar{x}_3\bar{x}_4\bar{x}_5) \qquad \text{by Th.5.3.6}$$

5.6 THE METHOD OF ORDERED PARTITION: BISWAS (1970)

Let the example of the previous section be solved by this method. However we must first define the ordered partitioned tabular form. We have already come across the partitioned tabular form while forming table T_0 in the Quine–McCluskey method of minimizing a Boolean function.

Definition 5.6.1 If in each partition of the partitioned tabular form the minterms appear from top to bottom of the partition in ascending order of their decimal designation, then the resulting tabular form is known as the *ordered partitioned tabular form.*

Writing the function of Example 5.5.1, in the ordered partitioned tabular form, we get the table of Fig.5-5. We now look for columns with an all 0 or all 1 entry within each partition. These are the columns which are to be doubly complemented. In the table of Fig.5-5, partition of row-weights 1 and 4 indicate that columns of x_3, x_4 and x_5 should be double complemented, whereas the partitions of row-weights 2 and 3 indicate that the x_1 and x_2 columns should be doubly complemented. These two indications are not contradictory, as by

	x_1	x_2	x_3	x_4	x_5	r_W	r'_W with $\bar{x}_1$ $\bar{x}_2$ x_3 x_4 x_5	
8	0	1	(0)	(0)	(0)	1	1	$N(1) = 5$ ✓
16	1	0	(0)	(0)	(0)	1	1	$\binom{5}{1} = 5$
3	(0)	(0)	0	1	1	2	4	
5	(0)	(0)	1	0	1	2	4	$N(4) = 5$ ✓
6	(0)	(0)	1	1	0	2	4	
25	(1)	(1)	0	0	1	3	1	$\binom{5}{4} = 5$
26	(1)	(1)	0	1	0	3	1	
30	(1)	(1)	1	0	0	3	1	
15	0	1	(1)	(1)	(1)	4	4	
23	1	0	(1)	(1)	(1)	4	4	
	$\frac{5}{5}$	$\frac{5}{5}$	$\frac{5}{5}$	$\frac{5}{5}$	$\frac{5}{5}$			

Figure 5-5

Th.5.3.6 if the function turns out to be a TSF with x_1, x_2, $\bar{x}_3$, $\bar{x}_4$ and $\bar{x}_5$ as variables of symmetry, then it can also be expressed as a TSF with $\bar{x}_1$, $\bar{x}_2$, x_3, x_4 and x_5 as variables of symmetry. Using the latter set, the new row-weights r'_w are as shown.

Since $\binom{5}{1} = 5$ and $\binom{5}{4} = 5$, the function is a TSF, with the designation,

$$f = S_{14}(\bar{x}_1\bar{x}_2x_3x_4x_5)$$

$$= S_{14}(x_1x_2\bar{x}_3\bar{x}_4\bar{x}_5) \qquad \text{by Th.5.3.6.}$$

There may be cases, where no partition may show a column with an all 0/1 entry. In such cases, we must look within the *sub-partitions* for an all 0/1 entry. A sub-partition is obtained by drawing a dotted line dividing the 0's from 1's under the x_1 column within a partition. This is demonstrated in the following example.

Example 5.6.1 Is the following function a TSF? If so, determine its identifying designation.

$$f(x_1x_2x_3x_4x_5x_6) = \vee\,(0, 3, 6, 9, 12, 15, 18, 24, 27, 30, 33, 36, 39, 45, 48, 51, 54, 57, 60, 63).$$

Solution: Write the function in its ordered partitioned tabular form (Fig.5-6). The partitions of row-weights of 0 and 6 have one row each. Therefore these are of no help to us. The partitions of row-weights 2 and 4 do not have any column with an all 0 or all 1 entry. So, we check the sub-partitions, and find that columns x_1, x_3 and x_5 are to be doubly complemented. As the new row-weight matches the binomial distribution, the function is a TSF. Its identifying designation is

$$f = S_3(\bar{x}_1x_2\bar{x}_3x_4\bar{x}_5x_6)$$

$$= S_3(x_1\bar{x}_2x_3\bar{x}_4x_5\bar{x}_6) \qquad \text{by Th.5.3.6.}$$

The method as outlined in this section virtually presents the columns to be detected in a pictorial form. In fact, if a computer program is written to produce the ordered partitioned tabular form of the function under test, the columns to be doubly complemented will be immediately visible.

Although this algorithm gives correct results in most cases, it needs minor modification to take care of a limited number of functions which otherwise escape detection. Such a case may arise where the function under test exhibits all 0 and/or all 1 columns in the subpartitions of its ordered partitioned tabular form. Consider the following example.

Example 5.6.2 Is the following function a TSF? If so, determine its a-numbers and variables of symmetry.

$$f(x_1x_2x_3x_4x_5x_6) = \vee\,(3, 12, 20, 24, 29, 30, 33, 34, 39, 43, 51, 60).$$

Solution: The ordered partitioned tabular form of the function is shown in

	x_1	x_2	x_3	x_4	x_5	x_6	r_W	r'_W with $\bar{x}_1$ x_2 $\bar{x}_3$ x_4 $\bar{x}_5$ x_6
0	0	0	0	0	0	0	0	3
3	0	0	0	0	1	1	2	3
6	0	0	0	1	1	0	2	3
9	0	0	1	0	0	1	2	3
12	0	0	1	1	0	0	2	3
18	0	1	0	0	1	0	2	3
24	0	1	1	0	0	0	2	3
33	1	0	0	0	0	1	2	3
36	1	0	0	1	0	0	2	3
48	1	1	0	0	0	0	2	3
15	0	0	1	1	1	1	4	3
27	0	1	1	0	1	1	4	3
30	0	1	1	1	1	0	4	3
39	1	0	0	1	1	1	4	3
45	1	0	1	1	0	1	4	3
51	1	1	0	0	1	1	4	3
54	1	1	0	1	1	0	4	3
57	1	1	1	0	0	1	4	3
60	1	1	1	1	0	0	4	3
63	1	1	1	1	1	1	6	3
	$\frac{10}{10}$	$\frac{10}{10}$	$\frac{10}{10}$	$\frac{10}{10}$	$\frac{10}{10}$	$\frac{10}{10}$		

$N(3) = 20$ ✓

$\binom{6}{3} = 20$

Figure 5-6

Section A of Fig.5-7. The table has only two partitions, one of row-weight 2, and the other of row-weight 4. None of these partitions has columns with all 0/all 1 entries. Hence we construct the subpartitions by drawing dotted lines separating the 0's and 1's of the x_1 column. The dotted line divides the partition into two-sub-partitions. The subpartition with smaller number of rows is called the *indicator subpartition*, as it is in this subpartition that the all 0/all 1 entries are looked for. When the two subpartitions have equal numbers of rows, either can be taken as an indicator subpartition. The indicator subpartitions of Fig.5-7 now indicate that columns x_1, x_2, x_3 and x_4 are to be doubly complemented. If this is done, the variables heading the columns become $\overline{x}_1$, $\overline{x}_2$, $\overline{x}_3$, $\overline{x}_4$, x_5 and x_6 and the new row-weights turn out to be as shown in section B of Fig.5-7. It can now be seen that the number of occurrence of rows with row-weight 2 does not satisfy the binomial distribution. According to our algorithm as expounded above the function is to be declared as not being totally symmetric.

It will now be interesting to investigate the consequence of doubly complementing the column x_1 in the tabular form of section A of Fig.5-7. On so doing we obtain the table as shown in Section A of Fig.5-8. In this table the indicator subpartitions have become partitions of row-weights 1 and 5. The all 0/all 1 entries in these partitions indicate that columns of $\overline{x}_1, x_2, x_3$ and x_4 are to be doubly complemented. After double complementation, the

	SECTION A							SECTION B		SECTION C	
	x_1	x_2	x_3	x_4	x_5	x_6	r_w	r_w with $\overline{x}_1\ \overline{x}_2\ \overline{x}_3\ \overline{x}_4\ x_5\ x_6$		r_w with $x_1\ \overline{x}_2\ \overline{x}_3\ \overline{x}_4\ x_5\ x_6$	
3	0	0	0	0	1	1	2	6		5	
12	0	0	1	1	0	0	2	2	$N(6) = 1 \stackrel{\checkmark}{=} \binom{6}{6}$	1	$N(5) = 6 \stackrel{\checkmark}{=} \binom{6}{5}$
20	0	1	0	1	0	0	2	2		1	
24	0	1	1	0	0	0	2	2		1	
33	1	0	0	0	0	1	2	4	$N(2) = 5 \neq \binom{6}{2}$	5	$N(1) = 6 \stackrel{\checkmark}{=} \binom{6}{1}$
34	1	0	0	0	1	0	2	4		5	
29	0	1	1	1	0	1	4	2		1	
30	0	1	1	1	1	0	4	2		1	
39	1	0	0	1	1	1	4	4		5	
43	1	0	1	0	1	1	4	4		5	
51	1	1	0	0	1	1	4	4		5	
60	1	1	1	1	0	0	4	0		1	
	$\frac{6}{6}$	$\frac{6}{6}$	$\frac{6}{6}$	$\frac{6}{6}$	$\frac{6}{6}$	$\frac{6}{6}$					

Figure 5-7

SECTION A							SECTION B	
$\bar{x}_1$	x_2	x_3	x_4	x_5	x_6	r_w	r_w with $\bar{\bar{x}}_1$ $\bar{x}_2$ $\bar{x}_3$ $\bar{x}_4$ x_5 x_6	
1	0	0	0	1	1	3	5	
1	0	1	1	0	0	3	1	$N(5) = 6 \stackrel{\checkmark}{=} \binom{6}{5}$
1	1	0	1	0	0	3	1	
1	1	1	0	0	0	3	1	
0	0	0	0	0	1	1	5	$N(1) = 6 \stackrel{\checkmark}{=} \binom{6}{1}$
0	0	0	0	1	0	1	5	
1	1	1	1	0	1	5	1	
1	1	1	1	1	0	5	1	
0	0	0	1	1	1	3	5	
0	0	1	0	1	1	3	5	
0	1	0	0	1	1	3	5	
0	1	1	1	0	0	3	1	

Figure 5-8

variables heading the table become $\bar{\bar{x}}_1$, $\bar{x}_2$, $\bar{x}_3$, $\bar{x}_4$, x_5 and x_6, and the row-weights become as shown in section B of Fig.5-8. The numbers of occurrences satisfy the binomial distribution. Hence the function is totally symmetric, and can be written as

$$f = S_{15}(x_1\bar{x}_2\bar{x}_3\bar{x}_4x_5x_6)$$
$$= S_{15}(\bar{x}_1x_2x_3x_4\bar{x}_5\bar{x}_6).$$

Thus it is seen that the conclusion derived from the tables of sections A and B of Fig.5-7 was erroneous. This will happen in cases where the ordered partitioned table exhibits only sub-partitions, and further the indicator sub-partitions become partitions when the x_1 column is double complemented. In such a case, let w_1 and w_2 ($w_1 < w_2$) be the weights of the two indicator sub-partitions, and therefore of the partitions also. It should be noted that because the partitions are ordered, w_1 is always less than w_2. The upper subpartition of the partition of weight w_1 will have all 0's in column x_1, whereas the lower subpartition will have all 1's in column x_1. Similarly the upper subpartition of the partition of weight w_2 will have all 1's in column x_1, and the lower subpartition will have all 0's in column x_1. Hence, after doubly complementing column x_1, the new weights of the partitions will be $w_1 + 1, w_1 - 1, w_2 + 1$, a $w_2 - 1$ (see Fig.5-8). Since w_1 is less than w_2, weights $w_1 - 1$ and $w_2 + 1$ will always be different from the others. The weights $w_1 + 1$ and $w_2 - 1$ can be equal when $w_2 - w_1 = 2$.

Thus, a function such as that of this example, which exhibits two partition at first, will show 4 or 3 partitions after x_1 column is doubly complemented. It should be observed here that such a situation is brought about by doubly complementing only the x_1 column, and not any other column. This is because all other columns have both 0's and 1's in the rows of the other sub-partition (non-indicator) of a partition. Hence we must modify the previous algorithm. A generalized algorithm for the detection and identification of totally symmetric functions can now be written as follows.

Step 1: Write the function in its ordered partitioned tabular form.

Step 2: Determine the ratio of the number of 1's to that of 0's of each column. If the ratio is the same for all columns, go to step 7, If not, go to step 3.

Step 3: Check if columns have two ratios, one of which is the reciprocal of the other. If yes, then doubly complement the columns with the reciprocal ratios, and go to step 4. If the columns exhibit more than two ratios, or two ratios which are not reciprocal to one another, then the function is not a TSF. Stop.

Step 4: Determine the weight of each row, and count the number of rows wi weight a. Let these be $N(a_1), N(a_2), \ldots, N(a_k)$.

Step 5: Check if for each $a_i, N(a_i) = \binom{n}{a_i}$. If this relation is satisfied for all a_i's then the function is a TSF. Go to step 6. If the relation fails for even a single a_i, then the function is not a TSF. Stop.

Step 6: Note the variables heading the columns of the tabular form. Let these be $x_1, x_2, \ldots, x_n$. Then the designation of the function is as follows

$$S_{a_1, a_2 \ldots a_k}(x_1 x_2 \ldots x_n). \text{ Stop.}$$

Step 7: Check if the ratio is unity. If yes, go to step 8. If not, go to step 4.

Step 8: Determine the weight of each row and count the number of rows with weight a. Let these be $N(a_1), N(a_2), \ldots, N(a_k)$.

Step 9: Check if for each $a_i, N(a_i) = \binom{n}{a_i}$. If this relation is satisfied for all a_i's, then the function is a TSF. Go to step 6. If not, go to step 10.

Step 10: Detect column/columns having all 0 and/or all 1 entries in one or more partitions. If such columns are detected, go to step 12. If not go to the next step.

Step 11: Detect column/columns having all 0 and/or all 1 entries in one or more indicator subpartitions. If such columns are detected go to next step.

If not, the function is not a TSF. Stop.

Step 12: In case more than one partition or indicator subpartition have indicated all 0 and/or all 1 columns, check that they are not contradictory. If not, go to the next step. If yes, the function is not a TSF. Stop.

Step 13: Doubly complement the detected columns, calculate the new row-weights. Check that the number of occurrences of rows of row-weight a equals $\binom{n}{a}$. If this is not satisfied by even a single a, then go to the next step. If yes, then the function is a TSF. Write its designation with a-numbers and variables of symmetry. Stop.

Step 14: Check if the all 0 and/or all 1 columns have been detected by partitions or subpartitions. If by partitions, then the function is not a TSF; stop. If by sub-partitions, then go to the next step.

Step 15: Doubly complement all detected columns except that of x_1. Calculate the new row-weights. Check that the number of occurrences of rows of row-weight a equals $\binom{n}{a}$, for all a's. If this is not satisfied by even a single a, then the function is not a TSF, stop. If yes the function is a TSF. Write its designation with a-numbers and variables of symmetry. Stop.

5.7 REALIZATION OF SYMMETRIC FUNCTIONS BY CONTACT NETWORKS

Like any other Boolean function, symmetric functions can also be realized by series – parallel contact networks after the MSP form of the function has been determined. However, often a more minimal circuit can be arrived at by pursuing a different approach in implementing a symmetric function. The basis of the design is a single input-multiple output starting network. As an example, for any three variable TSF with x_1, x_2 and x_3 as variables of symmetry, the starting network is as shown in Fig.5-9. It can be seen that transmission between the terminals **I** and **O** is 1, if and only if none of the three relays X_1, X_2 and X_3 operates. In this condition, the transmission between I and any other terminal is 0. Between I and **1**, the transmission is 1, if and only if any one of the three relays operates at a time. At this condition the transmission between **I** and any other terminal is 0. Hence the output at terminal **1** is $S_1(x_1x_2x_3)$. Similarly the outputs at terminals 2 and 3 can be easily verified to be $S_2(x_1x_2x_3)$ and $S_3(x_1x_2x_3)$ respectively. Thus any of the four ESFs of the three variable function can be obtained from one of the four output terminals. If the function to be implemented is, say, $S_{13}(x_1x_2x_3)$ then, it can be obtained by joining terminals 1 and 3, and then simplifying the network to minimize the number of contacts. The minimization can be effected a) by removing superfluous contacts; b) by reducing the number of contacts by the application of well-known theorems such as those shown in Fig.5-10; and c) by folding.

Definition 5.7.1 When two nodes a and b reach the output terminal *via* identical contact-configurations, then one of the configurations is eliminated and

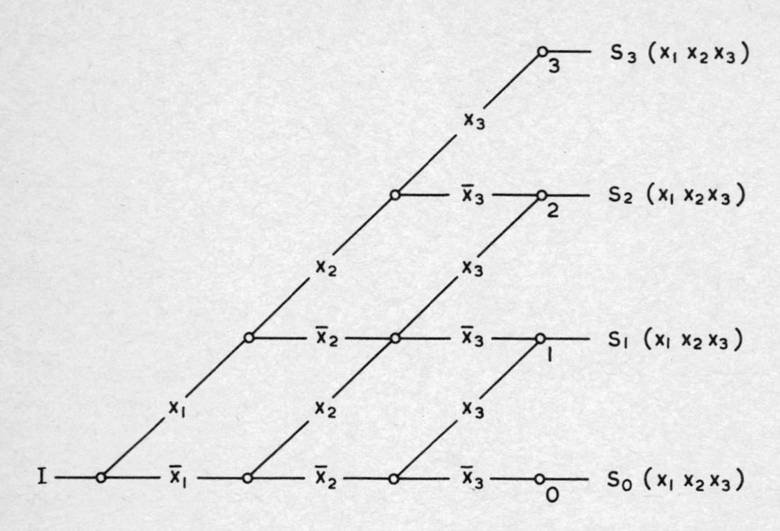

Figure 5-9 The starting network for the realization of any three-variable TSF with x_1, x_2 and x_3 as the variables of symmetry.

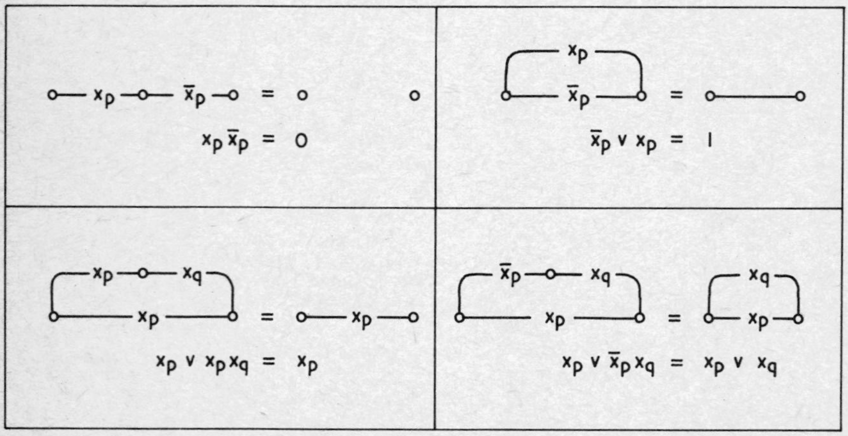

$x_p \bar{x}_p = 0$

$\bar{x}_p \vee x_p = 1$

$x_p \vee x_p x_q = x_p$

$x_p \vee \bar{x}_p x_q = x_p \vee x_q$

Figure 5-10 Theorems to reduce number of contacts.

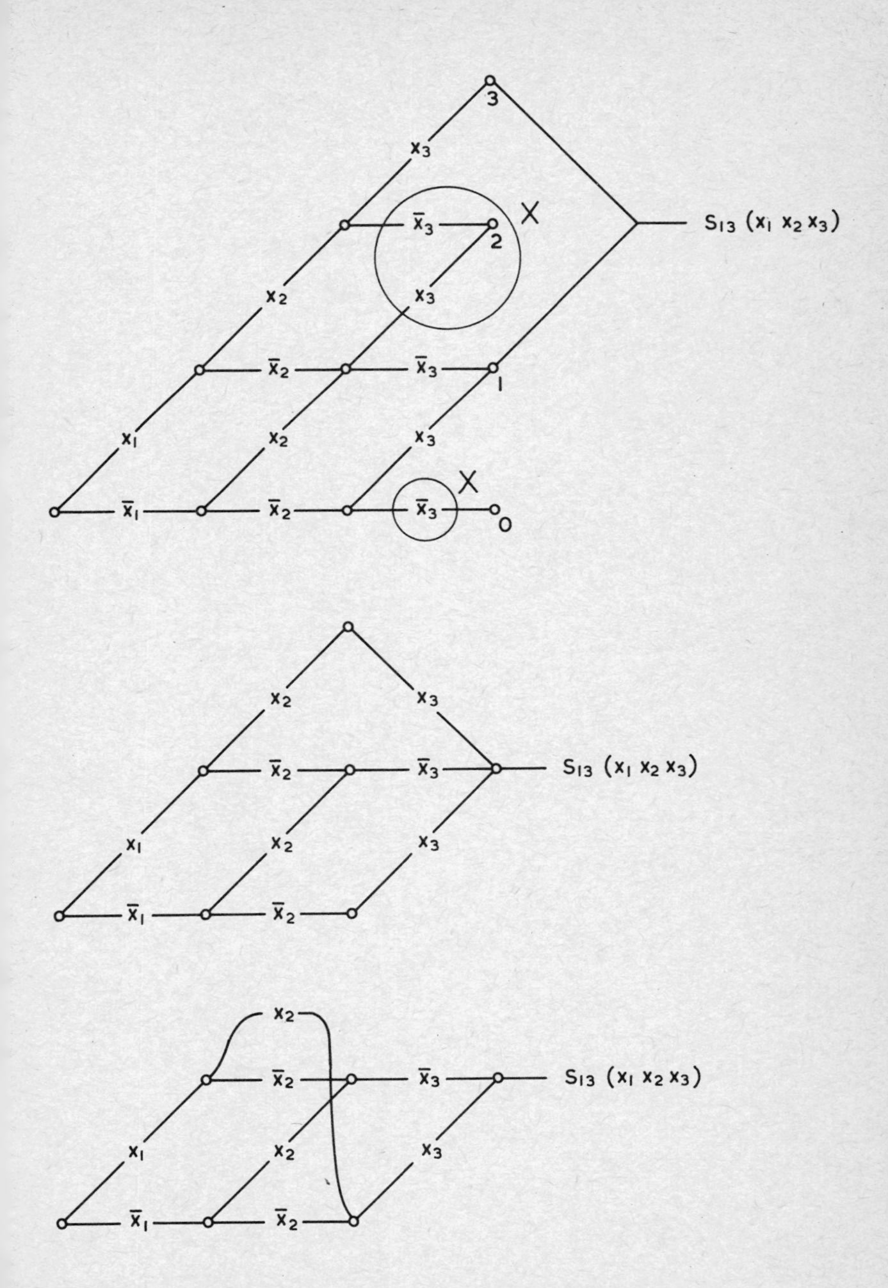

igure 5-11 Various steps in arriving at the minimal circuit for $S_{13}(x_1x_2x_3)$.

the node a is connected to node b by direct connection. This procedure is called *folding.*

The various steps to arrive at the minimal circuit for $S_{13}(x_1x_2x_3)$ are illustrated in Fig.5-11.

It can be easily seen that two or more symmetric functions with the same variables of symmetry but with disjoint sets of a-numbers can be realized from the same starting network, now functioning as a multiple output network. Also, if some of the variables of symmetry are complemented, then this must be reflected in the starting network by changing the true variables into complemented variables and *vice versa.*

Example 5.7.1 Realize a minimal network for $S_{01}(x_1x_2\bar{x}_3)$ and $S_3(x_1x_2\bar{x}_3)$.

Solution: The starting and the final networks are as shown in Fig.5-12.

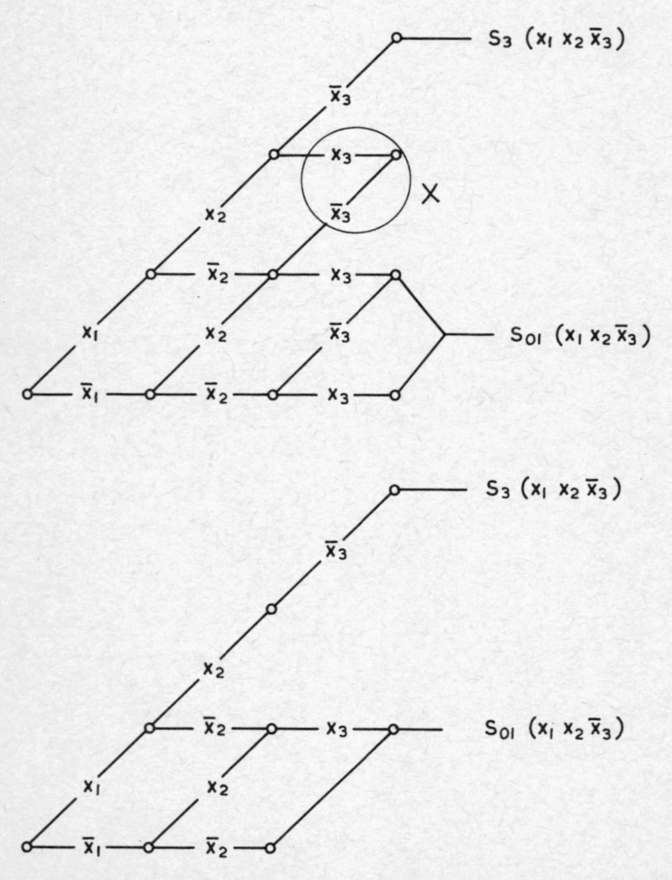

Figure 5-12 Starting and final networks to realize $S_{01}(x_1x_2\bar{x}_3)$ and $S_3(x_1x_2\bar{x}_3)$.

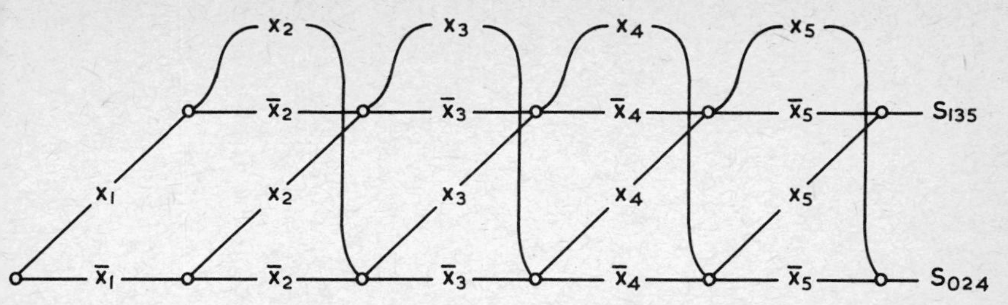

Figure 5-13 Folded network for the realization of $S_{024}(x_1x_2x_3x_4x_5)$ and $S_{135}(x_1x_2x_3x_4x_5)$.

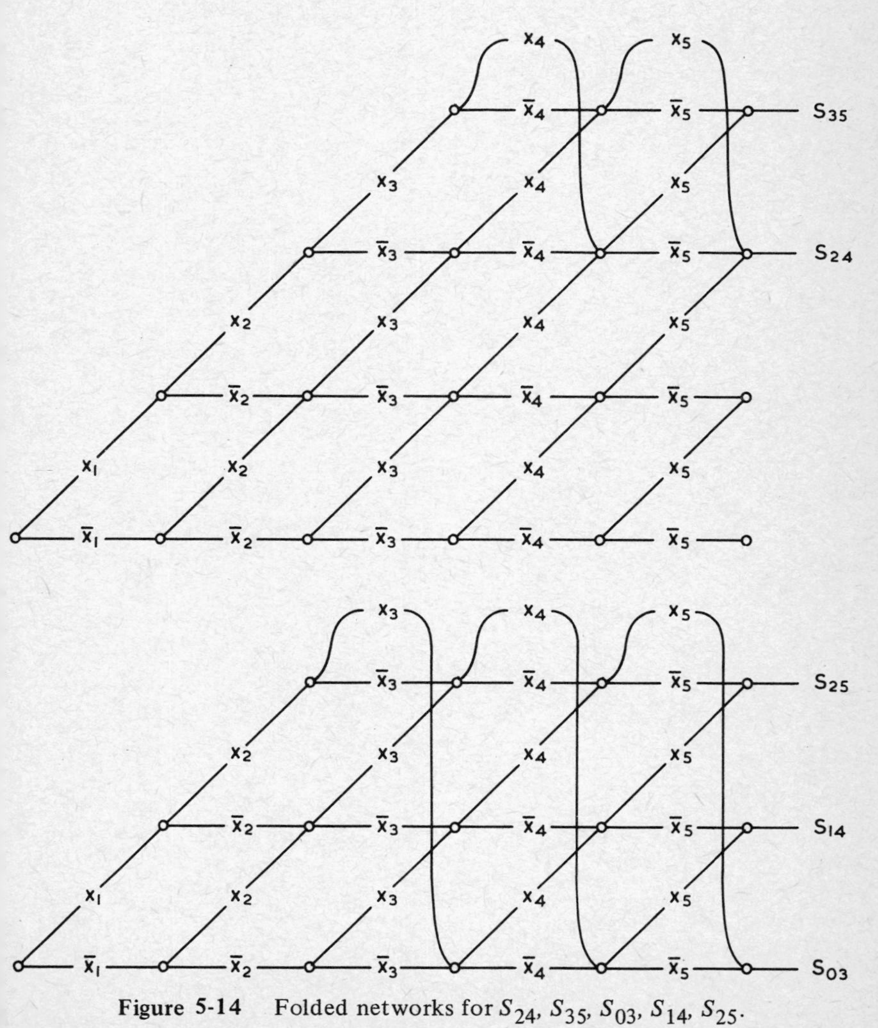

Figure 5-14 Folded networks for S_{24}, S_{35}, S_{03}, S_{14}, S_{25}.

Folded networks can be very conveniently used as a starting network for the synthesis of symmetric functions whose a-numbers form an arithmetic progression. Figure 5-13 shows the folded network whose two outputs realize the functions $S_{024}(x_1x_2x_3x_4x_5)$ and $S_{135}(x_1x_2x_3x_5)$. Figure 5-14 shows the folded networks for $S_{24}(x_1x_2x_3x_4x_5)$ and $S_{35}(x_1x_2x_3x_4x_5)$, and for $S_{03}(x_1x_2x_3x_4x_5)$, $S_{14}(x_1x_2x_3x_4x_5)$ and $S_{25}(x_1x_2x_3x_4x_5)$.

It must be observed that the progression of the a-numbers may start from any number but must terminate in the largest number nearest to n. Thus the functions $S_{0246}(x_1x_2x_3x_4x_5x_6)$ and $S_{246}(x_1x_2x_3x_4x_5x_6)$, can be realized by folded network. But no folded network exists for the functions $S_{024}(x_1x_2x_3x_4x_5x_6)$, and $S_{03}(x_1x_2x_3x_4x_5)$.

REFERENCES

Biswas, N. N.: "Special class of symmetric Boolean functions", *Electronics Letters, IEE (London)*, **5**, No.4, p.72 Feb.1969(a)

——: "Special class of symmetric Boolean functions: Proof and comment", *Electronics Letters, IEE (London),* **5**, No.19, pp.444–446, Sept.1969(b).

——: "On identification of totally symmetric Boolean functions", *IEEE Trans. Computers,* **C-19**, pp.645–648, July 70; **C-22**, pp.863–864, Sept.1973.

Caldwell, S. H.: "The recognition and identification of symmetric switching functions", *AIEE Trans. Communications and Electronics,* pp.142–146, May 1954.

Chu, Y.: *Digital Computer Design Fundamentals,* McGraw-Hill Book Co., New York, 1962.

Harrison, M. A.: *Introduction to Switching and Automata Theory,* McGraw-Hill Book Co., New York, 1965.

Marcus, M. P.: "The detection and identification of symmetric switching functions with the use of tables of combinations", *IRE Trans. Electronic Computers (Correspondence),* pp.237–239, Dec. 1956.

McCluskey, E. J.: "Detection of group invariance or total symmetry of a Boolean function", *Bell System. Tech. J.,* pp.1445–1453, Nov. 1956.

Miller, R. E.: *Switching Theory,* Vol.I, John Wiley and Sons, Inc., New York, 1965.

Povarov, G. N.: "A mathematical theory for the synthesis of contact networks with one input and k outputs". Proceedings International Symposium on the theory of Switching, April 2–5, 1957. *Ann. Computation Lab., Harvard University*, **30**, pp.74–94, 1959.

Shannon, C. E.: "A symbolic analysis of relay and switching circuits", *AIEE Trans.* pp.713–723, 1938.

PROBLEMS

5.1 Determine the variables of symmetry of the following totally symmetric functions:

$$f_1 = x_1 \vee \bar{x}_2 \vee x_3$$

$$f_2 = x_1\bar{x}_2x_3 \vee \bar{x}_1x_2\bar{x}_3$$

$$f_3 = (\bar{x}_1 \vee \bar{x}_2)(x_1 \vee x_2).$$

5.2 Express the functions of problem 5.1, in terms of a-numbers and variables of symmetry.

5.3 Determine the MSP forms of the following symmetric functions:

$$f_1 = S_{023}(x_1x_2x_3)$$

$$f_2 = S_{124}(x_1x_2x_3x_4)$$

$$f_3 = S_{34}(x_1\bar{x}_2x_3x_4)$$

$$f_4 = S_{023}(\bar{x}_1\bar{x}_2x_3x_4)$$

$$f_5 = S_{234}(\bar{x}_1\bar{x}_2\bar{x}_3\bar{x}_4).$$

5.4 Determine the number of 0's and 1's of each column of the following ESFs:

$$f_1 = S_1\ (x_1x_2x_3)$$

$$f_2 = S_2\ (x_1x_2x_3x_4)$$

$$f_3 = S_4\ (x_1x_2x_3x_4x_5)$$

$$f_4 = S_4\ (x_1x_2x_3x_4x_5x_6x_7)$$

$$f_5 = S_3\ (x_1x_2x_3x_4x_5x_6x_7x_8).$$

5.5 Determine the number of 0's and 1's of each column in the tabular form of the symmetric functions of problem 5.3.

5.6 If $f_1 = S_{23}(x_1x_2x_3x_4)$, $f_2 = S_{14}(x_1x_2x_3x_4)$ and $f_3 = S_{013}(x_1x_2x_3x_4)$, evaluate the following:

a) f_1f_2 b) f_2f_3 c) $f_1f_2f_3$ d) $f_2 \vee f_3$

e) $f_1 \vee f_2 \vee f_3$ f) $\bar{f}_1$ g) $\bar{f}_3$ h) $\bar{f}_1f_2$

i) $\bar{f}_2 \vee f_1$.

5.7 If $f_1 = S_{024}(x_1x_2x_3x_4)$ and $f_2 = S_{13}(\bar{x}_1\bar{x}_2\bar{x}_3\bar{x}_4)$, evaluate

a) f_1f_2 b) $f_1 \vee f_2$ c) $f_1\bar{f}_2$.

5.8 Determine if the following three variable functions are TSFs. If so, write them in the usual manner of designating TSFs.

a) 7; b) 26; c) 27; d) 36; e) 75; f) 177; g) 200; h) 226;

i) 350; j) 351; k) 357.

5.9 Are the following functions TSFs? If so, determine their identifying designations.

a) $f(x_1x_2x_3) = \vee\,(0, 1, 3, 4, 6, 7)$

b) $f(x_1x_2x_3x_4) = \vee\,(0, 3, 5, 6, 9, 10, 12, 15)$

c) $f(x_1x_2x_3x_4) = \vee\,(0, 4, 5, 6, 11, 12)$

d) $f(x_1x_2x_3x_4x_5) = \vee\,(1, 4, 5, 7, 8, 9, 11, 12, 14, 15, 16, 17, 19, 20, 22, 23, 24, 26, 27, 30)$

e) $f(x_1x_2x_3x_4x_5x_6) = \vee\,(2, 5, 8, 10, 11, 14, 16, 18, 19, 22, 24, 25, 27, 28, 30, 31, 33, 36, 37, 39, 42, 45, 50, 53, 58, 59, 62)$.

5.10 Detect and identify the following functions.

a) $\vee\,(1, 2, 4, 11, 13, 14)$

b) $\vee\,(0, 3, 5, 6, 9, 10, 12, 15)$

c) $\vee\,(3, 4, 8, 13, 14, 17, 18, 23, 27, 28)$

d) $\vee\,(2, 4, 7, 14, 22, 25, 38, 41, 49, 56, 59, 61)$.

5.11 Synthesize the following symmetric functions by contact network.

a) $S_{12}(x_1x_2x_3)$

b) $S_{23}(x_1\bar{x}_2x_3x_4)$

c) $S_{13}(x_1x_2\bar{x}_3x_4)$

d) $S_{02}(x_1x_2x_3x_4)$ and $S_{34}(x_1x_2x_3x_4)$

e) $S_{246}(x_1x_2x_3x_4x_5x_6)$

f) $S_{357}(x_1x_2x_3x_4x_5x_6x_7)$

g) $S_{036}(x_1x_2x_3x_4x_5x_6x_7)$.

Chapter 6

COMBINATIONAL CIRCUITS

6.1 INTRODUCTION

One of the major objectives of the logic designer is to design a circuit with the minimum cost. The minimization methods help achieve this object by removing redundant elements from the circuit. We have also seen how more economy can be introduced by special techniques if the Boolean function to be implemented turns out to be a symmetric one. In all these designs of combinational circuits, however, we have only encountered the AND, OR, and NOT operations. But these are by no means the only logical operations. In fact, it will be seen shortly that more economical circuits along with other advantages can be derived by using other types of logic gates.

6.2 FUNCTIONS OF TWO VARIABLES: LOGICAL OPERATIONS

It is now our purpose to show that besides the AND, OR and NOT logics, there are others which may serve the purpose as well or even better. A ready method of finding these operations is to take a close look at the various functions that may be possible of two variables. These are enumerated below in the form of truth table of two variables.

Functions f_0 through f_{15} are the sixteen different functions that are possible of two variables. If these are now written in terms of a and b, some significant logical operations can be found out. This has been done in Fig.6-2, where the

a	b	f_0	f_1	f_2	f_3	f_4	f_5	f_6	f_7	f_8	f_9	f_{10}	f_{11}	f_{12}	f_{13}	f_{14}	f_{15}
0	0	0	1	0	1	0	1	0	1	0	1	0	1	0	1	0	1
0	1	0	0	1	1	0	0	1	1	0	0	1	1	0	0	1	1
1	0	0	0	0	0	1	1	1	1	0	0	0	0	1	1	1	1
1	1	0	0	0	0	0	0	0	0	1	1	1	1	1	1	1	1

Figure 6-1 All functions of two variables.

Function	Boolean Expression	Operation	Symbol of Operation	Symbol of Logic Gate
f_0	0			
f_1	$\bar{a}\bar{b}$	NOR	$a \downarrow b$	a, b → [↓] → $\bar{a}\bar{b}$
f_2	$\bar{a}b$	INHIBIT		b, a (inhibiting) → [•] → $\bar{a}b$
f_3	$\bar{a}$			
f_4	$a\bar{b}$	INHIBIT		a, b (inhibiting) → [ ] → $a\bar{b}$
f_5	$\bar{b}$			
f_6	$\bar{a}b \vee a\bar{b}$	EXCLUSIVE OR	$a \circledvee b$	a, b → [ⓥ] → $\bar{a}b \vee a\bar{b}$
f_7	$\bar{a} \vee \bar{b}$	NAND	$a \mid b$	a, b → [|] → $\bar{a} \vee \bar{b}$
f_8	ab	AND	$a \bullet b$	a, b → [•] → ab
f_9	$\bar{a}\bar{b} \vee ab$	COINCIDENCE	$a \odot b$	a, b → [⊙] → $\bar{a}\bar{b} \vee ab$
f_{10}	b			
f_{11}	$\bar{a} \vee b$	INCLUSION		b, a (inhibiting) → [v] → $\bar{a} \vee b$
f_{12}	a			
f_{13}	$a \vee \bar{b}$	INCLUSION		a, b (inhibiting) → [v] → $a \vee \bar{b}$
f_{14}	$a \vee b$	OR	$a \vee b$	a, b → [v] → $a \vee b$
f_{15}	1			

Figure 6-2 Logical operations of two variables.

third column gives the names of the operations, and the fourth column the symb used.

Of the sixteen functions, f_0 and f_{15} are trivials, and the four functions f_3, f_5, f_{10} and f_{12} degenerate into functions of one variable. For all other functions except f_{11} and f_{13}, a logic circuit has been developed. Although, there is an inhibitor circuit, inhibition as such has not found much favor as a logic operator. So, among the logical operators, we have AND, OR, NAND, NOR, EXCLUSIVE–OR, and COINCIDENCE. We have already been familiar with the AND and OR operations. The rest of the chapter will be devoted to the other operators.

6.3 EXCLUSIVE–OR OPERATION

An EXCLUSIVE–OR function is defined by the truth table of Fig.6-3. The function is 1, when either a or b is 1, but not both. Hence the name EXCLUSIVE–OR. From the truth table,

$$f = \bar{a}b \vee a\bar{b} = (\bar{a} \vee \bar{b})(a \vee b).$$

Another feature of this gate is that it does not give an output when both the inputs are either 0 or 1. Only when the inputs are dissimilar, it produces an output. A close look at the table also reveals that the output function is identical with the sum modulo 2 function of two binary variables, as has been shown in the binary addition table in Chapter 0. For this reason an EXCLUSIVE–OR gate can also produce the Mod-2 sum.

a	b	f
0	0	0
0	1	1
1	0	1
1	1	0

$f = a \circledS b$

Figure 6-3 Truth Table for the EXCLUSIVE–OR operation.

Theorem 6.3.1 EXCLUSIVE–OR operation is associative.

Proof: $a \circledS (b \circledS c)$

$= \bar{a}(\bar{b}c \vee b\bar{c}) \vee a(\overline{\bar{b}c \vee b\bar{c}})$

$= \bar{a}\bar{b}c \vee \bar{a}b\bar{c} \vee a(\bar{b}\bar{c} \vee bc)$

$= (\bar{a}\bar{b} \vee ab)c \vee (\bar{a}b \vee a\bar{b})\bar{c}$

$= (\overline{a \circledS b})c \vee (a \circledS b)\bar{c}$

$= (a \circledS b) \circledS c.$ ▲

Theorem 6.3.2 EXCLUSIVE–OR operation is commutative.

Proof: $a \circledS b$

$= \bar{a}b \vee a\bar{b}$

$= \bar{b}a \vee b\bar{a}$

$= b \circledS a.$ ▲

Theorem 6.3.3 0, and not 1, is the identity element for EXCLUSIVE–OR operation.

Proof: $a \circledvee 0$

$= \bar{a}0 \vee \bar{0}a$

$= a.$

But, $a \circledvee 1$

$= \bar{a}1 \vee a\bar{1}$

$= \bar{a}.$ ▲

We already know that the INCLUSIVE–OR (or simply OR) and the AND operations form a Boolean Algebra. It will now be interesting to investigate if the two binary operations, EXCLUSIVE–OR and AND can form a Boolean Algebra. The following theorems are useful to have the answer.

Theorem 6.3.4 The AND operation is distributive over the EXCLUSIVE–OR operation. That is,

$a(b \circledvee c) = ab \circledvee ac.$

Proof: $ab \circledvee ac$

$= \overline{ab}\, ac \vee ab\, \overline{ac}$

$= (\bar{a} \vee \bar{b})ac \vee ab(\bar{a} \vee \bar{c})$

$= a\bar{b}c \vee ab\bar{c}$

$= a(\bar{b}c \vee b\bar{c})$

$= a(b \circledvee c).$ ▲

Theorem 6.3.5 The EXCLUSIVE–OR operation is not distributive over the AND operation, that is,

$a \circledvee bc \neq (a \circledvee b)(a \circledvee c).$

Proof: $a \circledvee bc$

$= \bar{a}bc \vee a\overline{bc}$

$= \bar{a}bc \vee a(\bar{b} \vee \bar{c})$

$= \bigvee (3, 4, 5, 6).$

But $(a \circledvee b)(a \circledvee c)$

$= (\bar{a}b \vee a\bar{b})(\bar{a}c \vee a\bar{c})$

$= (\bar{a}bc \vee a\bar{b}\bar{c})$

$= \bigvee (3, 4)$

$\therefore$ $a \circledvee bc \neq (a \circledvee b)(a \circledvee c).$ ▲

Theorem 6.3.5 shows that the two binary operations $\circledvee$ and . do not form a Boolean Algebra.

Following theorems bring out some other important properties of the EXCLUSIVE–OR operation.

Theorem 6.3.6 $a \circledvee \bar{a} = 1.$

Proof: $a \circledvee \bar{a} = \bar{a}\bar{a} \vee a\bar{\bar{a}}$

$= \bar{a} \vee a = 1.$ ▲

Theorem 6.3.7 $a \circledvee a \circledvee \ldots \ldots \circledvee a$ (number of a's is n) $= 0$ when n is even
$= a$ when n is odd.

Proof: $a \circledvee a = \bar{a}a \vee a\bar{a} = 0.$

When n is even,

$a \circledvee a \circledvee \ldots \circledvee a$

$= (a \circledvee a) \circledvee (a \circledvee a) \circledvee \ldots \circledvee (a \circledvee a)$

$= 0 \circledvee 0 \ldots \circledvee 0$

$= 0.$

When n is odd

$a \circledvee a \circledvee \ldots \circledvee a$

$= (a \circledvee a) \ldots (a \circledvee a) \circledvee a$

$= 0 \circledvee a$

$= a.$ ▲

6.4 COINCIDENCE OPERATION

The truth table for the COINCIDENCE operation is shown in Fig.6-4. In the equation form,

$$a \odot b = \bar{a}\bar{b} \vee ab$$

$$= (\bar{a} \vee b)(a \vee \bar{b}).$$

From the truth table it is apparent that it is complementary to the EXCLUSIVE–OR operation. The gate gives an output 1 when both the inputs are similar, that is, when both are 0's or both are 1's. Hence the name COINCIDENCE or EQUIVALENCE.

a	b	f
0	0	1
0	1	0
1	0	0
1	1	1

$f = a \odot b$

Figure 6-4 Truth Table for the COINCIDENCE operation.

Theorem 6.4.1 The EXCLUSIVE–OR and COINCIDENCE operations are duals of each other.

Proof: Let $f = a \circledvee b = \bar{a}b \vee a\bar{b}$

$\therefore f^D = (\bar{a} \vee b)(a \vee \bar{b})$

$= a \odot b.$ ▲

As a result of this duality, the theorems of EXCLUSIVE–OR operation have their duals with the EQUIVALENCE operation.

6.5 NAND OPERATION

NAND operation for two variables is defined by the following equation

$a \mid b = \overline{ab} = \bar{a} \vee \bar{b}.$

Its truth table is shown in Fig.6-5.

This operation was studied by Sheffer as a binary operation using the | (stroke) symbol. Hence this came to be known as *Sheffer's stroke function.* However, with the increase in its popularity as a logical operator, it became more commonly known as the NAND (a conjunction of NOT AND) operation.

From the truth table it can be seen that the logic does not produce an output when both the inputs are 1. For all other cases it produces an output. This definition is extended to cover an n variable case. Hence the NAND logic does not

a	b	f
0	0	1
0	1	1
1	0	1
1	1	0

$f = a \mid b = \bar{a} \vee \bar{b}$

Figure 6-5 Truth Table for the NAND Operation.

produce an output when all the inputs are 1. But even if only one input is 0 it produces an output, so that,

$a_1 \mid a_2 \mid \ldots \mid a_n = \bar{a}_1 \vee \bar{a}_2 \vee \ldots \vee \bar{a}_n.$

Theorem 6.5.1 The NAND operation is not associative.

$a \mid (b \mid c) \neq (a \mid b) \mid c.$

Proof: $a \mid (b \mid c)$

$= a \mid (\overline{bc})$

$= \bar{a} \vee \overline{\overline{bc}}$

$= \bar{a} \vee bc$

But $(a \mid b) \mid c$

$= \overline{ab} \mid c$

$= \overline{\overline{ab}} \vee \bar{c}$

$= ab \vee \bar{c}$

$\therefore$ $a \mid (b \mid c) \neq (a \mid b) \mid c.$ ▲

Theorem 6.5.2 The NAND operation is commutative. $a \mid b = b \mid a$.

Proof: $a \mid b$

$= \bar{a} \vee \bar{b}$

$= \bar{b} \vee \bar{a}$

$= b \mid a.$ ▲

Theorem 6.5.3 Neither 0 nor 1 is the identity element for the NAND operation.

$a \mid 0 = 1$ and $a \mid 1 = \bar{a}$.

Proof: $a \mid 0 = \bar{a} \vee \bar{0}$

$= \bar{a} \vee 1$

$= 1$

Also $a \mid 1 = \bar{a} \vee \bar{1}$

$= \bar{a} \vee 0$

$= \bar{a}.$ ▲

The second identity shows how a NAND gate can act as an inverter, if the logical 1 is available as an input.

Theorem 6.5.4 $a \mid a = \overline{a}$.

Proof: $a \mid a = \overline{a} \vee \overline{a}$

$= \overline{a}$. ▲

This theorem shows that the NAND operation is not idempotent. *It also shows how a NAND logic gate can act as an inverter even when the logical 1 is not available as an input.*

Theorem 6.5.5 The NAND operation is not distributive over the AND operation.

$a \mid (bc) \neq (a \mid b)\,(a \mid c)$.

Proof: $a \mid (bc) = \overline{a} \vee \overline{bc}$

$= \overline{a} \vee \overline{b} \vee \overline{c}$.

Again,

$(a \mid b)\,(a \mid c) = (\overline{a} \vee \overline{b})\,(\overline{a} \vee \overline{c})$.

$= \overline{a} \vee \overline{b}\overline{c}$

$\therefore\; a \mid (bc) \neq (a \mid b)\,(a \mid c)$. ▲

Theorem 6.5.6 The NAND operation is not distributive over the OR operation.

$a \mid (b \vee c) \neq (a \mid b) \vee (a \mid c)$.

Proof: $a \mid (b \vee c) = \overline{a} \vee \overline{b \vee c}$

$= \overline{a} \vee \overline{b}\overline{c}$

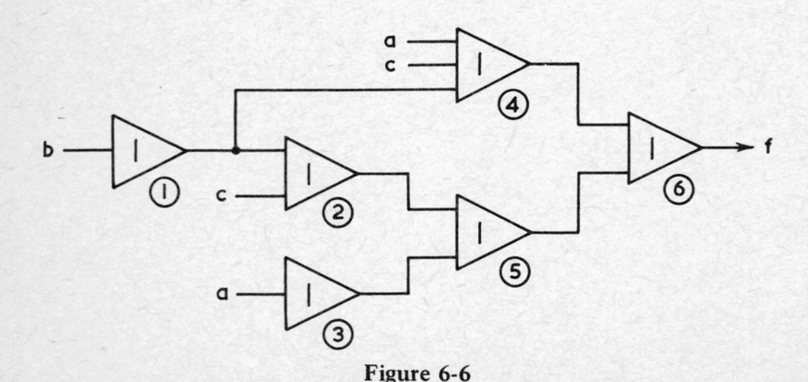

Figure 6-6

But

$$(a \mid b) \vee (a \mid c) = (\bar{a} \vee \bar{b}) \vee (\bar{a} \vee \bar{c})$$
$$= \bar{a} \vee \bar{b} \vee \bar{c}$$

$\therefore\ a \mid (b \vee c) \neq (a \mid b) \vee (a \mid c).$ ▲

Example 6.5.1 Evaluate f as given by the circuit of Fig.6-6.

Solution: Mark the outputs of each NAND block as shown.

Output at ① $= \bar{b}$

Output at ② $= \bar{\bar{b}} \vee \bar{c} = b \vee \bar{c}$

Output at ③ $= \bar{a}$

Output at ④ $= \bar{a} \vee \bar{c} \vee \overline{①} = \bar{a} \vee \bar{c} \vee b$

Output at ⑤ $= \overline{②} \vee \overline{③}$

$= \bar{b}c \vee a = a \vee \bar{b}c$

Output at ⑥ $= \overline{④} \vee \overline{⑤}$

$= a\bar{b}c \vee \bar{a}(b \vee \bar{c}) = f.$

6.6 NOR OPERATION

NOR operation for two variables is defined by the following equation:

$$a \downarrow b = \overline{a \vee b} = \bar{a}\bar{b}.$$

Its truth table is shown in Fig.6-7. The NOR operation was studied by the American Logician Pierce using the dagger (↓) symbol. Subsequently the Pierce function became more popularly known as the NOR (conjunction of NOT OR) function. It will be clear from the truth table that the logic produces an output when both the inputs are 0. In an n-variable case, the NOR gate produces an

a	b	f
0	0	1
0	1	0
1	0	0
1	1	0

$f = a \downarrow b = \bar{a}\bar{b}$

Figure 6-7 Truth Table for the NOR Operation.

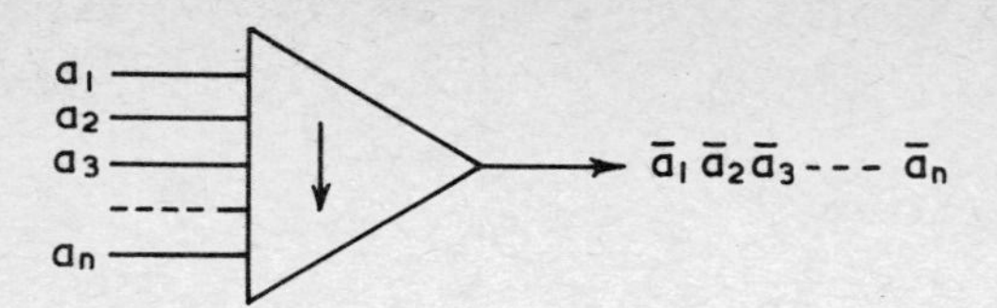

Figure 6-8 A NOR gate with n inputs.

output of 1 when all the inputs are zero. (Fig.6-8.) Therefore,

$$a_1 \downarrow a_2 \downarrow a_3 \downarrow \ldots \downarrow a_n = \bar{a}_1\bar{a}_2\bar{a}_3 \ldots \bar{a}_n.$$

Theorem 6.6.1 The NOR operation is the dual of the NAND operation.

Proof: Let $f = a \downarrow b = \bar{a}\bar{b}$

$\therefore\ f^D = \bar{a} \vee \bar{b} = a \mid b.$ ▲

As a consequence of this duality, all the identities of the NAND operation have their corresponding duals with the NOR operation.

Example 6.6.1 Evaluate f as given by the NOR circuit of Fig.6-9.

Solution: Number the outputs of the six NOR blocks as shown. Now,

Output at $\textcircled{1} = \bar{c}$

Output at $\textcircled{2} = \bar{b}\,\overline{\textcircled{1}} = \bar{b}c$

Output at $\textcircled{3} = \bar{a}\,\overline{\textcircled{2}} = \bar{a}(b \vee \bar{c})$

Output at $\textcircled{4} = \bar{a}\,\overline{\textcircled{3}} = \bar{a}(a \vee \bar{b}c) = \bar{a}\bar{b}c$

Output at $\textcircled{5} = \overline{\textcircled{3}}\,\overline{\textcircled{2}} = (a \vee \bar{b}c)\,(b \vee \bar{c})$

$= ab \vee a\bar{c} = a(b \vee \bar{c})$

Output at $\textcircled{6} = \overline{\textcircled{4}}\,\overline{\textcircled{5}} = (a \vee b \vee \bar{c})\,(\bar{a} \vee \bar{b}c)$

$= \bar{a}(b \vee \bar{c}) \vee a\bar{b}c.$

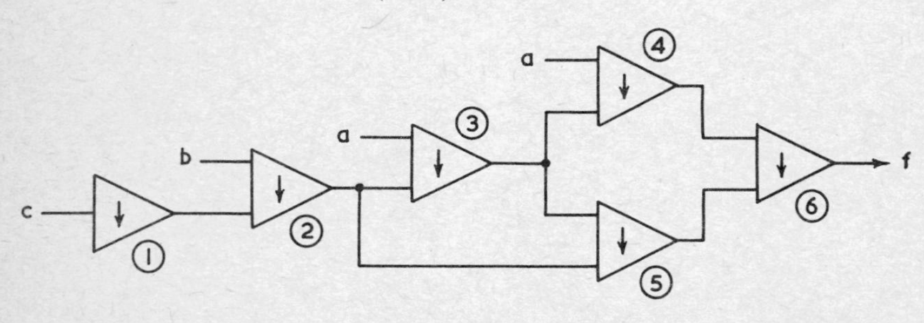

Figure 6-9

6.7 FUNCTIONALLY COMPLETE SETS

In the previous sections we have studied important properties of some of the logical operations. Similar studies can be made for the INHIBIT and INCLUSION logics too. It is interesting to investigate the minimum logical operations necessary to implement any Boolean function. Such a set of logical operation(s) which is (are) just sufficient to implement any Boolean function is called a *functionally complete minimal set.* In order to prove a set to be functionally complete, we must show that the set is capable of performing the AND, OR, and NOT operations.

a) AND and NOT: That this set can perform the OR operation is evident from the following:

$$a \vee b = \overline{\overline{a \vee b}} = \overline{\bar{a}\bar{b}}.$$

b) OR and NOT: Its capability of performing the AND operation can be shown as follows:

$$ab = \overline{\overline{ab}} = \overline{\bar{a} \vee \bar{b}}\,.$$

c) EXCLUSIVE–OR, and AND: Its capacity to perform the OR, and NOT operations can be shown by the following identities:

$$a \vee b = a \circledvee b \circledvee ab$$

$$\bar{a} = a \circledvee 1.$$

d) COINCIDENCE, and AND: It can perform the OR and NOT operations in this way:

$$a \vee b = a \odot b \odot ab$$

$$\bar{a} = a \odot 0.$$

e) NAND only: That the NAND gate alone can perform the AND, OR, and NOT operations will be apparent from the following identities:

$$ab = (a \mid b) \mid (a \mid b)$$

$$a \vee b = (a \mid a) \mid (b \mid b)$$

$$\bar{a} = a \mid a.$$

f) NOR only: That the NOR logic alone can perform the AND, OR, and NOT operations will be evident from the following identities:

$$ab = (a \downarrow a) \downarrow (b \downarrow b)$$

$$a \vee b = (a \downarrow b) \downarrow (a \downarrow b)$$

$$\bar{a} = a \downarrow a.$$

g) INHIBIT only: The following identities demonstrate that the INHIBIT logic by itself can form a functionally complete set:

$$1 \cdot \bar{a} = \bar{a}$$

$$\bar{a} \cdot \bar{b} = \bar{a}\bar{b}, \text{ and}$$

$$1 \cdot \overline{\bar{a}\bar{b}} = a \vee b.$$

This shows that the OR operation can be performed with the help of three INHIBIT gates. Again

$$a\bar{b} = a\bar{b}$$

$$a\overline{(a\bar{b})} = a(\bar{a} \vee b) = ab.$$

This shows that the AND operation can be obtained with two INHIBIT gates. The NOT operation can be performed by a single INHIBIT gate, since,

$$1 \cdot \bar{a} = \bar{a}.$$

h) INCLUSION only: The versatility of the INCLUSION logic is apparent from the following. For OR operation:

$$a \vee \bar{b} = a \vee \bar{b}$$

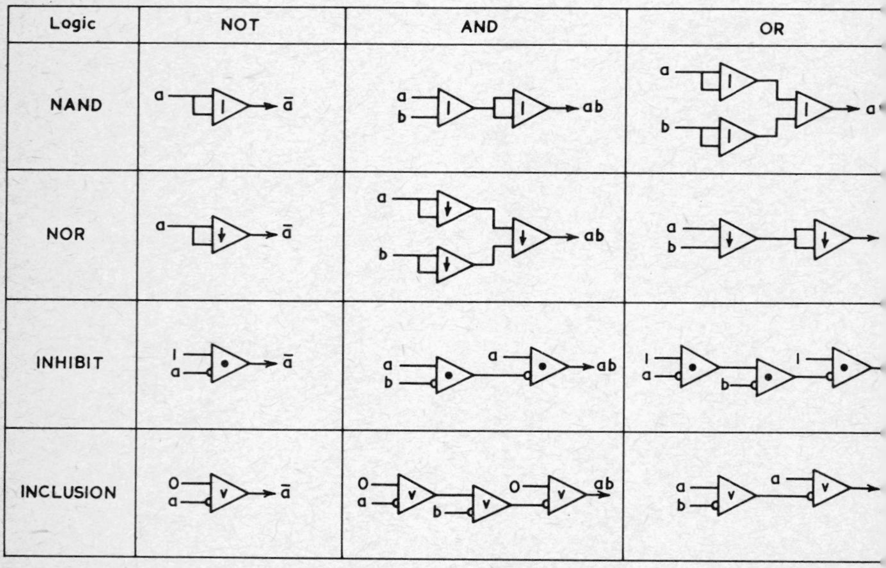

Logic	NOT	AND	OR
NAND			
NOR			
INHIBIT			
INCLUSION			

Figure 6-10 Versatility of the NAND, NOR, INHIBIT and INCLUSION logic gates.

$$a \vee \overline{a \vee \overline{b}} = a \vee \overline{a}b = a \vee b.$$

For AND operation:

$$0 \vee \overline{a} = \overline{a}$$

$$\overline{a} \vee \overline{b} = \overline{a} \vee \overline{b}, \quad 0 \vee \overline{\overline{a} \vee \overline{b}} = 0 \vee ab = ab.$$

For NOT operation:

$$0 \vee \overline{a} = \overline{a}.$$

Figure 6-10 shows the gate configurations for the last four logical operators. In the table of Fig.6-11 we have listed all these functionally complete sets in three groups. It can be easily seen that from the engineering point of view the members of group C are most promising, as in these only one type of gate without even any logical constant is adequate to implement any Boolean function. This is the reason why most of the present day logic circuits are made of NAND or NOR logic gates only.

	Group	Elements of functionally complete minimal set
A	Two elements only	AND, NOT OR, NOT EXCLUSIVE–OR, AND COINCIDENCE, AND
B	One element with one logical constant (0 or 1)	INHIBIT INCLUSION
C	One element only	NAND NOR

Figure 6-11 Functionally complete sets.

6.8 NAND AND NOR LOGIC: FURTHER THEOREMS

We have seen how only the NAND or only the NOR logic is sufficient to implement any Boolean function. Now, for *straightforward implementation* of a Boolean function in the AND–OR (SP), or in the OR–AND (PS) form, the following theorems are very useful.

Theorem 6.8.1 A Boolean expression expressed in the AND–OR form remains unaltered if the AND, and OR connectives are replaced by NAND connectives.

Proof: $ab \vee cd = \overline{\overline{ab \vee cd}}$

$= \overline{\overline{ab} \,.\, \overline{cd}}$

$= \overline{ab} \mid \overline{cd}$

$= (a \mid b) \mid (c \mid d).$ ▲

This also means that in a two-level AND–OR logic the output remains unaffected if the AND, and OR logics are simultaneously replaced by NAND logic blocks.

Theorem 6.8.2 If in a Boolean expression expressed in the OR–AND form, the OR and AND connectives are replaced by the NAND connectives, then the dual of the expression is obtained.

Proof: Let the expression be $(a \vee b)(c \vee d) = F$ say. Now,

$(a \mid b) \mid (c \mid d) = ab \vee cd$

$= F^D.$ ▲

These theorems help us implement a Boolean function by the NAND logics only. This is illustrated by the following examples.

Example 6.8.1 Implement the following function by only NAND logic assuming a) complemented variables are available, b) complemented variables are not available.

$f = \bar{a} \vee bc \vee \bar{b}\bar{c}.$

Solution: $f = \bar{a} \vee bc \vee \bar{b}\bar{c}$

$= \bar{a}\bar{a} \vee bc \vee \bar{b}\bar{c}.$

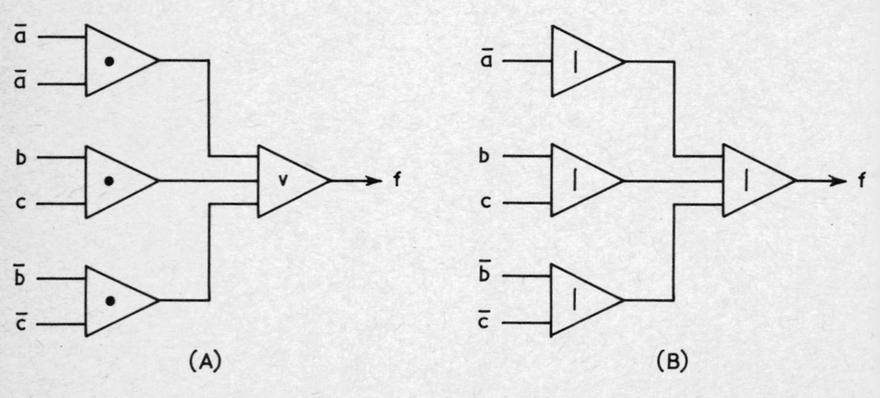

Figure 6-12 Implementation of $f = \bar{a} \vee bc \vee \bar{b}\bar{c}$ by the NAND gates.

The two-level implementation by the AND–OR logic gates is as shown in Fig.6-12(A). Applying theorem 6.8.1, the AND and OR gates are replaced by NAND gates, and the resulting circuit is as shown in Fig.6-12(B).

When complemented variables are available the function can be implemented by this circuit. When complemented variables are not available, $\overline{a}$, $\overline{b}$, and $\overline{c}$, are to be obtained by using NAND blocks as inverters. The resulting circuit becomes as shown in Fig.6.13(A). Knowing that the two NAND blocks in cascade is a short circuit, the circuit finally reduces to as that shown in Fig.6-13(B).

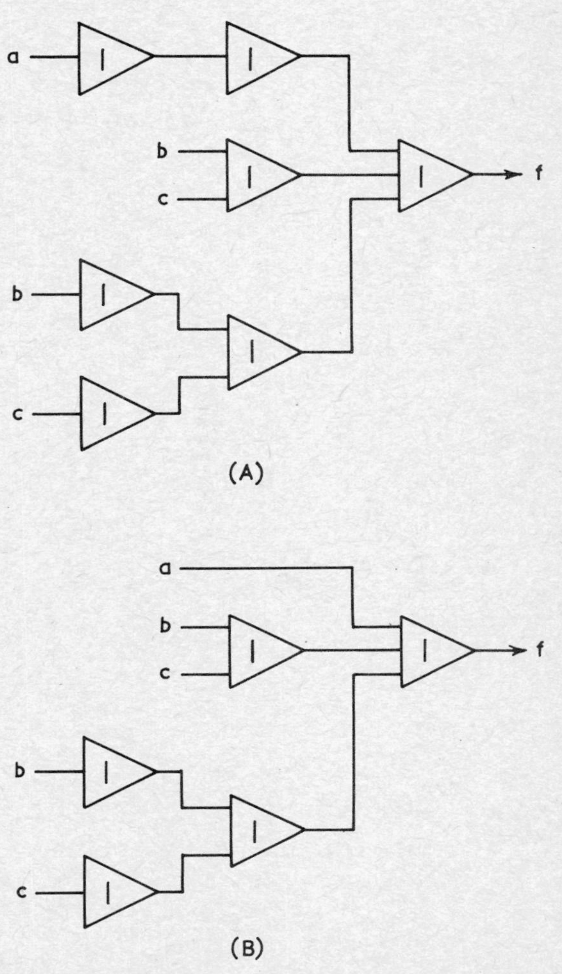

Figure 6-13 Implementation of the function of Fig.6-12 by NAND gates when complemented variables are not available.

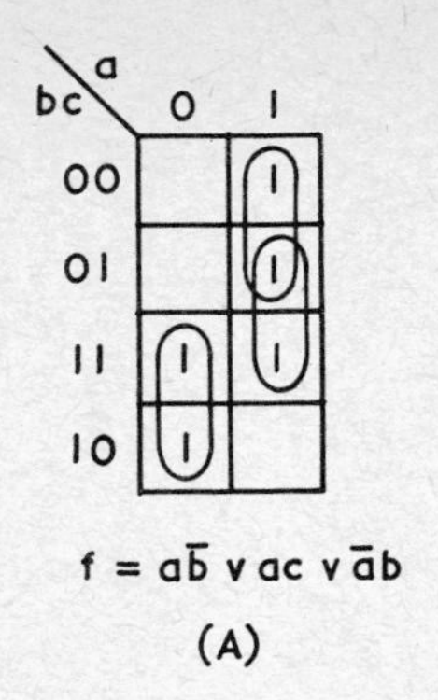

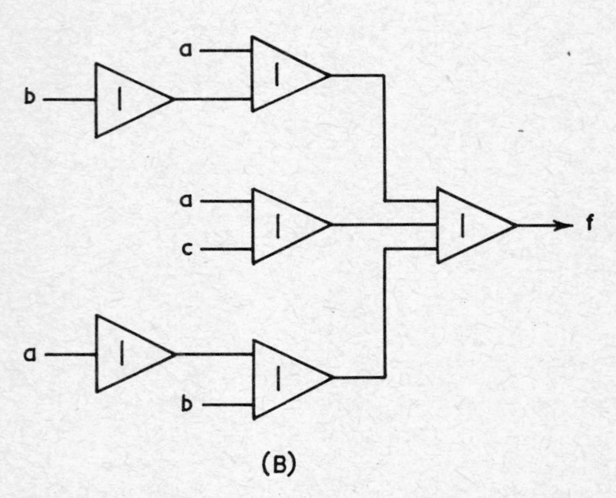

Figure 6-14 Implementation of the octal function 274 by NAND gates only. The inputs are unipolar.

Example 6.8.2 Implement the octal function 274 by NAND logics. The inputs are unipolar, that is, only a, b, and c, and not their complements, are available.

$(274)_8 = (10, 111, 100)_2$

$\therefore f = \vee\, (2, 3, 4, 5, 7).$

Simplifying on the V–K map (Fig.6-14(A))

$f = a\bar{b} \vee ac \vee \bar{a}b.$

Following the procedure as illustrated in the previous example, the final circuit is as shown in Fig.6-14(B).

Here we must make one important observation that these implementations may or may not be minimal. For example, whereas the 1st solution has turned out to be minimal, for the second example, there is a circuit which uses only five blocks instead of six. The circuit is given in Fig.6-15 and the reader may evaluate f, and verify the result.

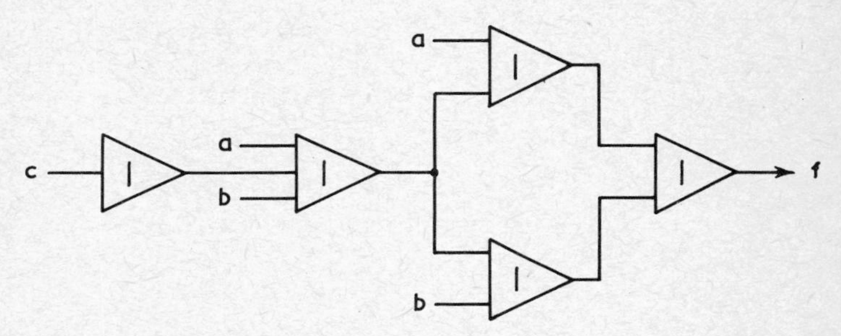

Figure 6-15 Implementation of the octal function 274 by only five NAND gates.

Similar to the NAND logic, the following theorems of NOR logic are useful in the straightforward implementation of a Boolean function. Proofs are similar to those of the NAND operation and are left to the reader.

Theorem 6.8.3 A Boolean expression expressed in the OR–AND form remains unaltered if the OR and AND connectives are replaced by NOR connectives.

Theorem 6.8.4 If in a Boolean expression expressed in the AND–OR form, the AND and OR connectives are replaced by NOR connectives, then the dual of the expression is obtained.

Example 6.8.3 Implement the function of example 6.8.2 by NOR logic only.

Solution: $f = (274)_8 = \vee\,(2, 3, 4, 5, 7)$.

In order to express it in the OR–AND form, find the MPS form on the map (Fig.6-16(A)). Hence $\overline{f} = \overline{a}\overline{b} \vee ab\overline{c}$

$\therefore\ f = (a \vee b)\,(\overline{a} \vee \overline{b} \vee c)$.

The resulting NOR circuit is as shown in Fig.6-16(B).

It will be seen later that the NOR circuit has *by chance* been identical with the minimal NOR circuit to implement this function.

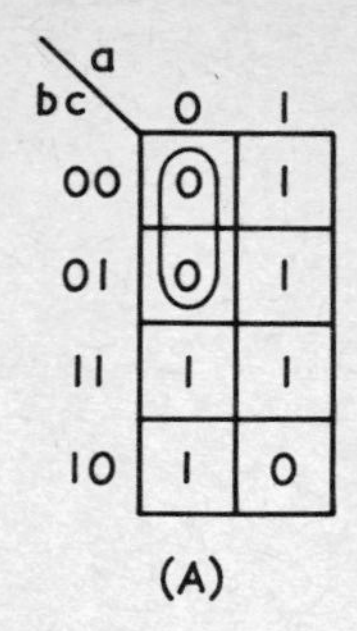

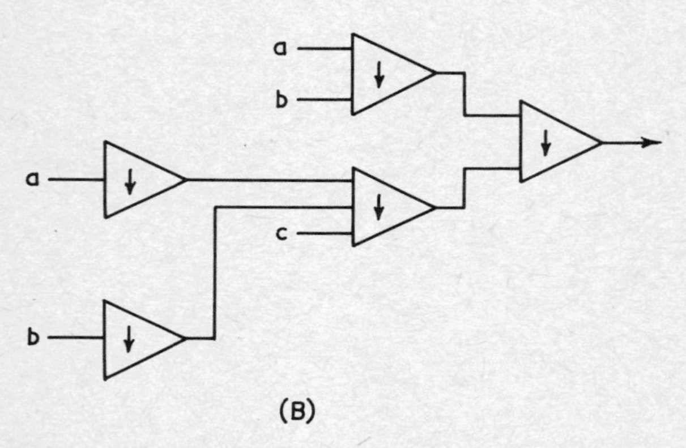

Figure 6-16 Implementation of the octal function 274 by NOR gates only. The inputs are unipolar.

6.9 MINIMAL NAND/NOR CIRCUITS

The minimization methods as discussed in Chapter 4 enable a circuit designer to implement a Boolean function with minimum number of AND and OR gates. By the procedures of the previous section the AND, OR gates can be replaced by the NOR or NAND gates. But does it ensure that the circuit has been implemented with the minimum number of NOR/NAND gates? To answer this let us try to implement the EXCLUSIVE–OR function by NAND gates only. We shall further assume that no complemented variables are available. With this constraint the NAND circuit implementing the function,

$$f = a \ ⓥ \ b = \bar{a}b \vee a\bar{b},$$

is as shown in Fig.6-17(A).

Again, the circuit shown in Fig.6-17(B) also realizes the same function but with one NAND gate less. From this experience we can conclude that a minimal

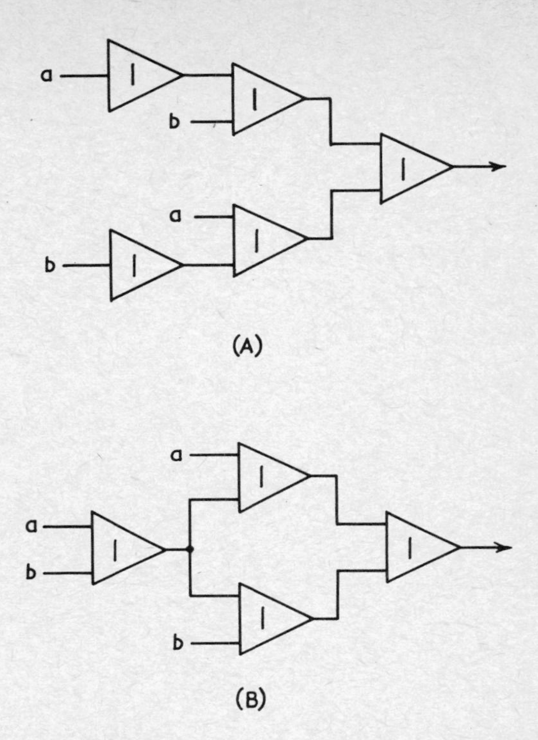

Figure 6-17 Realization of $f = a \ \textcircled{v}\ b$ by (A) five NAND gates, (B) four NAND gates.

AND–OR circuit does not necessarily lead to a minimal NAND/NOR circuit. Consequently, different techniques must be adopted to make a NAND/NOR circuit minimal. One approach is to devise a minimization procedure, for the NAND and NOR operation. Maley and Earle (1973) have described such a procedure for three variable functions. Hellerman (1963) of IBM has adopted a different approach for the design of minimal NAND/NOR circuits for three variable functions. He wrote a computer program which investigates all possible interconnections among the gates and variables, and then produces the circuit which is 'best'. It must be mentioned here that it is very difficult to have a unique definition of what is best. A circuit which is best for one application may only be a good one for another application. The criteria for the best circuit chosen by Hellerman are as follows:

1) The number of logic blocks of the circuit is least possible for performing the function.

2) The number of connections in the circuit (total number of inputs) is least possible, subject to the condition that the circuit satisfies the first condition.

3) The circuits satisfy certain reasonable restrictions on fan-in and fan-out.

Hellerman's catalog of three variable NOR and NAND logic circuits is reproduced in Appendix B. The procedure to design a circuit using this catalog has also been described in the Appendix.

Another catalog developed by Smith (1965) gives the minimal NOR/NAND circuits of three variables when the variables are available in both true and complemented forms.

6.10 MULTIPLE OUTPUT CIRCUITS

So far we have considered circuits which have only one output. In many instances there may be more than one output. Such circuits are called *multiple output* circuits. Take for example, the following circuit whose two outputs are given by f_1 and f_2

$$f_1 = \bar{a}bc \vee a\bar{b}c$$

$$f_2 = \bar{a} \vee \bar{b}c \vee b\bar{c}.$$

If the circuit is implemented as two single output circuits, then the implementation needs 6 gates as shown in Fig.6-18. However let us consider the three maps

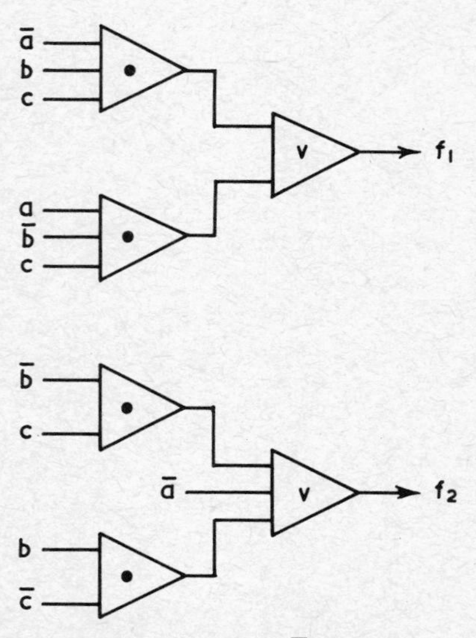

Figure 6-18 Implementation of $f_1 = \bar{a}bc \vee a\bar{b}c$ and $f_2 = \bar{a} \vee \bar{b}c \vee b\bar{c}$ with two single output circuits.

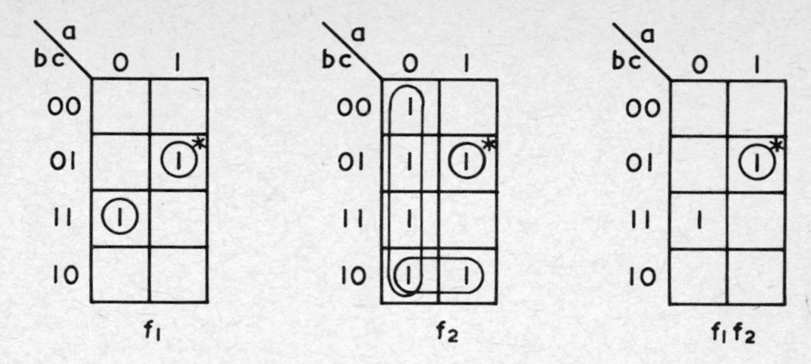

Figure 6-19 Plots of f_1, f_2 and $f_1 f_2$ on maps.

(Fig.6-19) showing f_1, f_2 and $f_1 f_2$. The product map is quite significant here. It shows that the minterm (5) can be a common gate between f_1 and f_2, although in f_1 it is a prime implicant, whereas, in f_2 it is simply an implicant. The choice of the implicant (5) in place of the prime implicant (1, 5) does not increase the number of gates to implement f_2 (Fig.6-20). On the other hand, the minterm (3) which is also common between f_1 and f_2 does not offer the prospect of reducing the number of gates as the implementation of f_2 will still require three more gates.

If any circuit has three outputs, so that three functions f_1, f_2, f_3 need to be considered, the procedure becomes more involved, as now the number of product maps to be examined increases to 4, namely those of $f_1 f_2$, $f_2 f_3$, $f_3 f_1$ and $f_1 f_2 f_3$. It is now very evident that the complexity of the design keeps on increasing as the number of outputs goes on increasing. Many authors [Bartee (1961), McCluskey (1965)] have expounded methods for speedy and efficient design of multiple output circuits.

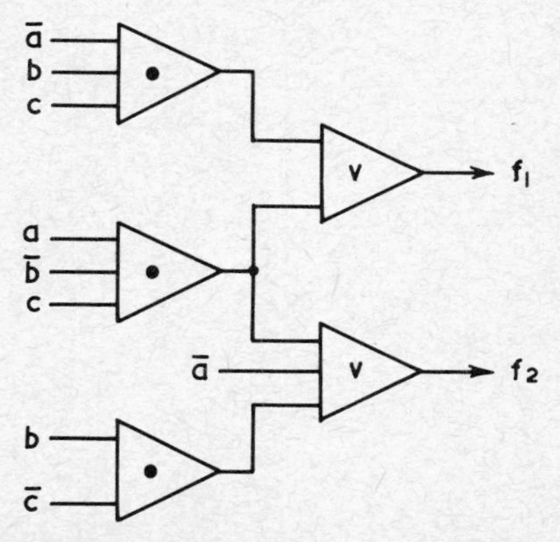

Figure 6-20 Implementation of f_1 and f_2 by a two-output circuit results in the saving of one gate.

6.11 UNIVERSAL LOGIC MODULES

With the advent of integrated circuits (ICs) the guiding philosophy in the design of combinational circuits underwent a major change. Prior to this the emphasis w in reducing the number of gates required to implement a given Boolean function, as the saving of even one gate meant the saving of one transistor, the costliest component of the design. But after ICs appeared on the scene, hundreds of NOR or NAND gates were available on a chip with as small an area as a square inch. Hence the emphasis shifted to exploit this easy availability of large number of gates to other advantageous feature of the design. This led to the development of universal logic modules (ULM). A ULM is a logic block which can implement any function of a certain number of variables by simply altering the interconnections. Research is also in progress to utilize the redundant gates to incorporate error detecting and correcting capabilities in the ULMs. This will lead to the development of *failsafe* and *failproof* ULMs.

REFERENCES

Bartee, T. C.: "Computer design of multiple-output logical networks", *IRE Trans. Electron Computers,* **EC-10**, No.2, pp.21–30, March 1961.

Chu, Y.: *Digital Computer Design Fundamentals,* McGraw-Hill Book Co., New York, 1962.

Forslund, D. C. and R. Waxman.: "The universal logic block (ULB) and its application to logic design", 7th Annual SYSWAT, pp.236–250, 1966.

Hellerman, L.: A Catalog of Three-Variable OR-invert and AND-Invert Logical Circuits; *IEEE Trans. Electronic Computers,* **EC-12**, pp.198–223, 1963.

Kohavi, Z.: *Switching and Finite Automata Theory,* McGraw-Hill Book Co., New York, 1970.

Krieger, M.: *Basic Switching Circuit Theory,* Macmillan, New York, 1967.

Maley, G. A. and J. Earle: *The Logic Design of Transistor Digital Computers,* Prentice-Hall, N.J. 1963.

McCluskey, E. J.: *Introduction to the Theory of Switching Circuits*, McGraw-Hill Book Co., New York, 1965.

Smith, R. A.: Minimal-variable NOR and NAND Logic Circuits, *IEEE Trans. Electronic Computers,* **EC-14**, pp.79–81, Feb.1965.

Yau, S. S. and C. K. Tang: Universal logic modules and their applications, *IEEE Trans. Electronic Computers,* **C-19**, No.2, pp.141–149, Feb. 1970.

PROBLEMS

6.1 Implement the following functions with two level AND, OR and NOT logics. Use minimum number of gates.

a) $f_1 = \vee\,(2, 3, 5, 7)$

b) $f_2 = \vee(0, 4, 6, 7)$

c) $f_3 = \vee(1, 2, 5, 9, 10)$

d) $f_4 = \vee(2, 3, 6, 12, 15)$

e) $f_5 = \vee(0, 6, 17, 19, 21)$

f) $f_6 = \vee(2, 9, 10, 12, 15, 16, 20, 30)$.

.2 Implement the above functions with multilevel AND, OR, and NOT logics.

.3 Implement the following functions with minimum number of AND, OR locks. Assume that complemented variables are available. If the gates are diode ates, count the number of diodes needed for each function.

a) $f_1 = ab \vee \bar{b}c \vee acd$

b) $f_2 = a \vee bc \vee d$

c) $f_3 = x \vee \bar{x}yz$

d) $f_4 = wx \vee \bar{w}y \vee \overline{xyz}$

e) $f_5 = (a \vee \bar{b})(b \vee c \vee \bar{a}d)$

f) $f_6 = (a \vee \bar{b})c \vee (\bar{a} \vee \bar{b})$.

.4 Prove the following theorems for EXCLUSIVE–OR operation.

a) $a \textcircled{v} ab = a\bar{b}$

b) $a \textcircled{v} ab = a \vee b$

c) $a \textcircled{v} (a \vee b) = \bar{a}b$

d) $a \textcircled{v} (\bar{a} \vee b) = \bar{a}\bar{b}$.

.5 Prove that the sum function of n variables is

$s = a_1 \textcircled{v} a_2 \textcircled{v} a_3 \textcircled{v} \ldots \textcircled{v} a_n$.

.6 Implement the following functions with the functionally complete set of an XCLUSIVE–OR and the AND operations.

a) $f = \bar{a}\bar{b} \vee ab$

b) $f = \bar{a}b \vee c$

c) $f = \vee(2, 3, 6, 7)$

d) $f = \vee(0, 1, 5, 6, 7)$.

7 Implement the functions of Prob.6.6 with the functionally complete set of OINCIDENCE, and AND logic.

6.8 Implement the following functions with INHIBIT logic only.

a) $f_1 = \vee(1, 2, 4, 6)$

b) $f_2 = \vee(7, 9, 10, 11, 12)$

c) $f_3 = \vee(0, 1, 2, 4, 6, 10, 12, 14)$.

6.9 Implement the following functions with INCLUSION logic only.

a) $f_1 = \vee(2, 3, 6, 7)$

b) $f_2 = \vee(0, 2, 3, 5, 8, 9)$

c) $f_3 = \vee(1, 4, 7, 10, 12, 13, 14, 15)$.

6.10 Prove the following:

a) An AND (OR) gate followed by an inverter is equivalent to a NAND (NOR) gate.

b) An OR (AND) gate preceded by an inverter at each input is equivalent to a NAND (NOR) gate.

c) A NAND (NOR) gate followed by an inverter is equivalent to an AND (OR) gate.

d) A NAND (NOR) gate preceded by an inverter at each input is equivalent to an OR (AND) gate.

e) A NAND (NOR) logic with the same literal at its inputs acts as an inverter.

f) Two NAND (NOR) gates in cascade act as a direct connection.

6.11 Implement the following functions with NAND logic only. The circuit need not be minimal.

a) $f_1 = \bar{a}b \vee a\bar{b}$

b) $f_2 = abc \vee \bar{a}\bar{b}\bar{c}$

c) $f_3 = a \vee \bar{b}c \vee \bar{a}d$

d) $f_4 = (a \vee \bar{b} \vee c)(\bar{a} \vee \bar{b} \vee \bar{c})$

e) $f_5 = (\bar{a} \vee \bar{b})(c \vee d)$

f) $f_6 = (a \vee b)(\bar{b} \vee c)(\bar{c} \vee d)$.

6.12 Implement the functions of Prob.6.11 with NOR logic only.

6.13 Evaluate f_1, f_2, and f_3, f_4, of the circuits of Fig.6-21 and express them in octal notation.

6.14 Implement the following octal functions with NAND logic only, using

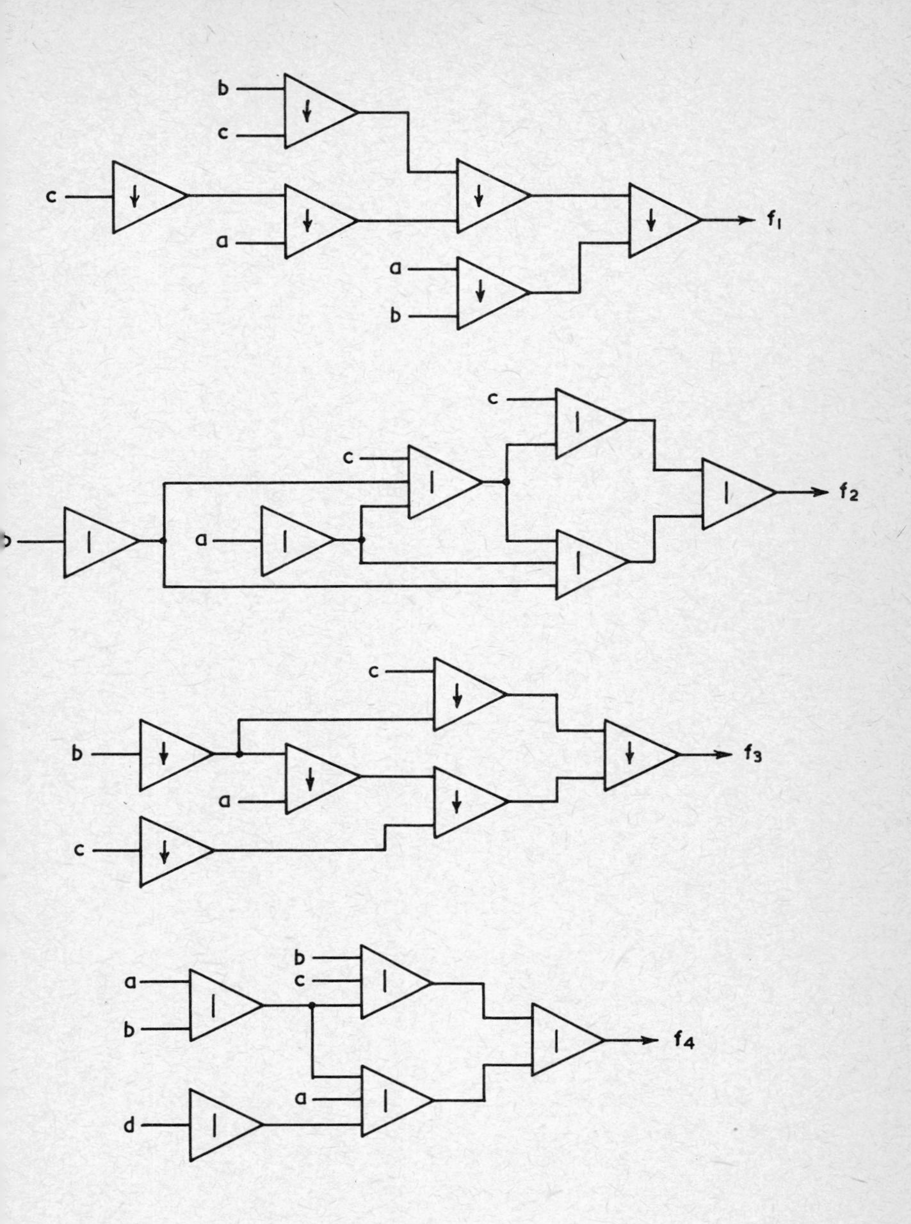

Figure 6-21

minimum number of gates. Assume that the variables are available only in true forms.

a) 51; b) 67; c) 231; d) 356.

6.15 Implement the functions of Prob.6.14 by NOR logic only, also using minin number of gates, and with the same assumption.

6.16 Repeat problems 6.14 and 6.15 assuming that variables are available in both true and complemented forms.

6.17 Design the following multiple output circuits

a) $f_1 = \bigvee (2, 3, 5, 6, 7)$

$f_2 = \bigvee (1, 5, 6, 7)$

b) $f_1 = \bar{a}b\bar{c} \vee ad \vee bd$

$f_2 = \bar{a}\,\bar{c}\,\bar{d} \vee \bar{a}\bar{b} \vee \bar{b}d.$

Chapter 7

THRESHOLD LOGIC

.1 INTRODUCTION

n the last chapter we have seen how combinational circuits can be designed with he help of NAND and NOR logic gates. In this chapter we shall study a more owerful logic known as threshold logic.

efinition 7.1.1 A *threshold gate* realizing a Boolean function $F = f(x_1 x_2 \ldots x_n)$ as n input lines, one for each of the variables $x_1, x_2, \ldots$ and x_n which can take n the value 0 or 1. Each input line has a *weight,* so that the weight of the x_i line a_i, where each a_i is a positive or negative real number. Each gate has also a iven *threshold* value t, which is also a positive or negative real number. For each ombination of the input variables $x_1, x_2, \ldots, x_n$ the corresponding value of the unction F is given by

$$F = 1 \text{ iff} \sum_{i=1}^{n} a_i x_i \geqslant t$$

nd

$$F = 0 \text{ iff} \sum_{i=1}^{n} a_i x_i < t,$$

here the Σ represents the usual arithmetic sum, and $a_i x_i$ is the arithmetic prouct of a_i and x_i. In this product x_i is considered to have the numerical value of or 1, when its logical value is 0 or 1 respectively. The a_i's are also called *coefficients.*

Let the threshold gate of Fig.7-1 have the values of a_i's and t as shown, then

$$\Sigma a_i x_i = x_1 + 2x_2 + 3x_3. \qquad (7.1)$$

n order to determine the Boolean function realized by this gate, the values of $a_i x_i$ for each combination of x_1, x_2, and x_3 are to be calculated. This is done in he table of Fig.7-2. Thus $F = 1$ for the minterms m_3, m_5, and m_7. Hence the

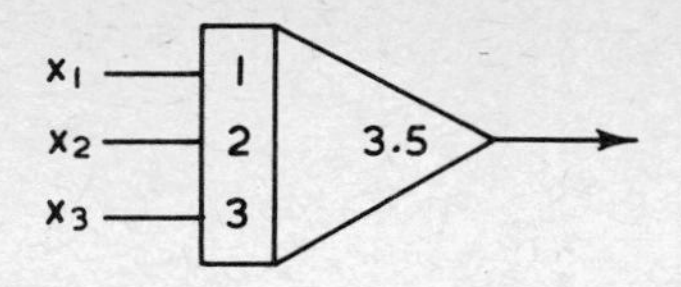

Figure 7-1 A threshold gate.

Boolean function realized by this gate is

$$F = f(x_1x_2x_3)$$
$$= \vee(3, 5, 7).$$

It can be seen that the gate realizes this and only this function. Again this gate is characterized by the set of weights a_i's and the threshold value t. Thus the Boolean function can also be written as

$$F = \langle x_1 + 2x_2 + 3x_3\rangle\, 3.5. \tag{7.2}$$

This is known as the separating function representation of the particular Boolean function.

Definition 7.1.2 Let $a_1, \ldots, a_n, t$ be the set of weights and the threshold value of a threshold gate realizing a Boolean function $f(x_1x_2 \ldots x_n)$, then

$$\langle a_1x_1 + a_2x_2 + \ldots + a_nx_n\rangle\, t$$

is the *separating function representation* of the given Boolean function.

7.2 LINEARLY SEPARABLE OR THRESHOLD FUNCTIONS

There are 2^{2^n} Boolean functions in an n-cube. Can all these functions be realized by *single* threshold gates? The answer to this question can be in the affirmative if each of the functions can have a separating function representation. Unfor-

m_i	x_1	x_2	x_3	Σa_ix_i	F
0	0	0	0	0	0
1	0	0	1	3	0
2	0	1	0	2	0
3	0	1	1	5	1
4	1	0	0	1	0
5	1	0	1	4	1
6	1	1	0	3	0
7	1	1	1	6	1

Figure 7-2 Determination of the function realized by the threshold gate of Fig. 7-1.

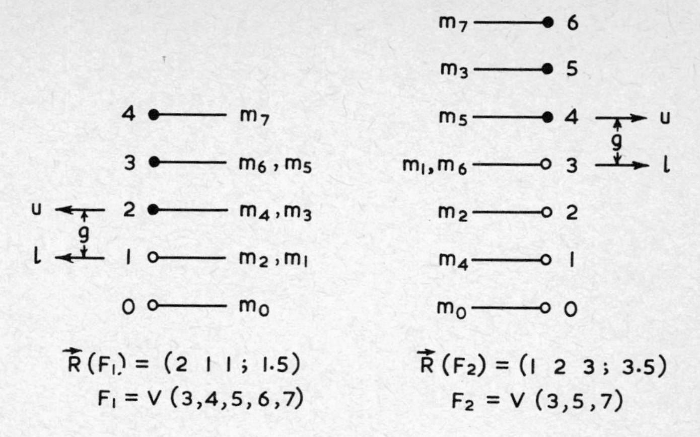

Figure 7-3 Mapping of minterms by weight vectors.

unately this is not the case. However, this is not a disadvantage exclusive to the hreshold logic, as other familiar logics such as the AND, OR, NOT, NAND, and NOR also cannot realize *all* functions by single gates. On the other hand, in nany cases where the NAND or NOR realizations may require a number of gates, he threshold logic may realize the function by only one gate. In fact, this articular merit of the threshold gate made it a superior and more powerful ogic than the more conventional NOR/NAND logic, and is responsible for the verwhelmingly large volume of research in this area.

Definition 7.2.1 A Boolean function which can be realized by a single threshold gate is known as a *threshold function* (TF) or a *linearly separable* (LS) function.

A few other definitions pertaining to the threshold functions can be given here.

Definition 7.2.2 The ordered set of weights $a_1, a_2, a_3, \ldots, a_n$ is known as the *weight vector*, and is written as

$$\mathbf{a} = (a_1 a_2 \ldots a_n).$$

The *realization vector* or the *weight-threshold vector* realizing a function F is written as

$$\mathbf{R}(F) = (a_1 a_2 \ldots a_n; t).$$

Definition 7.2.3 The ordered set of the logical values of the input variables $x_1, x_2, \ldots, x_n$ is known as the *input vector*, x.

It may be seen that the input vector can assume any of the 2^n values corresponding to the 2^n combinations of the input variables $x_1, x_2, \ldots, x_n$.

Let a linearly separable Boolean function, $F(x_1 x_2 \ldots x_n)$ be such that

$$F = \vee\, m_i,$$

and

$$\bar{F} = \vee m_j \quad m_j \neq m_i$$

where the m_i's and m_j's are the minterms. Further, let

$$\mathbf{R}(F) = (a_1 a_2 \ldots a_n; t).$$

Then it can be easily verified that

$$\mathbf{a} \cdot \mathbf{m}_i \geqslant t, \tag{7.3}$$

and

$$\mathbf{a} \cdot \mathbf{m}_j < t. \tag{7.4}$$

The . (dot) in Eqns.(7.3) and (7.4) indicates the dot or inner product of two vectors.

Now, for each minterm the dot product of the minterm and a weight vector has a value which is a real number. These real numbers can be arranged in their numerical order. This arrangement is known as the *mapping* of the minterms by the particular weight vector. Figure 7-3 shows how the 8 minterms of the 3 variable functions are mapped by two different weight vectors. If on the map on the right hand side the value of threshold t is so chosen that $3 < t \leqslant 4$, then the points 4, 5, and 6 will be separated from the points 0, 1, 2, and 3. Let the points 4, 5, 6, which are equal to or above the threshold value be represented by ● and others by ○. Since 4, 5, and 6 are the mappings of the minterms m_5, m_3, and m_7, these minterms have the value 1, and others 0 in the truth table representation of the Boolean function. Thus the map represents the Boolean function $F(x_1 x_2 x_3) = \vee(3, 5, 7)$. It will be apparent that the upper limit of the values that the threshold can have, (u) is the least value of the ● points, and the lower limit of t (l) must be greater than the highest value of the ○ points. The difference of these two values $u - l$ is called the *gap g*. Theoretically the value of the threshold t can be chosen anywhere within the gap. But to have higher tolerance and greater reliability of the gate circuit, a value midway between the gap is preferable. It can be easily verified that if t is chosen to be 1.5 in the weight vector on the left of Fig.7-3 then it separates the minterms m_7 m_5, m_4, and m_3 from m_0, m_1, and m_2. Hence

$$F_1 = \langle 2x_1 + x_2 + x_3 \rangle\, 1.5 = \vee(3, 4, 5, 6, 7) = x_1 \vee x_2 x_3.$$

Many times, instead of indicating the value of t, the values of u and l are written in the separating function. Thus for the above function,

$$F_1 = \langle 2x_1 + x_2 + x_3 \rangle\, 2{:}1.$$

In order to understand why a threshold function is called a linearly separable function, and why all functions are not LS it is necessary to visualize the geo-

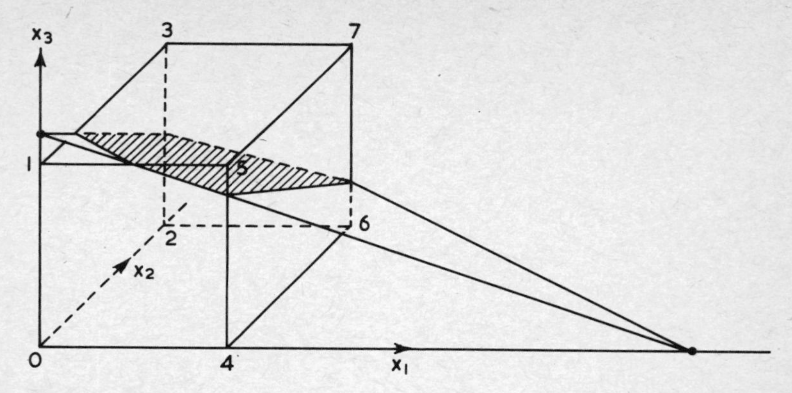

Figure 7-4 The plane $x_1 + 2x_2 + 3x_3 = 3.5$ separates the vertices 3, 7 and 5 from the vertices 1, 0, 2, 6 and 4.

metrical structure of a Boolean function as has already been discussed in Chapter 4. The 4 minterms of a 2 variable function can be represented as the 4 vertices of a square, and the 8 minterms of a 3 variable function can be represented as the 8 vertices of a cube as shown in Fig.4-2. In this geometrical representation, the separating function which separates the true vertices from the false vertices represents a hyperplane. For example, Fig.7-4 shows the plane, having the equation $x_1 + 2x_2 + 3x_3 = 3.5$, separating the true vertices 3, 5, and 7 from the false vertices, 0, 1, 2, 4 and 6. The plane becomes a straight line in a 2-cube, and a hyperplane in an n-cube ($n > 3$). Thus every threshold function is linearly separable in the sense that there exists a plane which separates its true vertices from the false vertices. It is now evident that all Boolean functions cannot be linearly separable. As an exercise the reader may try to draw a straight line in a 2-cube which will separate the pair of vertices 0 and 3 from the other pair 1 and 2. It will soon be apparent that there does not exist such a straight line. This leads to the conclusion that the two variable function

$$f = \vee\,(0, 3) = \bar{x}_1\bar{x}_2 \vee x_1x_2$$

is not an LS or a threshold function.

7.3 REALIZATION VECTOR: SEPARATING FUNCTIONS

We have seen before that if a NOR or a NAND circuit is designed to implement a certain Boolean function, the same circuit can implement other Boolean functions which can be derived from the original function by permuting and/or complementing some or all of the variables. Similarly, if the input variables of a threshold gate are permuted, or if one or more of them are complemented, the gate realizes another Boolean function which can be obtained from the original function by applying the same transformation as has been applied to the input variables of the gate. These properties of a threshold gate are stated in the following two theorems.

Theorem 7.3.1 If the realization vector, $(a_1 a_2 \ldots a_p \ldots a_q \ldots a_n; u:l)$ realizes the Boolean function $F_1 = f(x_1 x_2 \ldots x_p \ldots x_q \ldots x_n)$, then the realization vector, $(a_1 a_2 \ldots a_q \ldots a_p \ldots a_n; u:l)$ realizes the Boolean function $F_2 = f(x_1 x_2 \ldots x_q \ldots x_p \ldots x_n)$.

Proof: The proof is obvious, as the value of $\Sigma a_i x_i$ remains unchanged irrespective of the permutation of the variables. ▲

Theorem 7.3.2 If the separating function,

$$\langle a_1 x_1 + a_2 x_2 + \ldots + a_p x_p + \ldots + a_n x_n \rangle u:l$$

realizes the Boolean function $F_1 = f(x_1 x_2 \ldots x_p \ldots x_n)$ then the separating function $\langle a_1 x_1 + a_2 x_2 + \ldots + a_p \bar{x}_p + \ldots + a_n x_n \rangle u:l$ realizes the Boolean function $F_2 = f(x_1 x_2 \ldots \bar{x}_p \ldots x_n)$.

Proof: The validity of this theorem is also obvious. ▲

As a result of these two theorems, if we know that the Boolean function $F_1 = x_1 \vee x_2 \bar{x}_3 x_4$ has the realization

$$\langle 3x_1 + x_2 + \bar{x}_3 + x_4 \rangle 3:2$$

then, we can immediately write the realizations of all functions which are in the permutation class of F_1. Thus,

$$F_2 = x_2 \vee x_1 \bar{x}_3 x_4 = \langle 3x_2 + x_1 + \bar{x}_3 + x_4 \rangle 3:2$$

$$F_3 = x_4 \vee x_1 x_2 x_3 = \langle 3x_4 + x_1 + x_2 + x_3 \rangle 3:2.$$

Thus, one has to find the realization of only one member of each class. Usually the member whose MSP form has all variables in the uncomplemented or positive form is chosen to be the representative of the class.

In the above theorems, we have discussed two types of transformations which change both the separating function and the corresponding Boolean function that is realized by it. There are also certain transformations which change only the realization vector, whereas the Boolean function realized by it remains unaltered.

Definition 7.3.1 If two realization vectors realize the same Boolean function, then they are considered *equivalent*.

The following theorems show the transformations by which the realization vector changes to other equivalent forms.

Theorem 7.3.3 If the realization vector, $(a_1 a_2 \ldots a_n; t)$ realizes a Boolean function F, then the realization vector $(ba_1\ ba_2 \ldots ba_n; bt)$ also realizes the function F, where b is a real positive number.

Proof: Let $F = \vee\, m_i$

$\therefore\ \langle \Sigma a_i x_i \rangle t = \vee\, m_i$

then

$\mathbf{a} \,.\, \mathbf{m}_i \geqslant t$

$\therefore\ b\mathbf{a} \,.\, \mathbf{m}_i \geqslant bt$

as long as b is positive and real.

$\therefore\ F = \vee\, m_i = \langle \Sigma b a_i x_i \rangle bt.$ ▲

Theorem 7.3.4 If the separating function $\langle \Sigma a_i x_i \rangle t$ realizes the Boolean function F, then the separating function $\langle b + \Sigma a_i x_i \rangle b + t$ also realizes the same function, where b is a real number.

Proof: Let $\langle \Sigma a_i x_i \rangle t = \vee\, m_i$, then

$\mathbf{a} \,.\, \mathbf{m}_i \geqslant t$

$\therefore\ b + \mathbf{a} \,.\, \mathbf{m}_i \geqslant b + t,$

where b is any number. Hence

$F = \vee\, m_i = \langle b + \Sigma a_i x_i \rangle b + t.$ ▲

Theorem 7.3.5 In the separating function $\langle \Sigma a_i x_i \rangle t$ any x_i can be replaced by $1 - \bar{x}_i$.

Proof: The validity of this theorem follows from the fact that the arithmetic values of x_i and $(1 - \bar{x}_i)$ remain the same for the logical values of 0 and 1 for the variable x_i. ▲

The above theorems are very convenient to go from one form of separating function to another equivalent form, as will be evident from the following example.

Example 7.3.1 Find a separating function which is equivalent to

$\langle \tfrac{1}{6}x_1 - \tfrac{3}{4}x_2 + x_3 \rangle \tfrac{1}{2}$

but whose coefficients are neither fractional nor negative.

Solution: The least common multiple of 6, and 4, the denominators of the fractional coefficients is 12. Hence,

$$\begin{aligned}
\langle \tfrac{1}{6}x_1 - \tfrac{3}{4}x_2 + x_3 \rangle \tfrac{1}{2} &= \langle 12(\tfrac{1}{6}x_1 - \tfrac{3}{4}x_2 + x_3) \rangle 12(\tfrac{1}{2}) && \text{by Th.7.3.3} \\
&= \langle 2x_1 - 9x_2 + 12x_3 \rangle 6 \\
&= \langle 2x_1 - 9(1 - \bar{x}_2) + 12x_3 \rangle 6 && \text{by Th.7.3.5}
\end{aligned}$$

$$= \langle -9 + 2x_1 + 9\bar{x}_2 + 12x_3 \rangle 6$$

$$= \langle 2x_1 + 9\bar{x}_2 + 12x_3 \rangle 15 \qquad \text{by Th.7.3.4}$$

7.4 COMPLEMENTARY AND DUAL FUNCTIONS

It is known that if $F = \vee m_i$, then the complementary function of F, $\bar{F} = \vee m_j$, where $\{m_i\}$ and $\{m_j\}$ are two partitions of the set of all minterms. If F is an LS function, it has a separating function which separates the set of minterms $\{m_i\}$ from the set $\{m_j\}$. Therefore it can be expected that the same separating function will also realize the complementary function. Considering the geometrical picture of the Boolean function, it is apparent that the hyperplane whic separates the m_i vertices from the m_j vertices, can be made to represent either $\vee m_i$ or $\vee m_j$ if appropriate signs are assigned to its coefficients. The requirement can be specifically stated in the form of the following theorem.

Theorem 7.4.1 If the separating function $\langle \Sigma a_i x_i \rangle u : l$ realizes the Boolean function F, then the separating function $\langle \Sigma a_i \bar{x}_i \rangle \sigma - l : \sigma - u$ realizes $\bar{F}$, the complementary function of F, where $\sigma = \Sigma a_i$.

Proof: Let $F = \vee m_i$. Then $\bar{F} = \vee m_j$, where $\{m_i\}$ and $\{m_j\}$ are two partitions of the entire set of minterms. Since $\langle \Sigma a_i x_i \rangle u : l$ realizes F, for any m_i

$$\mathbf{a} \,.\, \mathbf{m}_i \geqslant u$$

and for any m_j

$$\mathbf{a} \,.\, \mathbf{m}_j \leqslant l.$$

The above two conditions can be re-written as

$$-\mathbf{a} \,.\, \mathbf{m}_i \leqslant -u$$

and

$$-\mathbf{a} \,.\, \mathbf{m}_j \geqslant -l.$$

Hence the separating function $\langle \Sigma - a_i x_i \rangle - l : -u$ realizes the function $\vee m_j$, that is, $\bar{F}$. Now,

$$\langle \Sigma - a_i x_i \rangle - l : -u = \langle \Sigma - a_i(1 - \bar{x}_i) \rangle - l : -u \qquad \text{by Th.7.3.5}$$

$$= \langle -\Sigma a_i + \Sigma a_i \bar{x}_i \rangle - l : -u$$

$$= \langle \Sigma a_i \bar{x}_i \rangle \sigma - l : \sigma - u \qquad \text{by Th.7.3.4.} \quad \blacktriangle$$

This theorem tells us that if a function F is LS, then its complement $\bar{F}$ is also LS, and the separating function of one can be derived from that of the other. This property is particularly useful in the synthesis of threshold functions, inasmuch as if one is successful in realizing a function F, then the realization of $\bar{F}$ follows automatically from that. Much to the advantage of designers of threshold logic, this property also holds good for the dual functions.

Theorem 7.4.2 If the separating function $\langle \Sigma a_i x_i \rangle u : l$ realizes the Boolean function F, then the dual function F^D is realized by the separating function

$$\langle \Sigma a_i x_i \rangle \sigma - l : \sigma - u,$$

where $\sigma = \Sigma a_i$.

Proof: Let $F = f(x_1 x_2 \ldots x_n)$, then $F^D = \bar{f}(\bar{x}_1 \bar{x}_2 \ldots \bar{x}_n)$. Now, since $f(x_1 x_2 \ldots x_n)$ is realized by $\langle \Sigma a_i x_i \rangle u : l$, $f(\bar{x}_1 \bar{x}_2 \ldots \bar{x}_n)$ must be realized by $\langle \Sigma a_i \bar{x}_i \rangle u : l$. Therefore, $\bar{f}(\bar{x}_1 \bar{x}_2 \ldots \bar{x}_n)$ must be realized by $\langle \Sigma a_i \bar{\bar{x}}_i \rangle \sigma - l : \sigma - u$ by Th.7.4.1. Hence F^D is realized by $\langle \Sigma a_i x_i \rangle \sigma - l : \sigma - u$. ▲

7.5 ISOBARIC FUNCTIONS

Definition 7.5.1 Two threshold functions are *isobaric* if they can be realized by the same weight vector but with different threshold values.

Definition 7.5.2 Two Boolean functions are *comparable* if one of them is implied by the other.

Thus F_1 and F_2 are comparable if

$$F_1 \geqslant F_2$$

or

$$F_2 \geqslant F_1.$$

Theorem 7.5.1 If $F_1(x_1 x_2 \ldots x_n)$ and $F_2(x_1 x_2 \ldots x_n)$ are two isobaric threshold functions, so that

$$F_1 = \langle \Sigma a_i x_i \rangle u_1 : l_1,$$

and

$$F_2 = \langle \Sigma a_i x_i \rangle u_2 : l_2$$

then F_1 and F_2 are comparable. Moreover,

$$F_1 > F_2 \text{ if } u_1 < u_2$$

and

$$F_1 < F_2 \text{ if } u_1 > u_2.$$

Proof: Let $F_1 = \vee\, m_p$, and $F_2 = \vee\, m_q$. Then, for any m_p,

$$\mathbf{a} \,.\, \mathbf{m}_p \geqslant u_1 \tag{7.5}$$

and, for any m_q,

$$\mathbf{a}\,.\,\mathbf{m}_q \geqslant u_2. \tag{7.6}$$

Equations (7.5) and (7.6) clearly indicate that if $u_1 < u_2$ then $\{m_p\} \supset \{m_q\}$, i.e. $F_1 > F_2$ and if $u_2 < u_1$ then $\{m_q\} \supset \{m_p\}$, i.e., $F_2 > F_1$.

Example 7.5.1 Let $F_1 = \langle 3a + b - c + d\rangle 2:1$ and $F_2 = \langle 3a + b - c + d\rangle 3:2$ find the DCFs of F_1 and F_2 and verify that $F_1 > F_2$

Solution: The DCFs of F_1 and F_2 can be determined from the table of combinations as follows:

m_i	a	b	c	d	$3a+b-c+d$	F_1	F_2
0	0	0	0	0	0	0	0
1	0	0	0	1	1	0	0
2	0	0	1	0	−1	0	0
3	0	0	1	1	0	0	0
4	0	1	0	0	1	0	0
5	0	1	0	1	2	1	0
6	0	1	1	0	0	0	0
7	0	1	1	1	1	0	0
8	1	0	0	0	3	1	1
9	1	0	0	1	4	1	1
10	1	0	1	0	2	1	0
11	1	0	1	1	3	1	1
12	1	1	0	0	4	1	1
13	1	1	0	1	5	1	1
14	1	1	1	0	3	1	1
15	1	1	1	1	4	1	1

So, $F_1 = \vee\,(5, 8, 9, 10, 11, 12, 13, 14, 15)$ and $F_2 = \vee\,(8, 9, 11, 12, 13, 14, 15)$
Hence, $F_1 > F_2$.

7.6 1-MONOTONICITY AND POLS FUNCTIONS

Any Boolean function of n variables can be expanded about any one of the variables. Thus the expansion of $F(x_1x_2x_3x_4)$ about the variable x_3 can be written as follows:

$$F(x_1x_2x_3x_4) = \bar{x}_3F(x_1x_20x_4) \vee x_3F(x_1x_21x_4),$$

where $F(x_1x_20x_4)$ and $F(x_1x_21x_4)$ are functions of the remaining variables x_1, x_2, and x_4.

The function $F(x_1x_20x_4)$ is called the reduced function of F when $x_3 = 0$, and the function $F(x_1x_21x_4)$ is called the reduced function of F when $x_3 = 1$.

The notations that will be used to denote the reduced functions are

$$F(x_1x_20x_4) = F_0^3$$

and

$$F(x_1x_21x_4) = F_1^3.$$

Reduced functions can also be evaluated by expanding about more than one variable. Thus the reduced function of F when $x_1 = 0$ and $x_2 = 0$ is

$$F(00x_3x_4) = F_{00}^{12}$$

and that of F, when $x_1 = 0$ and $x_2 = 1$, is

$$F(01x_3x_4) = F_{01}^{12}.$$

If

$$F(x_1x_2x_3x_4) = x_1x_2 \vee \bar{x}_3x_4$$

then

$$F_0^3 = x_1x_2 \vee x_4 \text{ and } F_1^3 = x_1x_2$$

$$F_{00}^{12} = \bar{x}_3x_4 \text{ and } F_{01}^{12} = \bar{x}_3x_4$$

$$F_{10}^{12} = \bar{x}_3x_4 \text{ and } F_{11}^{12} = 1.$$

In general, the reduced functions can be defined as follows.

Definition 7.6.1 A *reduced function* of the Boolean function $F(x_1x_2 \ldots x_n)$ is obtained by assigning zero and unity values to S_i and S_j, where S_i and S_j are two disjoint subsets of the variables. Either S_i or S_j can be the empty set.

The tabular forms of the reduced functions can also be written very easily from the tabular form of the given function, as can be seen in the Tables of Fig. 7-5, where

F

x_1	x_2	x_3	x_4
0	0	1	1
0	1	1	1
1	0	1	1
1	1	0	0
1	1	0	1
1	1	1	0
1	1	1	1

F_0^3

x_1	x_2	x_4
1	1	0
1	1	1

F_1^3

x_1	x_2	x_4
0	0	1
0	1	1
1	0	1
1	1	0
1	1	1

Figure 7-5 Tabular forms of reduced functions.

$F = x_1x_2 \vee x_3x_4 = \vee(3, 7, 11, 12, 13, 14, 15).$

Reduced functions of threshold functions exhibit many interesting properties.

Theorem 7.6.1 If the function $F(x_1x_2 \ldots x_n)$ is linearly separable, then the reduced functions F_0^k and F_1^k are isobaric LS functions.

Proof: Since $F(x_1x_2 \ldots x_n)$ is an LS function, let

$$F(x_1x_2 \ldots x_k \ldots x_n) = \langle a_1x_1 + a_2x_2 + \ldots + a_kx_k + \ldots + a_nx_n \rangle u : l \quad (7.7)$$

Putting $x_k = 0$ in both sides

$$F_0^k = \langle a_1x_1 + \ldots + a_{k-1}x_{k-1} + a_{k+1}x_{k+1} + \ldots + a_nx_n \rangle \, u : l. \quad (7.8)$$

Similarly, putting $x_k = 1$ in both sides of Eqn. (7.7) and by Th.7.3.4

$$F_1^k = \langle a_1x_1 + \ldots + a_{k-1}x_{k-1} + a_{k+1}x_{k+1} + \ldots + a_nx_n \rangle u - a_k : l - a_k. \quad (7.9)$$

Equations (7.8) and (7.9) show that F_0^k and F_1^k are isobaric LS functions. ▲

Theorem 7.6.2 The reduced functions F_0^k and F_1^k are related as follows:

$$F_0^k > F_1^k$$

if a_k is negative, and

$$F_0^k < F_1^k$$

if a_k is positive.

Proof: F_0^k and F_1^k have the same weight vector but differ in the threshold value the upper limits of which are u and $u - a_k$ for F_0^k and F_1^k respectively.

If a_k is positive, then $u > u - a_k$, and by Th.7.5.1, $F_0^k < F_1^k$.

If a_k is negative, then $u < u - a_k$, and by Th.7.5.1, $F_0^k > F_1^k$. ▲

Thus *the sign of a weight associated with a variable in the weight vector of an LS function depends on the direction of containment of the reduced functions with that variable equal to zero and unity.*

Theorems 7.6.1 and 7.6.2 show that all LS functions are 1-monotonic, where 1-monotonicity is defined as follows.

Definition 7.6.2 If the Boolean function $F = f(x_1x_2 \ldots x_n)$ is such that for all i's $(1 \leqslant i \leqslant n)$ the reduced functions F_0^i and F_1^i are comparable, then the function is called *1-monotonic.*

Thus all LS functions are 1-monotonic, but all 1-monotonic functions are not necessarily linearly separable. Hence, 1-monotonicity is a necessary but not a sufficient condition for linear separability. However, it has been shown that *all 1-monotonic functions of three or fewer variables are linearly separable.*

According to Th.7.6.1, the reduced functions F_0^i and F_1^i are isobaric LS functions. Hence, the reduced functions of say F_0^i, that is F_{00}^{ij} and F_{01}^{ij} are also (again by Th.7.6.1) isobaric LS functions. Thus F_0^k and F_1^k are also 1-monotonic unctions. Let the reduced functions F_0^k and F_1^k be called the *first generation* of educed functions and those like F_{00}^{ij} and F_{01}^{ij} be called the *second generation* of educed functions.

Definition 7.6.3 If all the reduced functions of the ith generation, $1 \leqslant i \leqslant k$, of a Boolean function are comparable, then the function is *k-monotonic*. When $k = n$, the function is called *completely monotonic* or simply *monotonic*.

It can now be seen that all LS functions are completely monotonic. However even complete monotonicity is not a sufficient condition for linear separability. This has been clearly demonstrated by Moore (1957) and Gableman (1961) who have reported monotonic functions of twelve and nine variables respectively which are not LS. However a stronger property than monotonicity, namely *asummability* is both a necessary and a sufficient condition for linear separability. These are, however, beyond the scope of this text. Readers desirous to know more about these topics are referred to the books dealing with threshold logic exclusively, some of which have been listed in the references at the end of this chapter.

Coming back to 1-monotonicity, we now proceed to show that it is just another name for unateness which may be defined as follows:

Definition 7.6.4 If the MSP form of a Boolean function contains each variable only in one form, either complemented or uncomplemented, then the given function is called a *unate* function. Thus

$$F_1 = x_1 \vee x_2 x_3$$

and

$$F_2 = x_1 \bar{x}_2 \vee x_3 x_4$$

are unate functions, whereas the function

$$F_3 = \bar{x}_1 x_2 \vee x_1 \bar{x}_2$$

is not unate.

A unate function is called *positive*, if all the variables in the MSP form appear in the true form, and *negative*, if they appear in the complemented form.

Theorem 7.6.3 A Boolean function is unate if, and only if, it is 1-monotonic. Moreover, in the MSP form of the function the variable $x_k (1 \leqslant k \leqslant n)$ is

a) positive if $F_0^k < F_1^k$ and conversely,

b) negative if $F_0^k > F_1^k$ and conversely, and

c) non-existent if $F_0^k = F_1^k$ and conversely.

Proof: Let the function $F(x_1 \ldots x_n)$ be 1-monotonic. Then the reduced unctions F_0^k and $F_1^k (1 \leqslant k \leqslant n)$ are comparable. Further let $F_0^k < F_1^k$. Both

F_0^k and F_1^k can be expressed as sums of minterms of variables

$$x_1 \ldots x_{k-1}, \quad x_{k+1} \ldots x_n.$$

Let the minterms subsuming F_0^k and F_1^k constitute the sets A and B respectively, so that,

$$F_0^k = \vee\,(\mathrm{A}) \text{ and } F_1^k = \vee\,(\mathrm{B}).$$

Since,

$$F_0^k < F_1^k$$

$$A \subset B$$

$$\therefore\ B = A \cup C, \text{ say.}$$

Now, the function F can be expressed as

$$\begin{aligned} F &= \bar{x}_k F_0^k \vee x_k F_1^k \\ &= \bar{x}_k (\vee(A)) \vee x_k (\vee(B)) \\ &= \bar{x}_k (\vee(A)) \vee x_k (\vee(A)) \vee x_k (\vee(C)) \\ &= (\vee(A)) \vee x_k (\vee(C)). \end{aligned}$$

Hence in the MSP form of F the variable x_k must appear in the x_k or positive form. Similarly it can be shown that if

$$F_0^k > F_1^k,$$

then the variable x_k appears in the MSP form in the $\bar{x}_k$ or negative form.

If $F_0^k = F_1^k$, then $A = B$ and

$$\begin{aligned} F &= \bar{x}_k (\vee(A)) \vee x_k (\vee(A)) \\ &= \vee\,(A) \end{aligned}$$

and the variable x_k does not appear in the MSP form. Arguing, in the reverse direction, the converse of the theorem can be proved. ▲

Theorem 7.6.4 In the tabular form of a unate function, the relation betwee the ratio of number of 1's to that of 0's, r_k, of a column headed by the variable x_k, and the form of x_k in the MSP form of the function, is as follows:

$r_k > 1$ if x_k is positive, and conversely,

$r_k < 1$ if x_k is negative, and conversely, and

$r_k = 1$ if x_k is non-existent, and conversely.

Proof: When x_k is positive,

$$F_0^k < F_1^k.$$

Also

$F_0^k = \vee(A), F_1^k = \vee(B)$ and $A \subset B$.

Therefore the number of elements in B is larger than that of A. Hence the number of rows in the tabular form of F_1^k is more than that in the tabular form of F_0^k. But all rows whose sum constitutes F_1^k have 1's under x_k, and all rows whose sum constitutes F_0^k have 0's under x_k. Hence, under the x_k column, the number of 1's is greater than that of 0's. Hence, $r_k > 1$. Conversely, a ratio greater than 1 under the x_k column implies that $B \supset A$, and therefore

$F_1^k > F_0^k$.

The two other relations can be proved by similar reasoning. ▲

COROLLARY **7.6.4A** In a positive unate function all r_k's are greater than 1.

As all LS functions are unate, there is a class of LS functions which are positive unate. Again, among these there are some whose r_k's are ordered, that is

$r_1 \geqslant r_2 \geqslant \ldots r_{n-1} \geqslant r_n$.

This leads us to the definition of a POLS function.

Definition 7.6.5 If the ratios of the number of 1's to that of 0's of the columns (r_k's) of an LS function are greater than 1, and if $r_1 \geqslant r_2 \ldots \geqslant r_{n-1} \geqslant r_n$, then the function is positive unate and has ordered ratios. Such a function will be called a *positive ordered linearly separable* (POLS) function. It is also called a *canonic threshold function.*

It is interesting to note that a POLS function may be considered the representative of its complementary symmetry class. Thus once the realization of a POLS function is known, the realization of any other function belonging to the class can be determined. Hence any method which is adequate for the realization of POLS functions will be good enough for the entire class of threshold functions.

Some interesting results regarding the co-efficients of the separating functions of POLS functions are now given by the following theorems.

Theorem 7.6.5 Let the ratios of the number of 1's to that of 0's of the columns i and j of a POLS function be r_i and r_j respectively, then if there exists a separating function of the POLS function having $a_i = a_j$, then $r_i = r_j$, and the function is symmetric in x_i and x_j.

Proof: Without any loss of generality, we let $i = 1$ and $j = 2$ so that $a_1 = a_2$.

Now let a minterm $\bar{x}_1 x_2 p$, where p is a product term of the variables x_3 through x_n be realized by the separating function, then:

$\mathbf{a} \cdot \bar{\mathbf{x}}_1 \mathbf{x}_2 \mathbf{p} = a_2 + N \text{ (say)} \geqslant u.$

For the minterm $x_1 \bar{x}_2 p$

$\mathbf{a} \cdot \mathbf{x}_1 \bar{\mathbf{x}}_2 \mathbf{p} = a_1 + N.$

But

$$a_1 = a_2,$$

therefore

$$\mathbf{a} \cdot \mathbf{x}_1\bar{\mathbf{x}}_2\mathbf{p} \geqslant u,$$

and the minterm $x_1\bar{x}_2p$ must also be realized by the same separating function.

Thus, for every row of the tabular form where a p has been preceded by 01 under the columns x_1 and x_2, there must be another row where the same p has been preceded by 10 under the columns x_1 and x_2. Hence, the number of rows with 01 under x_1x_2 is the same as the number of rows with 10 under x_1x_2. Other rows must have either 00 or 11 under x_1x_2. Hence $r_1 = r_2$.

Also the LS function can be written as:

$$F = \bar{x}_1\bar{x}_2F_a \vee (\bar{x}_1x_2 \vee x_1\bar{x}_2)F_b \vee x_1x_2F_c,$$

where F_a, F_b and F_c are functions of x_3 through x_n. Clearly F is symmetric in x_1 and x_2. ▲

Theorem 7.6.6 If in a positive unate LS function $r_i > r_j$, then $a_i - a_j \geqslant g$.

Proof: Here also we let $i = 1$ and $j = 2$ without any loss of generality.

Since $r_1 > r_2$, the number of rows with 10 under x_1x_2 is greater than those with 01 under x_1x_2. Therefore, there is at least one p, a product term of variables x_3 through x_n, for which

$$\mathbf{a} \cdot \bar{\mathbf{x}}_1\mathbf{x}_2\mathbf{p} = a_2 + N \text{ (say)} \leqslant l,$$

and

$$\mathbf{a} \cdot \mathbf{x}_1\bar{\mathbf{x}}_2\mathbf{p} = a_1 + N \geqslant u,$$

therefore,

$$a_1 - a_2 \geqslant u - l = g. \quad ▲$$

7.7 SYNTHESIS BY CHARACTERISTIC VECTOR: DERTOUZOS (1964)

Many authors have expounded methods for the testing and realization of linearly separable functions. In this section we shall discuss the method of characteristic vectors as given by Dertouzos. He has shown that the knowledge about the number of 1's and 0's in each column of the tabular form along with the number of rows is sufficient to test the linear separability, and to produce a *minimum integer*† *realization* of a function. In this method, a Boolean function $F(x_1 \ldots x_n)$ is characterized by $n + 1$ numbers b_n, $b_{n-1}, \ldots, b_1, b_0$. The vector $(b_nb_{n-1} \ldots b_0$

† a_i's are integers, and $\sum_{i=1}^{n} |a_i|$ is minimum.

is called the *characteristic vector* of the function. Let the method be illustrated by working out an example.

Example 7.7.1 Test if the following function is a threshold function. If yes, determine its separating function.

$f(x_1x_2x_3x_4) = \vee(2, 3, 9, 10, 11)$.

Solution: Write the function in its tabular form as shown in the Table of Fig.7-6.

	x_1	x_2	x_3	x_4
	0	0	1	0
	0	0	1	1
	1	0	0	1
	1	0	1	0
	1	0	1	1
N(1)	3	0	4	3
N(0)	2	5	1	2

Figure 7-6

Compute $N(1)$ and $N(0)$ for each column. Now, calculate all b_i's, where

$b_i = 2[N_i(1) - N_i(0)]$ for $1 \leqslant i \leqslant n$

and $b_0 = 2$ (number of rows) $- 2^n$. Note that b_0 and b_i's can be negative also. In the above example

$b_0 = 2(5) - 16 = -6$

$b_1 = 2(3 - 2) = 2$

$b_2 = 2(0 - 5) = -10$

$b_3 = 2(4 - 1) = 6$

$b_4 = 2(3 - 2) = 2.$

Consider only the magnitudes, and arrange $|b_i|$'s in the descending order of magnitude. For our example, it would be

$\lvert b_2\rvert$	$\lvert b_3\rvert$	$\lvert b_0\rvert$	$\lvert b_1\rvert$	$\lvert b_4\rvert$.
10	6	6	2	2

Looking into Dertouzos' table (Appendix C) the corresponding $|a_i|$'s are

3 2 2 1 1.

Restoring the order and sign as were those of $|b_i|$'s, the weight vector is

(1 −3 2 1).

The threshold value is given by

$$u : l = \tfrac{1}{2}(\sigma - a_0 + 1) : \tfrac{1}{2}(\sigma - a_0 - 1),$$

where

$$\sigma = \sum_{i=1}^{n} a_i.$$

In our example $\sigma = (1 - 3 + 2 + 1) = 1$ and $a_0 = -2$,

$$\therefore\ u : l = \tfrac{1}{2}(1 + 2 + 1) : \tfrac{1}{2}(1 + 2 - 1)$$

$$= 2 : 1.$$

∴ The realization for f is

$$f = \langle x_1 - 3x_2 + 2x_3 + x_4 \rangle 2 : 1.$$

7.8 MULTIGATE SYNTHESIS

It is obvious that Boolean functions which are not linearly separable would need more than one threshold gate for their synthesis. The procedure to carry out such synthesis is more involved and is beyond the scope of this text. However, we shall deal with the synthesis of symmetric functions requiring more than one gate. There are some symmetric functions which are linearly separable, and as such would need only one threshold gate.

Theorem 7.8.1 The totally symmetric function

$$S_{\{a \mid k \leqslant a \leqslant n\}}(x_1 x_2 \ldots x_n)$$

is linearly separable.

The proof of this theorem is left as an exercise to the reader. It may be observe that the TSF $S_{\{a \mid k \leqslant a \leqslant n\}}(x_1 x_2 \ldots x_n)$ is the sum of all minterms of weights $k, k + 1, \ldots, n$.

As it is a TSF,

$$a_1 = a_2 = \ldots = a_n = a \text{ say.}$$

To achieve minimum integer realization, make $a = 1$. If $m_{\min}(k)$ is the least minterm of weight k, then

$$u = \mathbf{a}.\mathbf{m}_{\min}(k)$$

$$= k \quad \because \mathbf{a} = (1\ 1 \ldots 1).$$

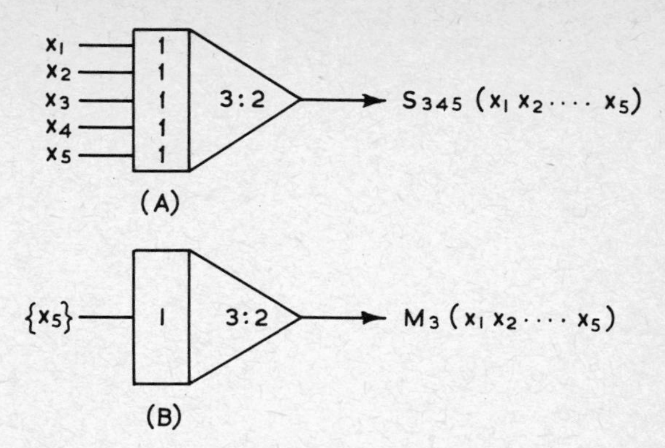

Figure 7-7 A single gate realization of the voting function $M_3(x_1x_2 \ldots x_5)$.

Thus the threshold gate realizing

$$S_{345}(x_1x_2 \ldots x_5)$$

is as shown in Fig.7-7(A).

It may be seen that if at any time any three of the input variables become 1 the output of the gate is 1. For this reason such a function and the gate are called *voting function* and *voting logic* respectively. These can be depicted very conveniently as shown in Fig.7-7(B). A convenient notation for a voting function is as follows:

$$S_{\{a \mid k \leqslant a \leqslant n\}}(x_1x_2 \ldots x_n)$$
$$= M_k(x_1x_2 \ldots x_n).$$

COROLLARY 7.8.1A *The totally symmetric function*

$$S_{\{a \mid 0 \leqslant a \leqslant k\}}(x_1x_2 \ldots x_n)$$

is linearly separable.

Proof: By Th.5.3.5

$$S_{\{a \mid k \leqslant a \leqslant n\}}(x_1x_2 \ldots x_n)$$
$$= \bar{S}_{\{a \mid (k+1) \leqslant a \leqslant n\}}(x_1x_2 \ldots x_n)$$
$$= \bar{M}_{k+1}(x_1x_2 \ldots x_n).$$

But

$$M_{k+1}(x_1x_2 \ldots x_n)$$

is an LS function. ∴ Its complement is also an LS function. ▲

Any other symmetric function which by itself is not a voting function or a complement of a voting function, can be expressed as a function of voting functions. Thus

$$\begin{aligned} &S_{14}(x_1x_2 \ldots x_5) \\ &\quad = S_{12345}(x_1x_2 \ldots x_5)\, \overline{S}_{2345}(x_1x_2 \ldots x_5) \\ &\quad \vee S_{45}(x_1x_2 \ldots x_5)\, \overline{S}_5(x_1x_2 \ldots x_5) \\ &\quad = M_1(x_1x_2 \ldots x_5)\, \overline{M}_2(x_1x_2 \ldots x_5) \\ &\quad \vee M_4(x_1x_2 \ldots x_5) \overline{M}_5(x_1x_2 \ldots x_5). \end{aligned}$$

At this point it may be verified that voting logic $M_{a_p}(x_1x_2 \ldots x_n)$ will pass the symmetric functions of a-numbers a_p and above. On the other hand the voting function $\overline{M}_{a_s}(x_1x_2 \ldots x_n)$ will stop all symmetric functions of a-numbers a_s and above. Thus a 'non-voting' TSF can be written as

$$F = M_{a_{p1}} \overline{M}_{a_{s1}} \vee M_{a_{p2}} \overline{M}_{a_{s2}} \vee \ldots \tag{7.10}$$

It has been shown by Minnick (1961) that such a function can be implemented by $n + 1$ threshold gates, where n is the number of $\overline{M}_{as}$ functions appearing in Eqn.(7.10).

Example 7.8.1 Implement the following function by a minimum number of threshold gates, $F = S_{14}(x_1x_2 \ldots x_5)$.

Solution:

$$\begin{aligned} F &= M_1(x_1x_2 \ldots x_5)\overline{M}_2(x_1x_2 \ldots x_5) \\ &\quad \vee M_4(x_1x_2 \ldots x_5)\, \overline{M}_5(x_1x_2 \ldots x_5) \\ &= M_{a_{p1}} \overline{M}_{a_{s1}} \vee M_{a_{p2}} \overline{M}_{a_{s2}}. \end{aligned}$$

The threshold gate network assumes the form as given in Fig.7-8. In the absence of β_1 and β_2, gate #0 will pass all a's which are 1 or more. This will require that

$$u_0 = a_{p1} = 1.$$

However gate #1 ensures that a-numbers 2, and 3 are stopped but 4 and above are passed. In order to achieve this, the value of β_1 should be such that

$$a_{p2} + \beta_1 = u_0 = a_{p1}$$

$$\therefore \beta_1 = a_{p1} - a_{p2} = 1 - 4 = -3.$$

Again gate #2 must stop a-number 5, but can pass a-number 6, since it does not exist. Hence, the value of β_2 is given by

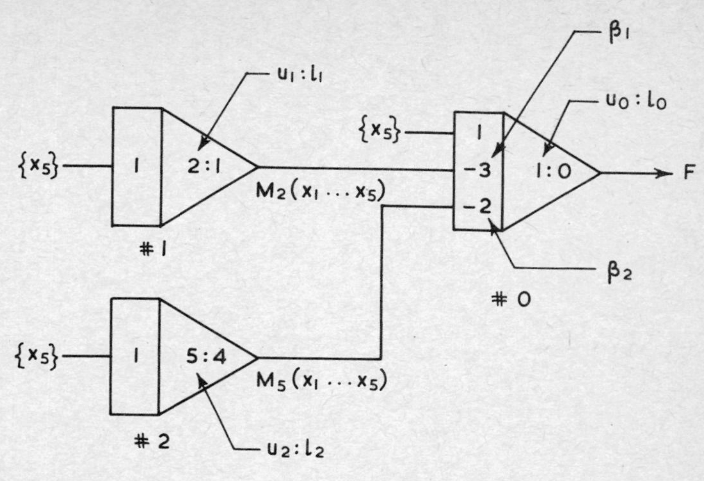

Figure 7-8 Realization of $S_{14}(x_1x_2 \ldots x_5)$ by the minimum number of threshold gates.

$$a_{p3} + \beta_1 + \beta_2 = u_0 = a_{p1}$$

$$\therefore \beta_2 = a_{p1} - a_{p3} - (a_{p1} - a_{p2})$$

$$= a_{p2} - a_{p3} = 4 - 6 = -2.$$

The various values of $u_1, u_2 \ldots u_m$, and $\beta_1, \beta_2 \ldots, \beta_m$ of the $m + 1$ number of threshold gates (Fig.7-9) realizing a symmetric non-voting function of the form as given in Eqn.(7.10), can be very conveniently calculated from the following two equations

$$u_0 = a$$

$$\begin{bmatrix} u_1 & \beta_1 \\ u_2 & \beta_2 \\ \cdot & \cdot \\ \cdot & \cdot \\ u_m & \beta_m \end{bmatrix} = \begin{bmatrix} a_{s1} & a_{p1} - a_{p2} \\ a_{s2} & a_{p2} - a_{p3} \\ \cdot & \cdot \\ \cdot & \cdot \\ a_{sm} & a_{pm} - a_{p(m+1)} \end{bmatrix}$$

In the above example

$a_{p1} = 1, a_{p2} = 4$, and $a_{p3} = 6$ (by implication)

$a_{s1} = 2, a_{s2} = 5.$

Hence, $u_0 = a_{p1} = 1$, and

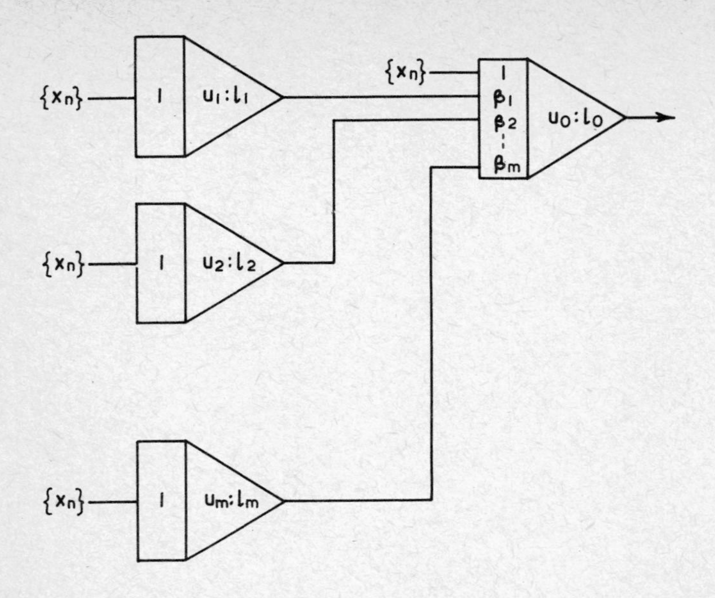

Figure 7-9 $(m + 1)$ number of threshold gates realizes a symmetric non-voting function.

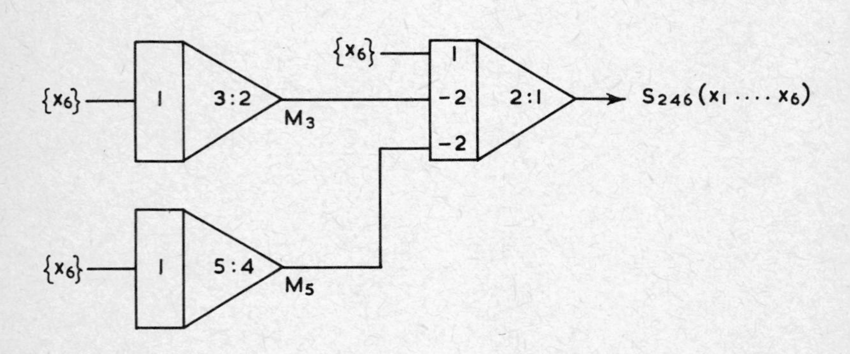

Figure 7-10 Multigate realization of $S_{246}(x_1 \ldots x_6)$.

$$\begin{bmatrix} u_1 & \beta_1 \\ u_2 & \beta_2 \end{bmatrix} = \begin{bmatrix} 2 & 1-4 \\ 5 & 4-6 \end{bmatrix} = \begin{bmatrix} 2 & -3 \\ 5 & -2 \end{bmatrix}$$

Example 7.8.2 Implement by threshold gates the following function

$F = S_{246}(x_1 x_2 \ldots x_6)$.

Solution:

$$F = S_{246}(x_1 \ldots x_6)$$
$$= M_2\overline{M}_3 \vee M_4\overline{M}_5 \vee M_6.$$

Here

$a_{p1} = 2, \quad a_{p2} = 4, \quad a_{p3} = 6$

and

$a_{s1} = 3, \quad a_{s2} = 5.$

Number of gates required = 2 + 1 = 3.

$$u_0 = a_{p1} = 2$$

$$\begin{bmatrix} u_1 & \beta_1 \\ u_2 & \beta_2 \end{bmatrix} = \begin{bmatrix} 3 & 2-4 \\ 5 & 4-6 \end{bmatrix} = \begin{bmatrix} 3 & -2 \\ 5 & -2 \end{bmatrix}$$

The network is as shown in Fig.7-10.

REFERENCES

Chow, C. K.: "Boolean functions realizable with single threshold devices", *Proc. IRE,* **39**, pp.370–371, Jan.1961.

—— : "On the characterization of threshold functions", *SCTLD*, pp.34–38, Sept.1961.

Coates, C. L., R. B. Kirchner and P. M. Lewis, II: "A simplified procedure for the realization of linearly separable switching functions", *IRE Trans. Electronic Computers,* **EC-11**, pp.447–458, Aug.1962.

Dertouzos, M. L.: "An approach to single-threshold element synthesis", *IEEE Trans. Electronic Computers,* **EC-13**, pp.519–528, Oct.1964.

—— : *Threshold Logic: a Synthesis approach,* MIT Press, Cambridge, Mass., USA, 1965.

Elgot, C. C.: "Threshold functions realizable by single threshold organs", *SCTLD,* 1960, pp.225–245, Sept.1961.

Gableman, I. J.: "The synthesis of Boolean functions using a single threshold element", *IEEE Trans. Electronic Computers,* **EC-11**, pp.639–642, Oct.1962.

Gaston, C. A.: "A simple test for linear separability", *IRE Trans. Electronic Computers,* **EC-12**, pp.134–135, April 1963.

Ghosh, S., D. Basu and A. K. Choudhury: "Multigate synthesis of general Boolean

functions by threshold logic elements", *IEEE Trans. Computers,* **C-18**, pp.451–456, May 1969.
Lewis, P. M. and C. L. Coates: *Threshold Logic,* John Wiley and Sons, Inc., New York, 1967.
McNaughton, R.: "Unate truth functions", *IRE Trans. Electronic Computers,* **EC-10**, pp.1–6, 1961.
Minnick, R. C.: "Linear input logic", *IRE Trans. Electronic Computers,* **EC-10**, pp.6–16, 1961.
Muroga, S., I. Toda, and S. Takasu: "Theory of majority decision elements", *J. Franklin Inst.* **271**, pp.376–418, 1961.
Paull, M. C. and E. J. McCluskey, Jr.: "Boolean functions realizable with single threshold devices", *Proc. IRE,* **48**, pp.1335–1337, 1960.
Roy, P. K. S.: "Synthesis of symmetric switching functions using threshold logic elements", *IEEE Trans Electronic Computers,* **EC-16**, pp.359–364, June 1967.
Sheng, C. L.: "A method for testing and realization of threshold functions", *IEEE Trans. Electronic Computers,* **EC-13**, pp.232–239, June 1964.
Sheng, C. L.: "Compound synthesis of threshold logic network for the realization of general Boolean functions", *IEEE Trans. Electronic Computers,* **EC-14**, pp.798–814, Dec.1965.
—— : *Threshold Logic,* Academic Press, 1969.
Sheng, C. L. and H. R. Hwa: "Testing and realization of threshold functions by successive higher ordering of incremental weights", *IEEE Trans. Electronic Computers,* **EC-15**, pp.212–220, 1966.
Torng, H. C.: "An approach for the realization of linearly separable switching functions", *IEEE Trans. Electronic Computers,* **EC-15**, pp.14–20, Feb.1966.
Winder, R. O,: "Single stage threshold logic", *SCTLD*, pp.321–332, Sept. 1961.

PROBLEMS

7.1 Determine the MSP forms of f_1 through f_6 (see Fig.7-11).

$$f_1 = \langle x_1 + 2x_2 - 3x_3 \rangle\, 2.5$$

$$f_2 = \langle -x_1 + 2x_2 + \bar{x}_3 - x_4 \rangle\, 3$$

$$f_3 = \langle 2x_1 - 2x_2 + 3x_3 + 3x_4 \rangle\, 5.2.$$

7.2 Plot the following functions on the appropriate n-cubes and ascertain their linear separability.

a) $f_1 = x_1 \vee \bar{x}_2$

b) $f_2 = \bar{x}_1\bar{x}_2 \vee x_1x_2$

c) $f_3 = x_1x_2 \vee \bar{x}_3$

d) $f_4(x_1x_2x_3) = \vee\,(0, 1, 2, 6)$

e) $f_5(x_1x_2x_3) = \vee\,(3, 5, 6, 7)$

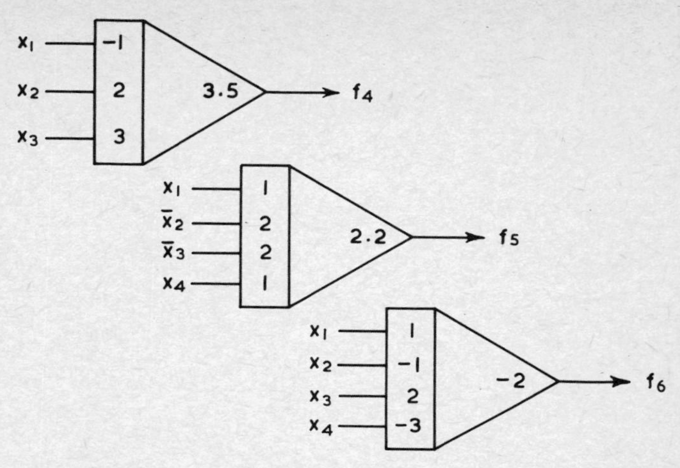

Figure 7-11

f) $f_6(x_1x_2x_3) = \vee\,(3, 4, 5, 6, 7)$.

7.3 Redesign the following threshold gates, so that the threshold values will be positive and the weights integers.

a) $\langle \frac{1}{2}x_1 - 2x_2 + x_3 \rangle 2$

b) $\langle -x_1 - x_2 + 2x_3 \rangle - 1.5$

c) $\langle \frac{1}{4}x_1 + \frac{1}{3}x_2 - x_3 - x_4 \rangle 3$

d) $\langle 3x_1 - x_2 + x_3 - \frac{1}{2}x_4 \rangle - 2.5$.

7.4 Find the realizations of the complementary and dual functions of the following:

a) $f_1 = \langle 2x_1 + 3x_2 + 4x_3 \rangle 4:3$

b) $f_2 = \langle -x_1 + x_2 - 2x_3 + 4x_4 \rangle 2:1$

c) $f_3 = \langle 2x_1 - x_2 + x_3 + 3x_4 \rangle 3:2$

d) $f_4 = \langle 2x_1 - x_2 + x_3 + 2x_4 + 3x_5 \rangle 3:2$.

7.5 Show that for a self-dual threshold function

$$u + l = \sum_{i=1}^{n} a_i.$$

7.6 Which of the functions of problem 7.4 are self-dual? Determine their MSP forms.

7.7 Given $F(x_1x_2x_3x_4) = \langle 3x_1 + 2x_2 - 2x_3 + x_4 \rangle 3:2$, determine the DCFs of F, F_0^2 and F_1^2. Also determine the realizations of F_0^2 and F_1^2.

7.8 Given $F = x_1\bar{x}_2 \vee x_2x_3 \vee \bar{x}_1x_4x_5$, determine F_0^1, F_{01}^{23}, F_{010}^{134}, and F_{0110}^{1235}.

7.9 Given $F_1 = \langle 3x_1 - 2x_2 + 3x_3 - x_4 + x_5 \rangle 3:2$, and $F_2 = \langle 3x_1 + 2\bar{x}_2 + 3x_3 + \bar{x}_4 - \bar{x}_5 \rangle 2:1$, show that $F_1 < F_2$.

7.10 Determine without finding the MSP forms if the following functions are unate.

a) $f_1(x_1x_2x_3) = \vee\,(0, 1, 2, 4, 7)$

b) $f_2(x_1x_2x_3) = \vee\,(3, 4, 5, 6, 7)$

c) $f_3(x_1x_2x_3x_4) = \vee\,(2, 5, 8, 9, 12, 14)$

d) $f_4(x_1x_2x_3x_4) = \vee\,(6, 7, 10, 11, 12, 13, 14, 15)$

e) $f_5(x_1x_2x_3x_4x_5) = \vee\,(1, 3, 5, 6, 7, 14, 15, 17, 19, 20, 21, 22, 23, 28, 29, 30, 31)$

f) $f_6(x_1x_2x_3x_4x_5) = \vee\,(1, 3, 7, 8, 9, 10, 11, 12, 13, 14, 15, 17, 19, 23, 25, 27, 31)$.

7.11 Verify your findings in problem 7.10 by determining the MSP forms.

7.12 Prove that the MSP form of a unate function is unique, that is, all its prim implicants are essential prime implicants. Is the converse true? If not, give a counter example.

7.13 Test if the following functions are LS, and determine minimum integer realizations for those functions which are.

$f_1 = x_1 \vee \bar{x}_2\bar{x}_3$

$f_2 = x_1x_2 \vee x_3x_4$

$f_3 = x_1x_2 \vee x_1x_2x_4$

$f_4 = x_1x_2\bar{x}_3 \vee \bar{x}_3x_4x_5 \vee x_1x_5$

$f_5(x_1x_2x_3) = \vee\,(1, 2, 3, 6, 7)$

$f_6(x_1x_2x_3x_4) = \vee\,(6\text{–}15)$

$f_7(x_1x_2x_3x_4) = \vee\,(0, 2\text{–}7, 12, 13, 15)$

$f_8(x_1x_2x_3x_4) = \vee\,(0, 1, 5, 6, 8, 9, 10\text{–}13, 15)$.

7.14 Show that in no case a threshold gate realization of a Boolean function needs more gates than the corresponding realization by AND, OR, and NOT gates.

Hint: Show that any AND, OR, and NOT gate can be replaced by a single threshold gate.

7.15 Implement the following symmetric functions with minimum number of threshold gates.

a) $S_{01}(x_1x_2x_3x_4)$

b) $S_{34}(x_1x_2x_3x_4)$

c) $S_{125}(x_1x_2x_3x_4x_5)$

d) $S_{145}(x_1x_2x_3x_4x_5x_6)$

e) $S_{2356}(x_1x_2x_3x_4x_5x_6x_7)$.

7.16 Implement the *odd parity function* $S_{135}(x_1x_2 \ldots x_6)$ by minimum number of threshold gates.

7.17 Enumerate all two variable LS functions.

7.18 Give one example for each of the following types of functions: a) symmetric and LS, b) symmetric and not LS, c) not symmetric and LS, d) not symmetric and not LS.

7.19 Show that if $F(x_2x_3 \ldots x_n)$ is LS, so are

a) x_1F b) $\bar{x}_1F$ c) $x_1 \vee F$

d) $\bar{x}_1 \vee F$ and e) $\bar{x}_1F \vee x_1F^D$.

7.20 Show that the EPI's of a unate function intersect in a common implicant, which is called the *center of gravity*.

7.21 Find the centers of gravity of the unate functions of Prob.7.10.

7.22 Show that every LS function is dual-comparable.

Chapter 8

LOGIC OF FLIP-FLOPS

8.1 INTRODUCTION

The circuits and devices which constitute the hardware of a digital computer, not only should be able to distinguish between the 0 and 1 of a Boolean variable, but also are many times required to store this information. These devices and circuits are known as memory elements. By the very nature of their function, they have two stable states. Such bistable circuits can be made in a variety of ways. They may consist of electromagnetic relays, magnetic cores, vacuum tubes, transistors, cyrotrons, etc. As an example, a typical circuit is described in this section. While all these bistable circuits, more commonly known as flip-flops, may differ in their make-up and composition, they exhibit certain definite logical properties. In fact, all flip-flops can be classified into a few types according to their logical behavior. This has been done in this chapter, and subsequent sections bring out the various types of flip-flops that can be used in the design of sequential circuits.

Figure 8-1 shows the circuit of an S–C (set and clear†) flip-flop having two transistors. The circuit has two stable states. In one of these states the transistor T_a is on, i.e., conducting, and the transistor T_b is off, i.e., non-conducting. In this state of the circuit, neglecting the very small voltage drop across the collector and emitter of transistor T_a which is in saturation, the point a can be considered to be at ground potential. This makes the base current of transistor T_b through resistor R_1 to be zero. Hence transistor T_b does not conduct. At this time the potential of point b is $V_{cc} - R_b I_2$ where I_2 is the base current of transistor T_a. This current is small in magnitude. Hence the point b will be at a high potential relative to ground. Let this be E volts. The circuit will come to this state whenever a pulse or a steady voltage is incident on the S wire. This state is called the 1 state of the flip-flop. Output 1 (E volts) is available at point b (terminal q), and output 0 (O volts) is available at point a (terminal $\overline{q}$). If now a pulse or a steady voltage is applied on wire C (assuming that the pulse or steady

† In the literature $S-C$ flip-flop is also referred to as $S-R$ (set and reset) flip-flop.

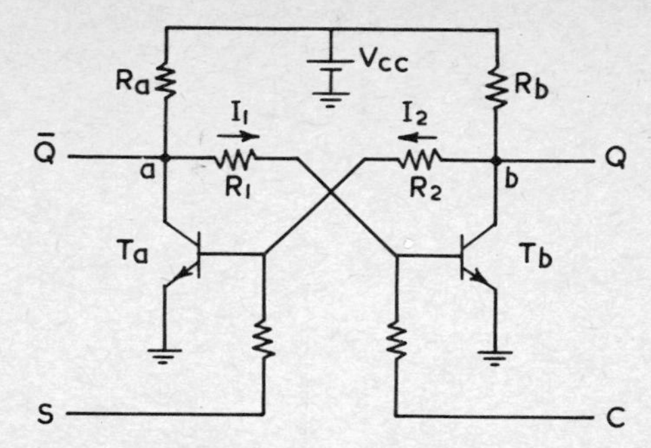

Figure 8-1 $S-C$ flip-flop circuit.

oltage on wire S has ceased to exist), base current will flow through transistor $_b$ bringing it to the on or conducting condition. This will ground the point b, hich results in cutting off the base current of transistor T_a. Consequently, ansistor T_a is made off or non-conducting. This is the other stable state, and this state, the potential of point b becomes 0 and that at point a becomes E olts. This state is called the 0 state of the flip-flop. At this time, the logical gnal on terminal q is 0 and on terminal $\overline{q}$ it is 1. It is obvious that if pulses or eady voltages occur simultaneously on the S and C lines, the circuit does not inction properly. Thus the restriction that S and C cannot be 1 (high) simul- ineously is stipulated as a required condition in $S-C$ flip-flop. The circuit as escribed here is simply for the purpose of introducing the reader to the under- ing principle of operation. For more details on this and other types of circuits, ooks dealing in circuits, such as those listed in the references, may be studied.

Like the $S-C$ flip-flop, other types of flip-flops are also capable of assuming two istinct states, the 0 state and 1 state. If a flip-flop is in 1 state, then it is not in state, and *vice-versa.* Like the $S-C$ flip-flop, there are other flip-flops which ave two input lines. Again there are flip-flops with only one or three input nes. Each of these input lines can be either 1 or 0 at a particular time. These gnals on the input lines as well as the state in which the flip-flop is in at that me (known as the present state) determine the next state of the flip-flop. he state of a flip-flop appears on a different line, known as the output line or mply the output of the flip-flop. More often there are two output lines, one f which gives the true state of the flip-flop, and the other gives the comple- ientary state. Now, we can define a few terms which are very useful in the udy of flip-flop logics.

efinition 8.1.1 In a sequential circuit employing flip-flops, the state of a ip-flop is to be observed at certain intervals of time. Then, it is customary to ill the state in the present interval as the *present state* (denoted by q) of the ip-flop, and the state in the interval immediately following the present interval the *next state* (denoted by Q).

Definition 8.1.2 The next state of a flip-flop is a function of the signals on its input lines (i.e., of its input variables) as well as its present state. The Boolean equation expressing this relation is known as the *characteristic function* of the flip-flop.

It will be seen later that each type of flip-flop is typified by its characteristic function.

8.2 SET AND CLEAR ($S-C$) FLIP-FLOP

The circuit of the $S-C$ flip-flop can be represented by a block diagram with two input lines, the S line and the C line and two output lines (Fig.8-2). We have also seen that when the S line is 1, the FF is set to 1 irrespective of its previous state, and when the C line is 1, the FF is cleared to 0, irrespective of its previous state. The state of the FF remains unchanged when both S and C are 0. Moreover, the S and C lines are not allowed to be 1 simultaneously, that is

$$SC = 0. \tag{8.1}$$

The truth table, its plotting and simplification are shown in Fig.8-2. The characteristic function is found to be

$$Q = S \vee \bar{C}q. \tag{8.2}$$

Let the usefulness of the characteristic function be illustrated by working out the following example.

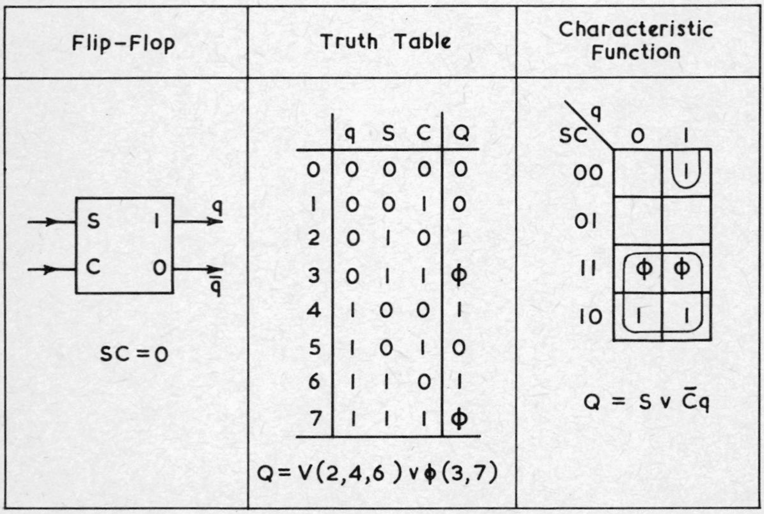

Flip-Flop	Truth Table	Characteristic Function
Inputs S, C; outputs 1 → q, 0 → $\bar{q}$ SC = 0	see truth table below $Q = \vee(2,4,6) \vee \phi(3,7)$	see map below $Q = S \vee \bar{C}q$

Truth table:

	q	S	C	Q
0	0	0	0	0
1	0	0	1	0
2	0	1	0	1
3	0	1	1	ϕ
4	1	0	0	1
5	1	0	1	0
6	1	1	0	1
7	1	1	1	ϕ

Characteristic function map:

SC \ q	0	1
00		1
01		
11	ϕ	ϕ
10	1	1

Figure 8-2 $S-C$ flip-flop.

Example 8.2.1 Design of modulo-5 counter which counts from 0 through 4 and then recycles. The decimal numbers are to be displayed in the binary code.

Solution: Since the display of 5 binary numbers requires three bits, the counter requires 3 flip-flops, one for each bit.

Let y_1, y_2 and y_3 be the three flip-flops. We must now write a flow table where the present states and the next states of the counter in terms of the flip-flops and the input (x) are shown. As will be apparent from the flow table (Fig.8-3), the next state of any particular flip-flop (shown by the corresponding capital letter) depends on the previous states of all flip-flops and the input. Thus, the next state of y_1 flip-flop, Y_1 is a function of y_1, y_2, y_3 and x. The Boolean equation which expresses Y_1 in terms of y_1, y_2, y_3 and x, is called the *transition function* of the flip-flop. The transition functions of the three flip-flops have also been computed in Fig.8-3.

Let us decide to implement this counter by S–C FFs, that is, the three FFs required will be of S–C type.

Confining our attention to the y_1 FF only for the time being, we can see that the next state of this FF must be a function of the present states of the three FFs and the input variable x. This function is given by the transition function of y_1 FF,

Count state	Decimal Cell	Present state				Next state			Transition functions
		x	y_1	y_2	y_3	Y_1	Y_2	Y_3	
C_0	0	0	0	0	0	0	0	0	
C_1	1	0	0	0	1	0	0	1	
C_2	2	0	0	1	0	0	1	0	$Y_1(xy_1y_2y_3)$
C_3	3	0	0	1	1	0	1	1	$= \vee (4, 11)$
C_4	4	0	1	0	0	1	0	0	$\vee\, \phi(5, 6, 7, 13, 14,$
Not used	5	0	1	0	1	ϕ	ϕ	ϕ	$15)$
	6	0	1	1	0	ϕ	ϕ	ϕ	
	7	0	1	1	1	ϕ	ϕ	ϕ	$Y_2(xy_1y_2y_3)$
C_0	8	1	0	0	0	0	0	1	$= \vee (2, 3, 9, 10)$
C_1	9	1	0	0	1	0	1	0	$\vee\, \phi(5, 6, 7, 13, 14$
C_2	10	1	0	1	0	0	1	1	$15)$
C_3	11	1	0	1	1	1	0	0	
C_4	12	1	1	0	0	0	0	0	$Y_3(xy_1y_2y_3)$
Not used	13	1	1	0	1	ϕ	ϕ	ϕ	$= \vee (1, 3, 8, 10)$
	14	1	1	1	0	ϕ	ϕ	ϕ	$\vee\, \phi(5, 6, 7, 13, 14,$
	15	1	1	1	1	ϕ	ϕ	ϕ	$15)$

Figure 8-3 Flow table of the modulo 5 counter of Example 8.2.1.

$$Y_1(xy_1y_2y_3) = \vee\,(4, 11) \vee \phi\,(5, 6, 7, 13, 14, 15). \tag{8.3}$$

Again, the next state of the y_1 *FF* is also a function of its present state and its two input lines as given by its characteristic function, that is,

$$Y_1 = S \vee \bar{C}y_1. \tag{8.4}$$

Equations (8.3) and (8.4) show that S and C must themselves be functions of x, y_1, y_2 and y_3 so that both these equations are simultaneously satisfied. Such functions are known as *excitation functions* or *input equations.* The immediate problem, therefore, is to derive these excitation functions. It is obvious that the excitation functions must satisfy the characteristic function. Again, in order that the design be most economical, excitation functions should contain the least number of literals. It must be mentioned that the nature of the problem remains the same if we decide to implement the counter by some other type of flip-flop. Since each of the excitation functions, is a function of x and the present states of the *FF*s, it can be plotted on a map. The object of the designer is to produce excitation function maps which will be most economical to implement. A criterion which is often accepted to meet this requirement is that the map must produce a minimal sum-of-product (MSP) form having the least number of literals. Such a map will be called the best rendering or solution of the excitation function. This can be achieved if the S and C maps satisfy the following theorem.

Theorem 8.2.1 The map of an excitation function gives the best solution if (a) the characteristic function of the particular flip-flop and the transition function are simultaneously satisfied, and (b) don't-care terms are maximized.

Proof: That condition (a) is necessary is obvious. But it is not sufficient. Condition (b) must also be fulfilled as only then the chances for the map to produce an MSP form with the least number of literals are most favorable. ▲

COROLLARY 8.2.1A All don't care terms are to be plotted as such on all maps.

Proof: The validity of this corollary follows as a direct consequence of Th.8.2.1(b). ▲

Reverting back to the problem of finding the functions connecting S and C with x, y_1, y_2 and y_3 it becomes a problem of finding the particular maps for S and C which will satisfy Th.8.2.1. Now, the Y_1 map has a 1, or a 0, or a ϕ (don't-care term) on each of its cells. The question is what must be plotted on those cells of S map, corresponding to which there are 1's on the Y_1 map, and what must be plotted on those cells for which there are 0's or ϕ's. The following theorems and corollaries now establish the algorithm setting forth the rules for plotting 1's and 0's of the Y_1 map on the corresponding cells of the S and C maps. Although the theorems are stated and proved for the y_1 *FF*, it can be

asily seen that the theorems are valid for the other *FF*'s too. This generaliza-
ion will be done while writing the algorithm in a tabular form, in which the
lip-flop has been referred to as a *q FF*.

Theorem 8.2.2 On the $\overline{y}_1$ half of *S* map 1's and 0's are plotted as such.

Proof: The maps of *S*, and *C* should be such that

$$Y_1 = S \vee \overline{C}y_1.$$

Of these the plots of Y_1 and y_1 are known as shown in Fig.8-4. The plots of *S* nd $\overline{C}$ are unknown. Now, 1's of the $\overline{y}_1$ half cannot be realized from $\overline{C}y_1$, as he y_1 map has 0's on all cells on the $\overline{y}_1$ half. Therefore, these 1's must be lotted as such on the *S* map. To satisfy theorem 8.2.1(a), 0's of $\overline{y}_1$ half must lso be plotted as such on the *S* map. ▲

COROLLARY 8.2.2A 1's of $\overline{y}_1$ half must be plotted as 0's on *C* map.

Proof: In *S–C FF SC* = 0. Hence, since on the $\overline{y}_1$ half 1's are plotted as 1's n the *S* map, they must be plotted as 0's on the *C* map. ▲

Theorem 8.2.3 On the $\overline{y}_1$ half of *C* map, 0's are plotted as ϕ's.

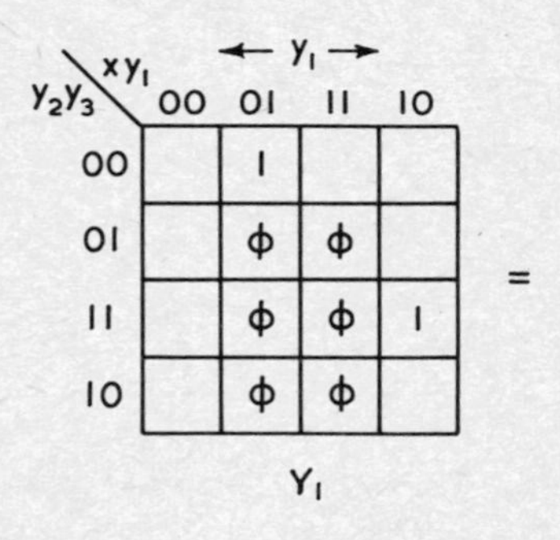

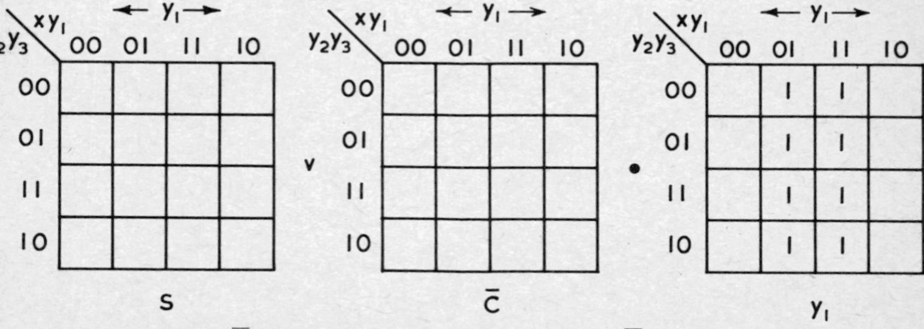

Figure 8-4 $Y_1 = S \vee \overline{C}y_1$ plotted on maps. The *S* and $\overline{C}$ functions are not known.

Proof: Since $\overline{C}$ map is multiplied by y_1 which has 0's on all cells on the $\overline{y}_1$ half, 0's of Y_1 map of $\overline{y}_1$ half can be plotted as a 0, or a 1, or a ϕ on the $\overline{C}$ map, as in all cases both the characteristic equation and the constraint $SC = 0$ are satisfied. But to satisfy Th.8.2.1(b), they must be plotted as ϕ's. ▲

Theorem 8.2.4 1's of y_1 half are plotted as 0's on the C map and as ϕ's on the S map.

Proof: It is evident from Fig.8-4 that the 1's of y_1 half of Y_1 map can be realized by plotting them as such on the S map, or on the $\overline{C}$ map, or on both. However, if they are plotted as such on S map, then the corresponding cells of C map must have 0's (since $SC = 0$). Hence the corresponding cells of $\overline{C}$ map must have 1's. This means that both S and $\overline{C}$ map must have 1's on the corresponding cells. On the other hand, if the 1's are realized by plotting them as such on the $\overline{C}$ map, they become 0's on the C map, and the corresponding cells of the S map can be don't-cares. The second alternative fulfils the requirement of Th.8.2.1(b). Hence the theorem is proved. ▲

Theorem 8.2.5 0's of y_1 half must be plotted as such on the S map and as 1's on the C map.

Proof: Fig.8-3 shows that 0's of y_1 half of Y_1 map must be plotted as such on the S map and also on the $\overline{C}$ map to satisfy the characteristic equation. Since, plotting a 0 on $\overline{C}$ map is equivalent to plotting a 1 on the C map, the validity of the theorem is established. ▲

The algorithm for deriving the input equation as developed in the above theorems can now be written in a tabular form as shown in Fig.8-5. Following these the input equations for the S and C inputs of the y_1 FF turn out to be (Fig.8-6)

$$S_1 = xy_2y_3 \text{ and } C_1 = xy_1. \tag{8.5}$$

Following the same rules the input equations of the y_2 and y_3 FF's are also found to be as follows: (Fig.8-7)

FF	On map	On $\overline{q}$ half: For 0	On $\overline{q}$ half: For 1	On q half: For 0	On q half: For 1
$S-C$	S	0	1	0	ϕ
	C	ϕ	0	1	0

On both maps ϕ's are plotted as ϕ's

Figure 8-5 Rules for plotting on the S and C maps.

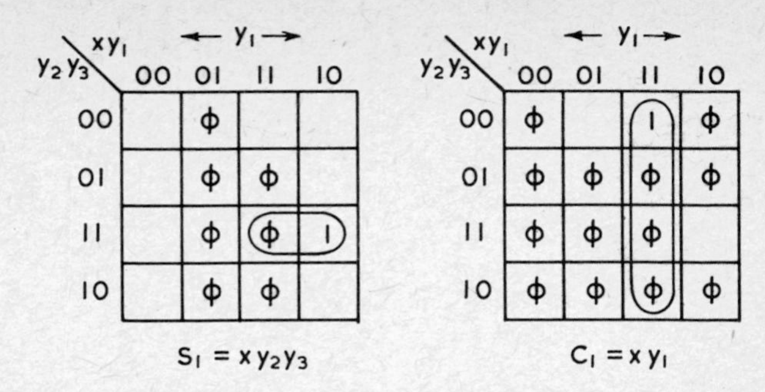

Figure 8-6 Excitation functions of y_1 ($S-C$ type) *FF*.

$$S_2 = x\bar{y}_2y_3 \text{ and } C_2 = xy_2y_3 \tag{8.6}$$

$$S_3 = x\bar{y}_1\bar{y}_3 \text{ and } C_3 = xy_3. \tag{8.7}$$

'he logical implementation of these equations with gates is shown in Fig.8-8.

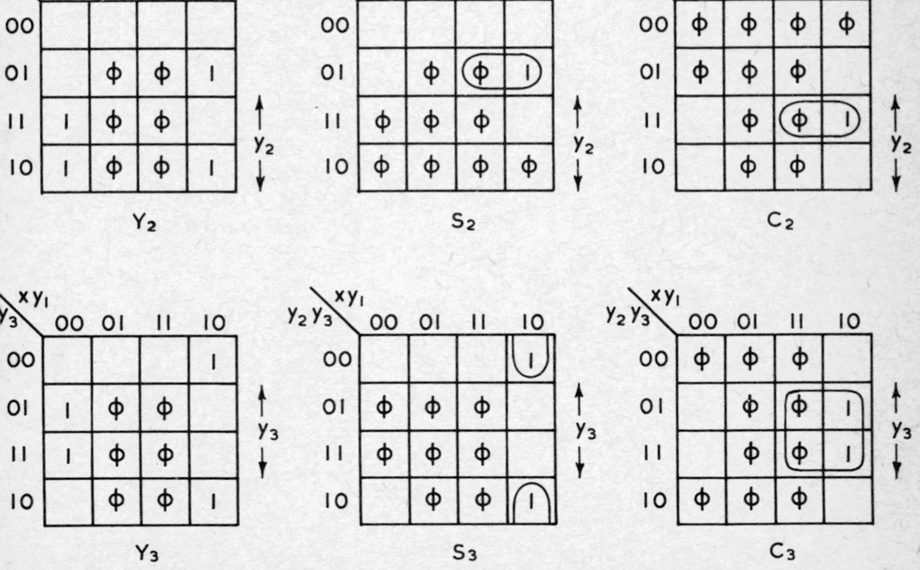

Figure 8-7 Excitation functions of the y_2 and y_3 ($S-C$ type) *FF*s.

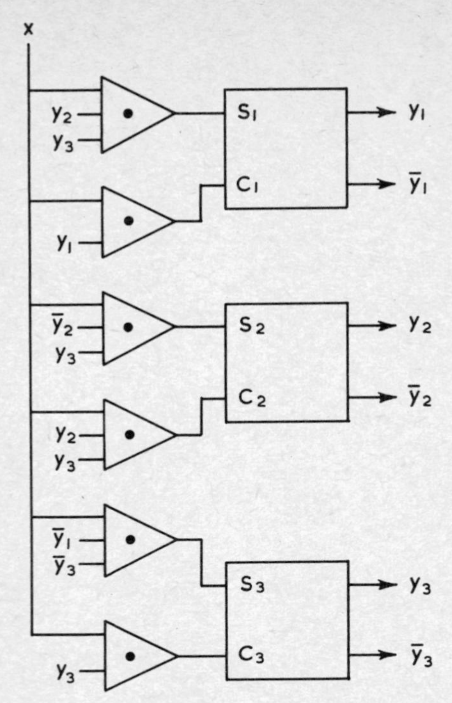

Figure 8-8 Implementation of the Mod-5 counter by $S-C$ flip-flops.

8.3 THE $J-K$ FLIP-FLOP

The $J-K$ flip-flop also has two input lines. The J and K correspond to the S and C inputs respectively of the $S-C$ flip-flop, that is, a 1 on J sets the FF to 1, and a 1 on K clears it to 0, irrespective of the previous states. However, here the J and K lines are allowed to become 1 simultaneously, and under this condition the state of the FF changes. The truth table, its plotting and simplification are shown in Fig.8-9. The characteristic function is found to be,

$$Q = J\bar{q} \vee \bar{K}q. \tag{8.8}$$

The algorithm for $J-K$ can be derived by solving for the y_1 FF of the counter for the previous example. Since, for $J-K$ FF the characteristic function is

$$Q = J\bar{q} \vee \bar{K}q,$$

for y_1 FF then,

$$Y_1 = J\bar{y}_1 \vee \bar{K}y_1. \tag{8.9}$$

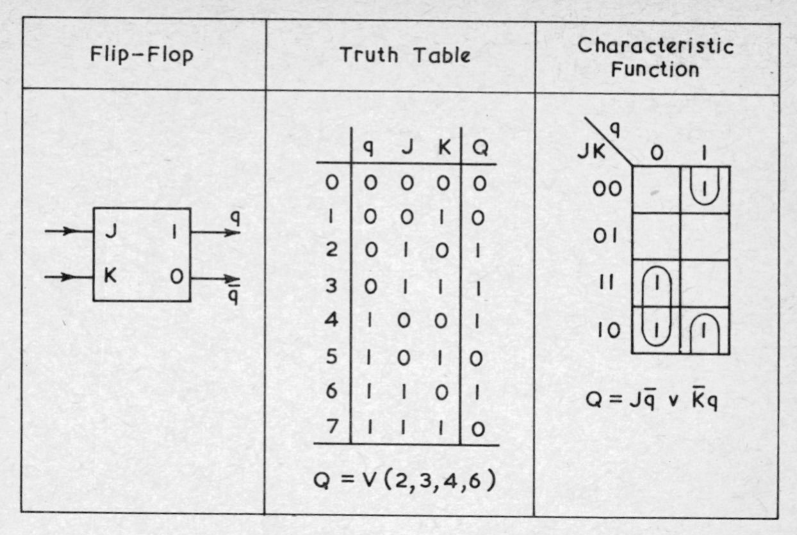

Figure 8-9 *J–K* flip-flop.

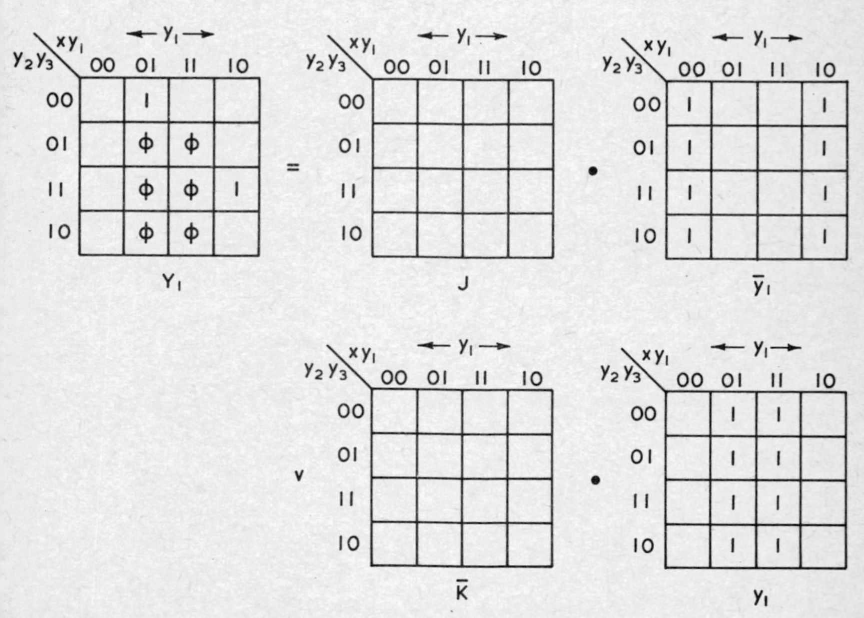

Figure 8-10 $Y_1 = J\bar{y}_1 \vee \bar{K}y_1$ plotted on maps. The J and $\bar{K}$ functions are not known.

Figure 8-10 shows this equation in the form of maps. Rules for plotting on the J and K maps are given by the following theorems.

Theorem 8.3.1 On the $\bar{y}_1$ half of the J map, 1's and 0's are plotted as such.

Proof: Since in the characteristic equation of the J–K *FF*, $\bar{K}$ is multiplied by y_1, which has 0's on cells of $\bar{y}_1$ half, 1's of this half must be obtained from the product of the J and $\bar{y}_1$ maps. Thus, 1's of $\bar{y}_1$ half must be plotted as such on the J map. Further, as the $\bar{y}_1$ map has 1's on all cells on the $\bar{y}_1$ half, the 0's of this half must also be plotted as such on the J map. ▲

Theorem 8.3.2 On the y_1 half of J map, 1's and 0's are plotted as ϕ's.

Proof: Since the J map is multiplied by the $\bar{y}_1$ half, having 0's on all cells in the y_1 half, 1's and 0's should be plotted as ϕ's to satisfy Th.8.2.1(b). ▲

Theorem 8.3.3 On the y_1 half of K map, 1's are plotted as 0's and 0's as 1's.

Proof: Since in the characteristic function of the J–K *FF*, J is multiplied by $\bar{y}_1$, which has 0's on all cells in the y_1 half, the 1's of this half must be obtained from the product of the $\bar{K}$ and the y_1 maps. Now, a 1 plotted on $\bar{K}$ map becomes a 0 on the K map. Hence on the y_1 half of K map, 1's are to be plotted as 0's. Again, as the y_1 map has 1's on all cells in the y_1 half, 0's of this half must be plotted as such on the $\bar{K}$ map, and consequently as 1's on the K map. ▲

Theorem 8.3.4 On the $\bar{y}_1$ half of K map, 1's and 0's are plotted as ϕ's.

Proof: The validity of this theorem follows from similar arguments as given for Th.8.3.2 above. ▲

Following the results of these theorems the algorithm for the J–K *FF* is as shown in the table of Fig.8-11.

		On $\bar{q}$ half		On q half	
FF	On map	For 0	For 1	For 0	For 1
J–K	J	0	1	ϕ	ϕ
	K	ϕ	ϕ	1	0

On both maps ϕ's are plotted as ϕ's.

Figure 8-11 Rules for plotting on the J and K maps.

8.4 THE $T-G$ FLIP-FLOP

The $T-G$ *FF* also has two input lines, the T line, and the G line. When the G line is 1, the *FF* assumes the state as dictated by the input on the T line. When the G line is 0, the *FF* does not respond to the T line, and its state remains unchanged. Its truth table, plotting, and the characteristic function are shown in Fig.8-12. Its characteristic function turns out to be:

$$Q = TG \vee \bar{G}q. \tag{8.10}$$

Flip-Flop: inputs T, G; outputs q, $\bar{q}$

Truth Table

	G	T	q	Q
0	0	0	0	0
1	0	0	1	1
2	0	1	0	0
3	0	1	1	1
4	1	0	0	0
5	1	0	1	0
6	1	1	0	1
7	1	1	1	1

$Q = \vee(1,3,6,7)$

Characteristic Function

Tq \ G	0	1
00		
01	1	
11	1	1
10		1

$Q = TG \vee \bar{G}q$

Figure 8-12 $T-G$ flip-flop.

To find its algorithm let us implement the counter by $T-G$ *FF*'s. Then, to solve for T and G of the y_1 *FF*, they can be plotted as shown in Fig.8-13. Here,

$$Y_1 = TG \vee \bar{G}y_1. \tag{8.11}$$

The following five theorems now establish the algorithm for the $T-G$ flip-flop.

Theorem 8.4.1 On the $\bar{y}_1$ half of the T and G maps, 1's are to be plotted as such.

Proof: Since the $\bar{y}_1$ half of the y_1 map (Fig.8-13) has 0's on all cells, the product of the $\bar{y}_1$ half of the $\bar{G}$ and y_1 maps will be always 0. Hence, the only way Th.8.2.1(a), regarding the 1's of $\bar{y}_1$ half of Y_1, is satisfied is by plotting them as such on the $\bar{y}_1$ half of the T and G maps. ▲

Theorem 8.4.2 On the $\bar{y}_1$ half of the T and G maps, a 0 is plotted either on

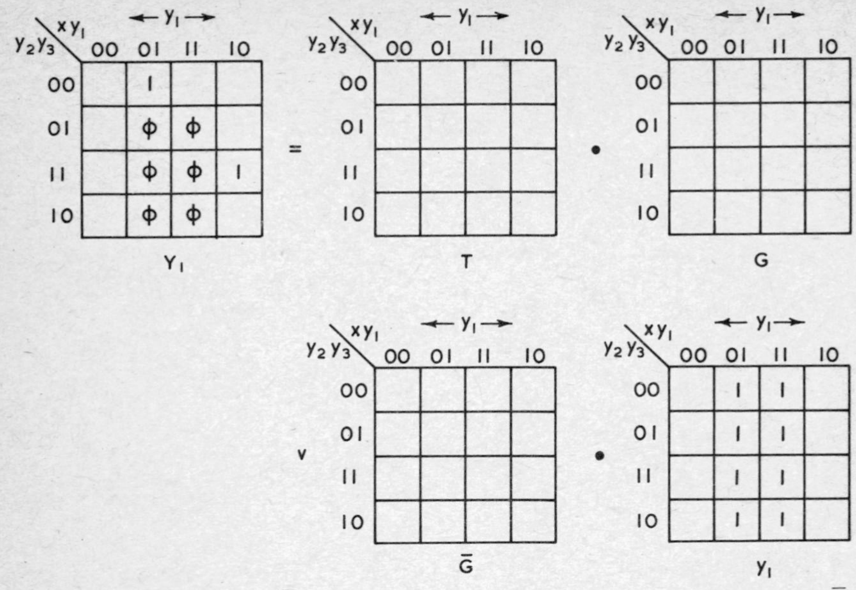

Figure 8-13 $Y_1 = TG \vee \bar{G}y_1$ plotted on maps. The *T*, *G* (and consequently $\bar{G}$) functions are not known.

the *T* map or the *G* map, and the corresponding cell of the other map is plotted with a ϕ.

Proof: For all cells with 0's on the $\bar{y}_1$ half of the Y_1 map, the products of the corresponding cells of the *T* and the *G* maps must be 0's. This can be achieved b plotting a 0 on any of the maps, and as required by Th.8.2.1(b), the corresponding cell of the other map should be a ϕ. ▲

According to this theorem, the 0's of the $\bar{y}_1$ half can be plotted in three different ways. Two straightforward ways are to plot all 0's of this half as such on either the *T* or the *G* map, and as ϕ's on the other map. The third way (which may be called the *mixed way*) is to plot some of the 0's as 0's on the *T* map and as ϕ's on the *G* map. Obviously, this arrangement is the most flexible and by a judicious plotting, best solutions of the excitation functions can be obtained.

Theorem 8.4.3 On the y_1 half, 0's are to be plotted as 1's on the *G* map.

Proof: For a cell with a 0 on the y_1 half of Y_1 map, the products of the corresponding cells of the $\bar{G}$ and the y_1 maps must also be 0's. Since the y_1 map has all 1's on the y_1 half, the $\bar{G}$ map must then have 0's on the corresponding cells. Since a 0 on the $\bar{G}$ map becomes a 1 on the *G* map, the theorem is proved. ▲

Theorem 8.4.4 On the y_1 half, 0's are to be plotted as 0's on the T map.

Proof: The product of the T and G maps for a cell with a 0 on the y_1 half of Y_1 map must be 0. Since, by Th.8.4.3 such a cell on the G map has a 1, the corresponding cell on the T map must have a 0. ▲

Theorem 8.4.5 On the y_1 half, 1's are to be plotted as 0's on the G map, and as ϕ's on the T map.

Proof: 1's of the y_1 half of Y_1 map can be realized in two alternative ways. (a) Plotting them as 1's on both the T and G maps. The corresponding cells of the $\overline{G}$ map then must have 0's. Alternatively, (b) Plotting them as 1's on the $\overline{G}$ map. The y_1 half of y_1 map already has 1's on all cells. The corresponding cells of G map then have 0's and ϕ's can be plotted on the corresponding cells of the T map. As alternative (b) satisfies Th.8.2.1(b), the validity of the theorem is established. ▲

The algorithm derived in the above theorems can now be summarized as in the table of Fig.8-14. Following the table and carrying out the three different

	On $\overline{q}$ half		On q half	
On map	For 0	For 1	For 0	For 1
T	$0/\phi$	1	0	ϕ
G	$\phi/0$	1	1	0

On both maps ϕ's are plotted as ϕ's

Figure 8-14 Rules for plotting on the T and G maps.

ways of plotting as mentioned under Th.8.4.2, the T and G maps appear as in Fig.8-15(A), (B), and (C). The three sets of input equations are:

$$T_1 = xy_2y_3 \qquad G_1 = x \tag{8.12}$$

$$T_1 = y_2 \qquad G_1 = xy_1 \vee xy_2y_3 \tag{8.13}$$

$$T_1 = y_2y_3 \qquad G_1 = x. \tag{8.14}$$

Figure 8-15 and Eqns (8.12), (8.13), and (8.14) illustrate the fact that the best solution is obtained by *mixed plotting.* Hence the third set, Eqn.(8.14), having the smallest cost function, is chosen.

However, in order that the algorithm can be programmable, a systematic procedure must be established to obtain the excitation functions as given by Eqn.(8.14). Such a procedure is described in the following steps:

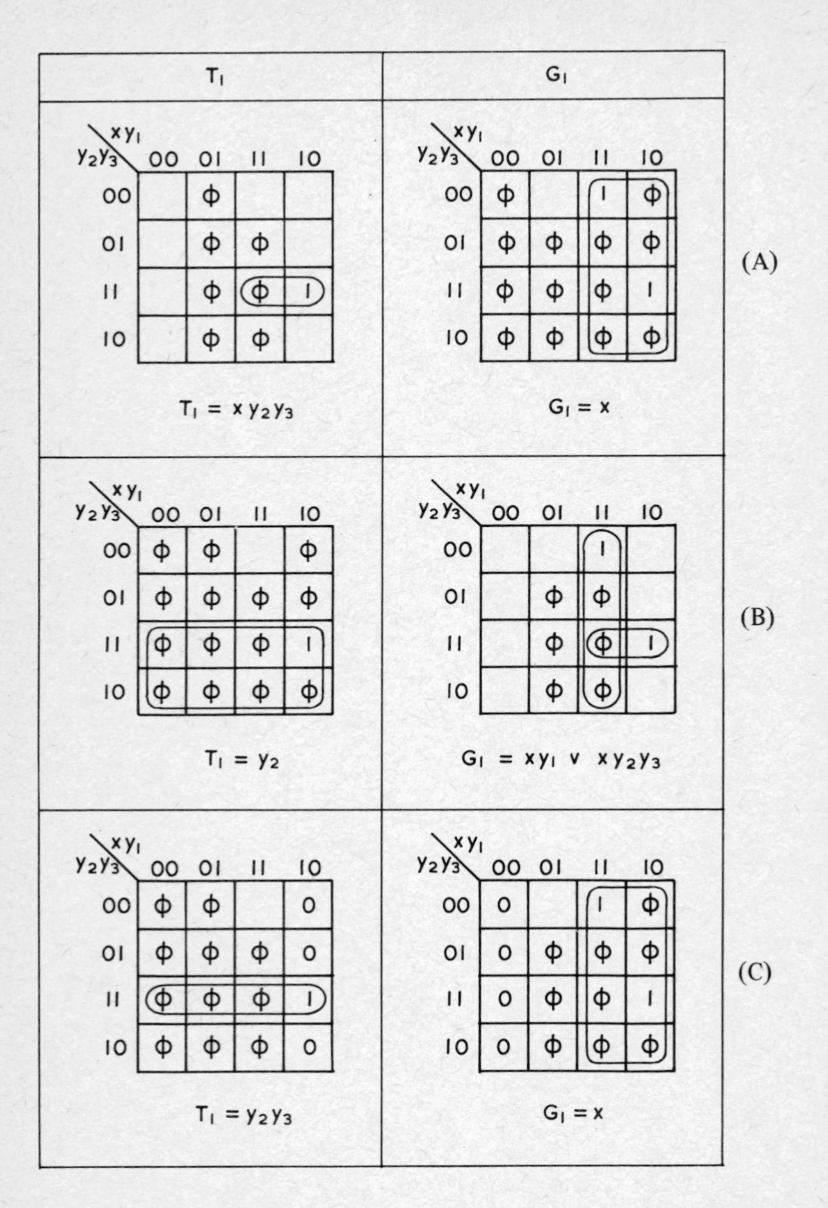

Figure 8-15 Excitation functions of the y_1(T–G type) FF. (A) and (B) are two alternative solutions due to straightforward plotting. (C) is the solution due to mixed plotting.

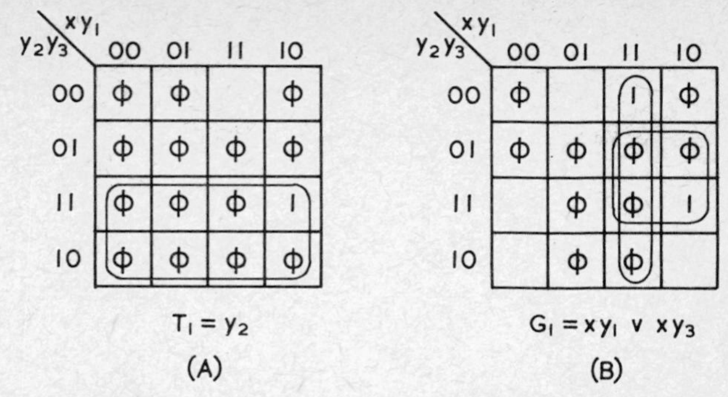

Figure 8-16 Excitation functions as given by steps 1 and 2.

Step 1: Complete plottings of the T_1 map by plotting all 0's of the $\bar{y}_1$ half as ϕ's. Obtain the minimized value of T_1 (Fig.8-16(A)). As a result of this minimization, some of the ϕ's of the $\bar{y}_1$ half are taken as 1's, leaving the rest to be taken as 0's.

Step 2: Complete plottings of the G_1 map by plotting 0's of the $\bar{y}_1$ half as 0's on cells where ϕ's have been taken as 1's on the T_1 map, and as ϕ's on cells where they (ϕ's) have been taken as 0's (Fig.8-16(B)). Obtain the minimized value of G_1. In our example, these turn out to be:

$$T_1 = y_2 \qquad G_1 = xy_1 \vee xy_3. \tag{8.15}$$

Step 3: Repeat the procedure of step 1 on the G_1 map in place of the T_1 map, and obtain the minimized value of G_1 (Fig.8-17(A)).

Step 4: Repeat the procedure of step 2 on the T_1 map in place of the G_1 map, and obtain the minimized value of T_1 (Fig.8-17(B)).

In our example, these two steps produce the following input equations:

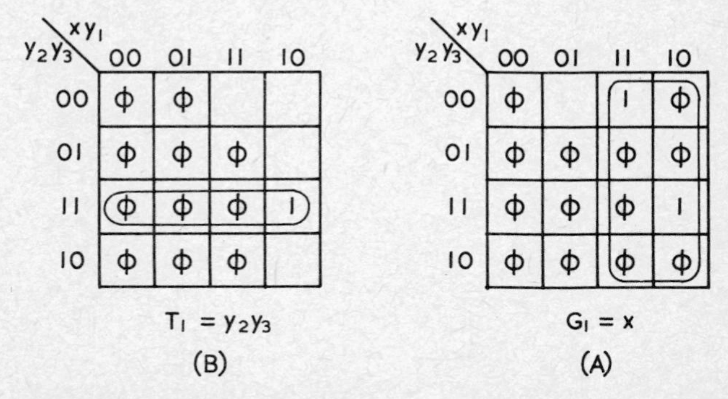

Figure 8-17 Excitation functions as given by steps 3 and 4.

$$T_1 = y_2 y_3 \qquad G_1 = x. \tag{8.16}$$

It is obvious that steps 1 and 2 allow us to have the best solution for T_1 first and then as good a solution for G_1 as may be compatible with that of T_1. On the other hand, steps 3 and 4 render the best solution for G_1 first, and the best compatible solution for T_1 next. The design can now be completed with the following final step.

Step 5: Among the two sets of input equations, select the one having the least cost function.

In our example, Eqn.(8.16) (rather than (8.15)) gives the required solution. This is the same as that given by Eqn.(8.14).

The mixed plotting of the T–G *FF* as has been expounded here brings out an interesting point. Many designers consider the unique algorithm of the S–C, and the J–K *FF*'s as a distinct advantage over the T–G *FF*. By now, it is apparent, however, that the 'ambiguous' algorithm of the T–G flip-flop is really an inherent strength, and there exists a systematic rather than a cut-and-try method to obtain the best possible solutions for the excitation functions.

8.5 THE DELAY (*D*) ELEMENT

The Delay element has only one input line and is not a flip-flop in the sense that it is not a bistable circuit. It simply stores the 0 or 1 fed to its input and passes the same on to the output line after a specified amount of delay. Consequently its next state equals the signal on its input line at the present time. Hence, its truth table is as shown in Fig.8-18. After plotting the truth table on the map and simplifying, the characteristic function of the D element turns out to be

$$Q = D. \tag{8.17}$$

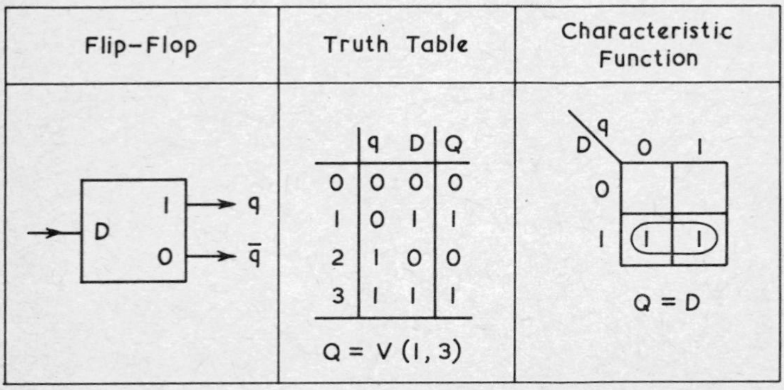

Flip-Flop

D 1 → q
0 → $\bar{q}$

Truth Table

	q	D	Q
0	0	0	0
1	0	1	1
2	1	0	0
3	1	1	1

Q = V (1, 3)

Characteristic Function

D \ q	0	1
0		
1	1	1

Q = D

Figure 8-18 The D element.

Hence for the y_1 *FF* of the counter

$$Y_1 = D_1. \tag{8.18}$$

Thus the plot of y_1 is also the plot for D_1. Hence, no special rule for plotting on the D map is called for.

8.6 THE TRIGGER (T) FLIP-FLOP

The T flip-flop also has only one input line. When a 1 arrives at the input, the state of the *FF* changes. For a 0 input, on the other hand, the state remains unchanged. The truth table and the derivation of the characteristic function are shown in Fig.8-19. The characteristic function for this *FF* is,

$$Q = T\bar{q} \vee \bar{T}q. \tag{8.19}$$

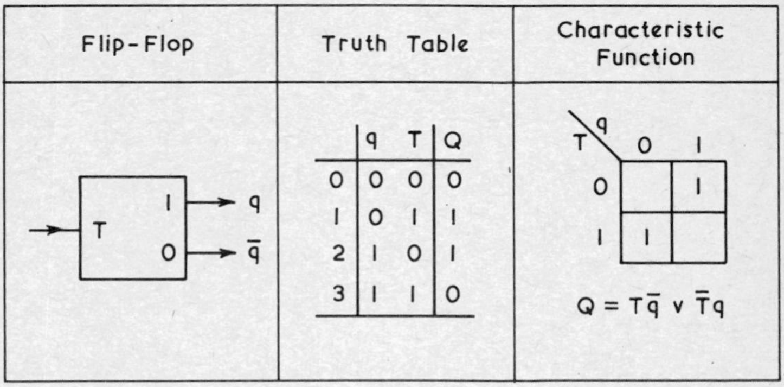

Flip-Flop | Truth Table | Characteristic Function

Flip-Flop: input T; outputs 1 → q, 0 → $\bar{q}$

Truth Table:

	q	T	Q
0	0	0	0
1	0	1	1
2	1	0	1
3	1	1	0

Characteristic Function:

T \ q	0	1
0		1
1	1	

$Q = T\bar{q} \vee \bar{T}q$

Figure 8-19 The T flip-flop.

Let the counter of the previous example be implemented by T *FF*'s, that is, the *FF*'s, y_1, y_2, and y_3 will all be of the Trigger type. For T *FF*, the characteristic function is,

$$Q = T\bar{q} \vee \bar{T}q.$$

For y_1 *FF*, then,

$$Y_1 = T\bar{y}_1 \vee \bar{T}y_1. \tag{8.20}$$

This equation is plotted on maps in Fig.8-20. The algorithm for plotting on the T and $\bar{T}$ maps is revealed by the following theorems.

Theorem 8.6.1 On the $\bar{y}_1$ half of the T map, 1's and 0's are plotted as such.

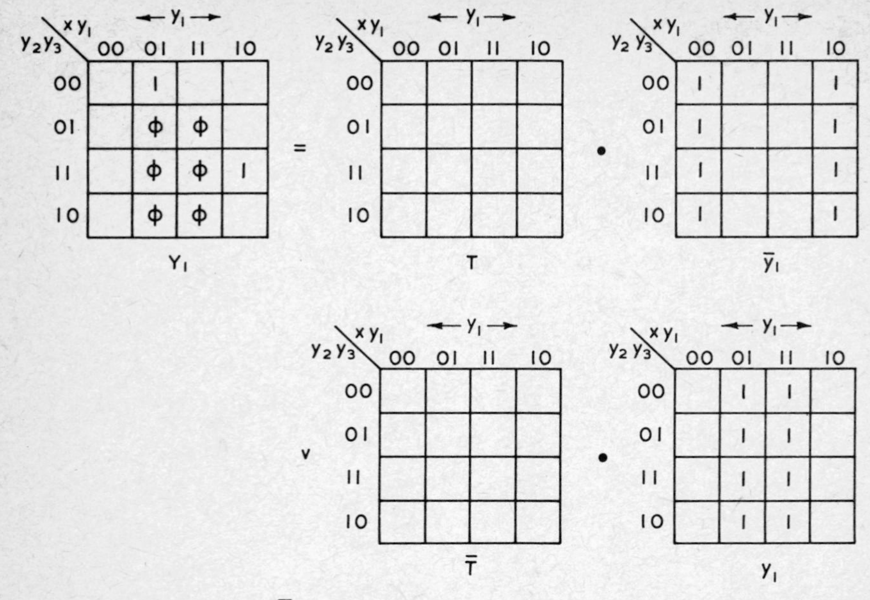

Figure 8-20 $Y_1 = T\bar{y}_1 \vee \bar{T}y_1$ plotted on maps. The T (and consequently $\bar{T}$) functions are not known.

Proof: It is evident from Fig.8-20 that 1's of the $\bar{y}_1$ half cannot be realized fro the product of $\bar{T}$ and y_1 maps. Therefore, 1's must be realized from the product T and $\bar{y}_1$ maps. Hence they must be plotted as such on the T map. Again, a 0 must be plotted as such on the T map to satisfy Th.8.2.1(a). ▲

Theorem 8.6.2 On the y_1 half of the T map 1's are plotted as 0's and 0's as 1's.

Proof: It is clear from Fig.8-20 that 1's of the y_1 half cannot be realized from the product of the T and the $\bar{y}_1$ maps, and have to be realized from the product of the $\bar{T}$ and y_1 maps. Hence, they are to be plotted as such on the $\bar{T}$ map. Again to satisfy the characteristic equation, 0's of the y_1 half must also be

		On $\bar{q}$ half		On q half	
FF	On map	For 0	For 1	For 0	For 1
T	*T*	0	1	1	0

On both halves ϕ's are plotted as ϕ's

Figure 8-21 Rules for plotting on the T map.

plotted as such on the $\overline{T}$ map. Since, a 1 and a 0 on the $\overline{T}$ map are equivalent to a 0 and 1 respectively on the T map, the theorem is proved. ▲

These rules are now tabulated in the generalized form in the table of Fig.8-21. Following this algorithm the input equation for T_1 turns out to be (Fig.8-22)

$$T_1 = xy_1 \vee xy_2y_3. \tag{8.21}$$

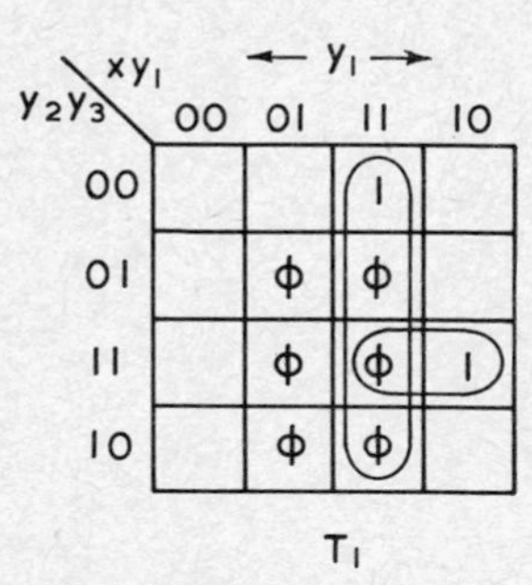

Figure 8-22 Excitation function of the y_1 (T type) FF.

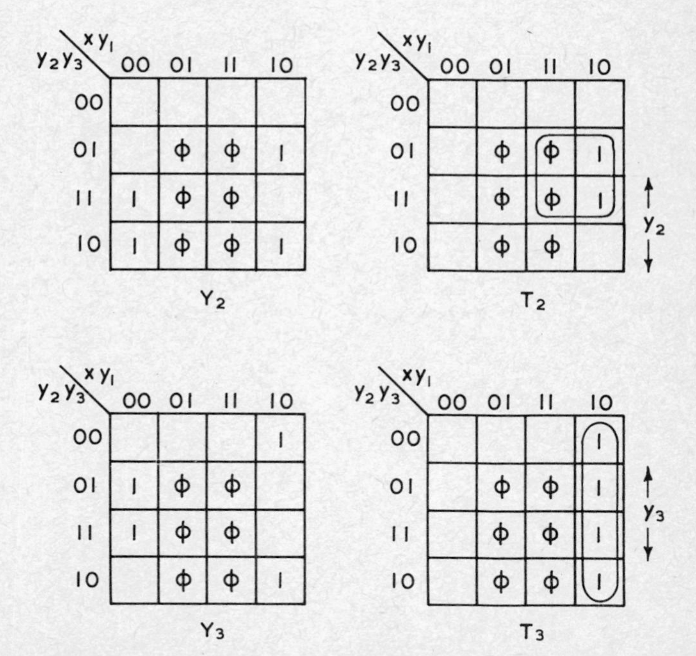

Figure 8-23 Excitation functions of the y_2 and y_3 (T type) FFs.

In a similar manner, T_2 and T_3 can be found out as shown in Fig.8-23. These are

$$T_2 = xy_3 \tag{8.22}$$

$$T_3 = x\bar{y}_1. \tag{8.23}$$

Block diagrams to implement these equations are shown in Fig.8-24.

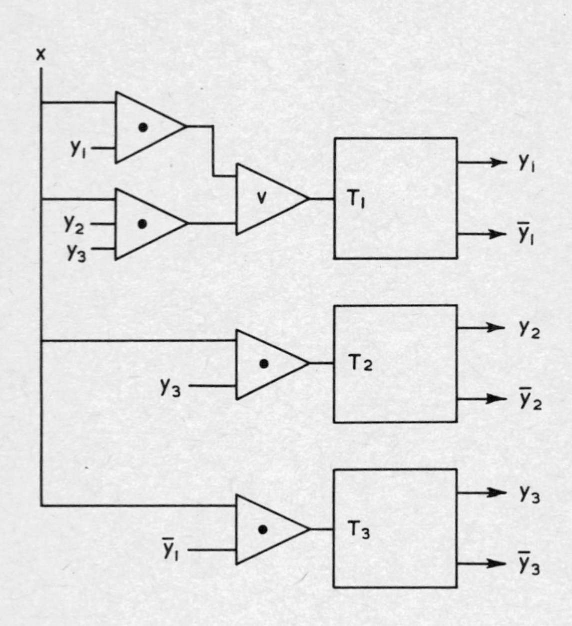

Figure 8-24 Implementation of the counter by trigger flip-flops.

8.7 THE *S–C–T* FLIP-FLOP

The *S–C–T* flip-flop has three input lines, *S, C,* and *T*. When *S* is 1, the *FF* is set to 1. When *C* is 1, the *FF* is cleared to 0, and when *T* is 1, the *FF* changes state. No two of the *S, C,* and *T* lines are allowed to be 1 simultaneously, which means

$$SC = CT = TS = 0. \tag{8.24}$$

The truth table, and the derivation of the characteristic function are shown in Fig.8-25. The latter turns out to be,

$$Q = S \vee T\bar{q} \vee \overline{C}\overline{T}q. \tag{8.25}$$

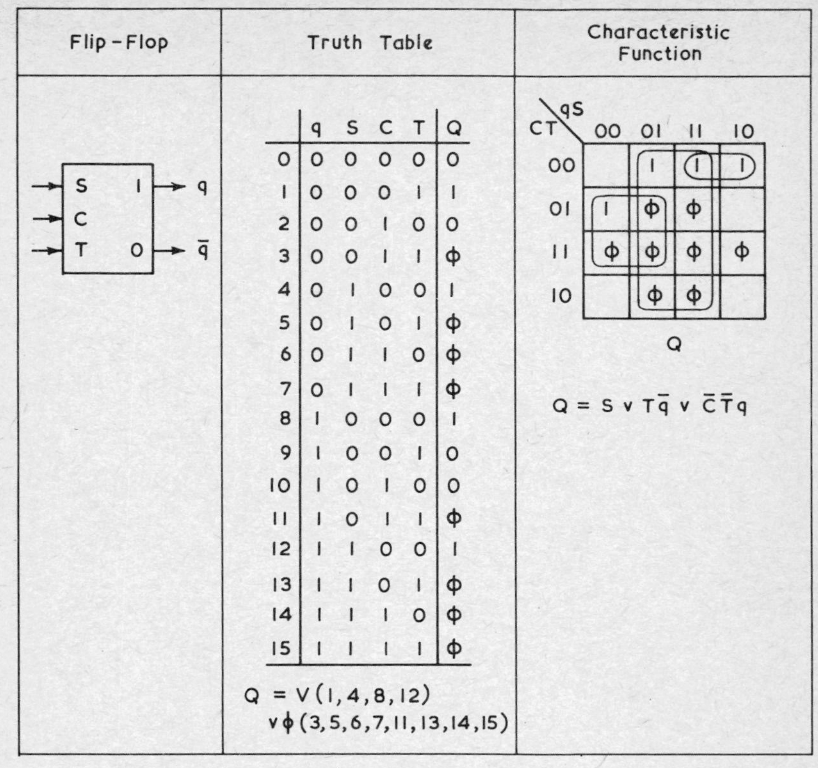

Figure 8-25 The $S-C-T$ flip-flop.

To find out the rules of this FF, let us solve for the y_1 FF of the counter in question. Since the characteristic function of the FF is,

$$Q = S \vee T\bar{q} \vee \bar{C}\bar{T}q,$$

for y_1 FF then,

$$Y_1 = S \vee T\bar{y}_1 \vee \bar{C}\bar{T}y_1. \tag{8.26}$$

This equation is plotted on maps as shown in Fig.8-26.

Following theorems establish the algorithm for the $S-C-T$ FF.

Theorem 8.7.1 1's of the $\bar{y}_1$ half are plotted as $\bar{S}$ on T, and as $\bar{T}$ on S.

Proof: The 1's of Y_1 for the $\bar{y}_1$ half must be plotted on the $\bar{y}_1$ half of either S or T as the $\bar{C}$ and $\bar{T}$ maps are multiplied by y_1 which has 0's on the $\bar{y}_1$ half. Since S and T cannot be 1 simultaneously, where there is a 1 on S, there must be a 0 on the corresponding cell of T and *vice-versa.* This can be indicated by plotting $\bar{T}$ on S and $\bar{S}$ on T on the $\bar{y}_1$ halves of S and T. ▲

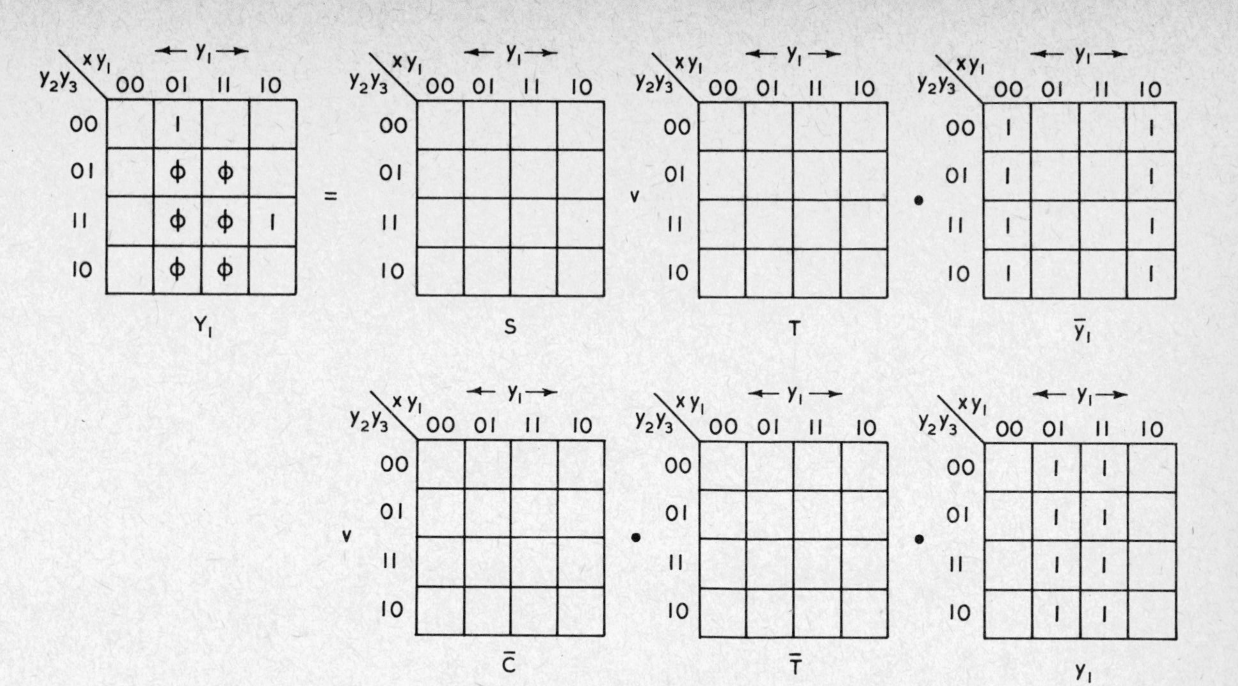

Figure 8-26 $Y_1 = S \vee T\bar{y}_1 \vee \bar{C}\bar{T}y_1$ plotted on maps. The S, T (and consequently $\bar{T}$) and $\bar{C}$ functions are not known.

COROLLARY 8.7.1A 1's of $\overline{y}_1$-half are plotted as 0's on C map.

Proof: Since on these cells either S or T is 1, and Eqn.8.24 must also be satisfied, there must be 0's on the corresponding cells on C. ▲

Theorem 8.7.2 All 0's of Y_1 map are plotted as such on the S map.

Proof: Figure 8-26 shows that this theorem must be valid to satisfy the characteristic function. ▲

Theorem 8.7.3 0's of the $\overline{y}_1$ half are plotted as 0's on the T map.

Proof: In Fig.8-26, T map has been multiplied by $\overline{y}_1$ which has 1's on all cells of $\overline{y}_1$ half. Hence 0's of this half must be plotted as such in order to satisfy the characteristic equation. ▲

Theorem 8.7.4 0's of $\overline{y}_1$ half are plotted as ϕ's on the C map.

Proof: 0's of the $\overline{y}_1$ half are plotted as such on both the S and T maps. Again the $\overline{C}$ map has been multiplied by y_1 which has 0's on all cells of $\overline{y}_1$ half. Hence 0's can be plotted on the $\overline{C}$ map either as 0's, 1's or ϕ's. As required by Th. 8.2.1(b), 0's must be plotted as ϕ's. ▲

Theorem 8.7.5 0's of the y_1 half are plotted as $\overline{C}$ and $\overline{T}$ on the T and C maps respectively.

Proof: By Th.8.7.2 these 0's are plotted as such on the S map. Therefore, these cells on the C and the T maps can have 1's in only one of the maps but not on both (since $CT = 0$). The choice of 1 on the T map and 0 on the C map is allowed since the product of T and $\overline{y}_1$ maps, and the product of the $\overline{C}$ and $\overline{T}$ maps render the final products as 0's on these cells. Again, the alternative of 0 on the T map and 1 on C (therefore 0 on $\overline{C}$ map) is also allowed because of the term $\overline{C}\overline{T}y_1$ in the characteristic function. ▲

Theorem 8.7.6 1's of the y_1 half are plotted as ϕ's on the S map, and as 0's on the C and T maps.

Proof: The theorem is proved by following similar reasonings as given in Th.8.2.5. 1's of the y_1 half cannot be realized from the product of the T and $\overline{y}_1$ maps. They can be realized either from the S map, or from the product of the $\overline{C}$, $\overline{T}$ and y_1 maps. Second alternative must be chosen as then the corresponding cells of the S map may have ϕ's satisfying Th.8.2.1(b). ▲

These rules for the S–C–T FF as well as those for other FF's are given in a single table in the next section.

Now, to determine the S, C, and T input equations of the y_1 FF, they are plotted on the maps as shown in Fig.8-27. Those cells on which there are $\overline{T}$, $\overline{R}$ and $\overline{S}$, they can be reckoned either as a 0 or a 1 depending on which one gives the best solution. The best solution from Fig.8-27 is

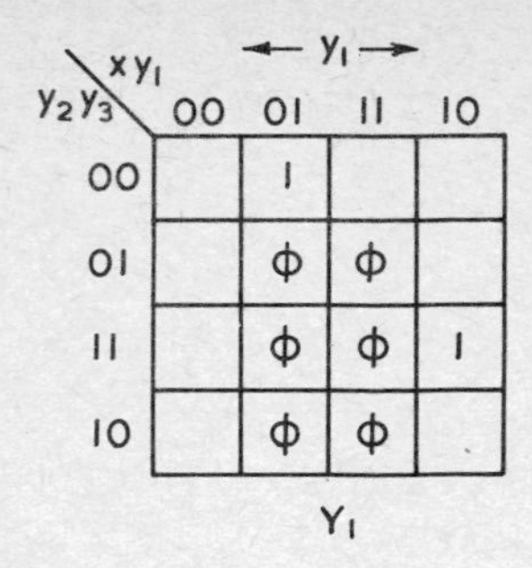

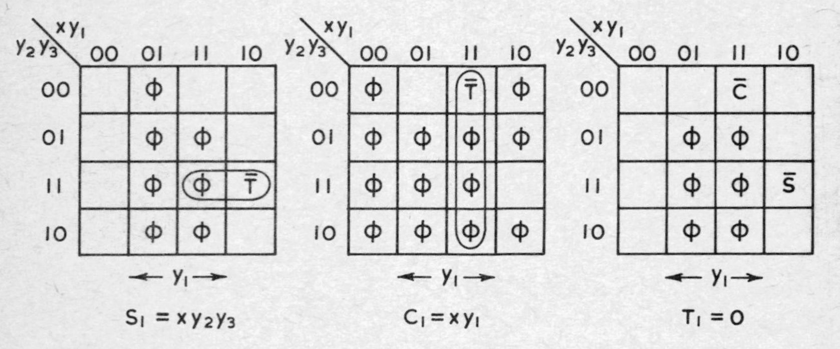

Figure 8-27 Excitation functions of the y_1 $(S-C-T)$ type FF.

$$S_1 = xy_2y_3 \qquad C_1 = xy_1 \text{ and } T_1 = 0. \tag{8.27}$$

Hence the 11th cell of the S_1 map has been thought of having a 1. This makes th 11th cell on T_1 map which has $\bar{S}$ on it a 0. Also the 12th cell of the C_1 map has also been considered to have a 1, making the 12th cell of the T_1 map to be a 0.

8.8 EXCITATION FUNCTION

During the analysis as outlined above, some rules have been framed for deriving the excitation functions by plotting the transition functions on individual maps. For ready reference these rules are summarized in a tabular form in the table of Fig.8-28. By following these rules, maps for input lines can be obtained very easily from the map of the transition function. This not only saves time, but alsc renders the derivation almost effortless.

8.9 SOME EXAMPLES

Example 8.9.1: Interconversion of flip-flops: Convert an $S-C$ flip-flop into (a) (b) a $J-K$ FF with the help of additional gates.

FF	On map	On $\bar{q}$ half For 0	For 1	On q half For 0	For 1
T	T	0	1	1	0
$S-C$	S	0	1	0	ϕ
	C	ϕ	0	1	0
$J-K$	J	0	1	ϕ	ϕ
	K	ϕ	ϕ	1	0
$T-G$	T	$0/\phi$	1	0	ϕ
	G	$\phi/0$	1	1	0
$S-C-T$	S	0	$\bar{T}$	0	ϕ
	C	ϕ	0	$\bar{T}$	0
	T	0	$\bar{S}$	$\bar{C}$	0

On all maps ϕ's are plotted as ϕ's.

Figure 8-28 Rules to derive excitation function maps.

Solution: a) The characteristic function of a T FF is,

$Q = T\bar{q} \vee \bar{T}q.$

Maps of its application equation, S and C are shown in Fig.8-29(A). Solving for S and C, the input equations become,

$S = T\bar{q}$

$C = Tq.$

The resulting circuit is now shown in Fig.8-29(B).

b) The application equation of the $J-K$ FF is,

$Q(qJK) = \vee\,(2, 3, 4, 6).$

The solutions for S and C, and the resulting circuit are shown in Fig.8-30.

Example 8.9.2: Analysis of a counter: The circuit of a counter is shown in Fig.8-31. Find its displaying sequence (assume binary coding), and write the flow table.

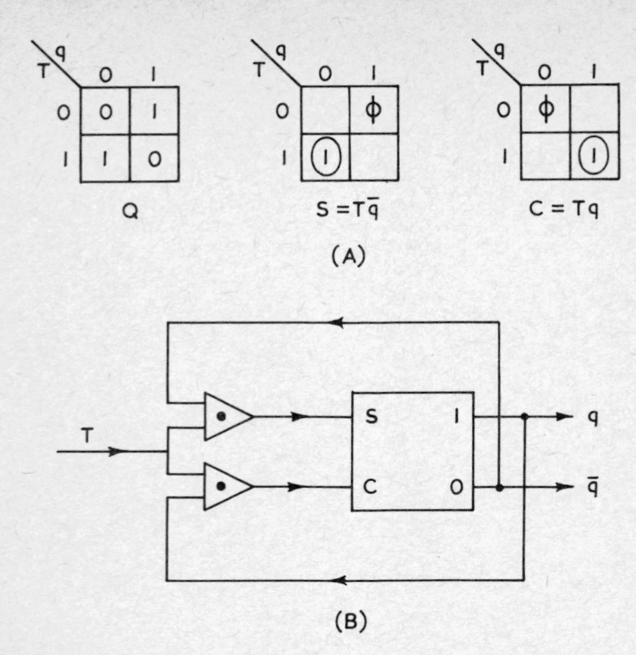

Figure 8-29 (A) Derivation of the S and C functions. (B) A T FF from an S–C FF.

Solution: From the figure it is clear that the counter has been implemented by three Trigger FF's a, b, and c. Their input equations are,

$$T_a = a \vee b$$

$$T_b = \bar{a} \vee c$$

$$T_c = b.$$

Since, the characteristic function of a T FF is $Q = T\bar{q} \vee \bar{T}q$, the application equations of the three FF's can be found out as follows:

$$a_+ = T_a\bar{a} \vee \bar{T}_a a$$

$$= (a \vee b)\bar{a} \vee (\bar{a}\bar{b})a = \bar{a}b$$

$$b_+ = T_b\bar{b} \vee \bar{T}_b b$$

$$= (\bar{a} \vee c)\bar{b} \vee (a\bar{c})b$$

$$= \bar{a}\bar{b} \vee \bar{b}c \vee ab\bar{c}$$

$$c_+ = T_c\bar{c} \vee \bar{T}_c c$$

$$= b\bar{c} \vee \bar{b}c.$$

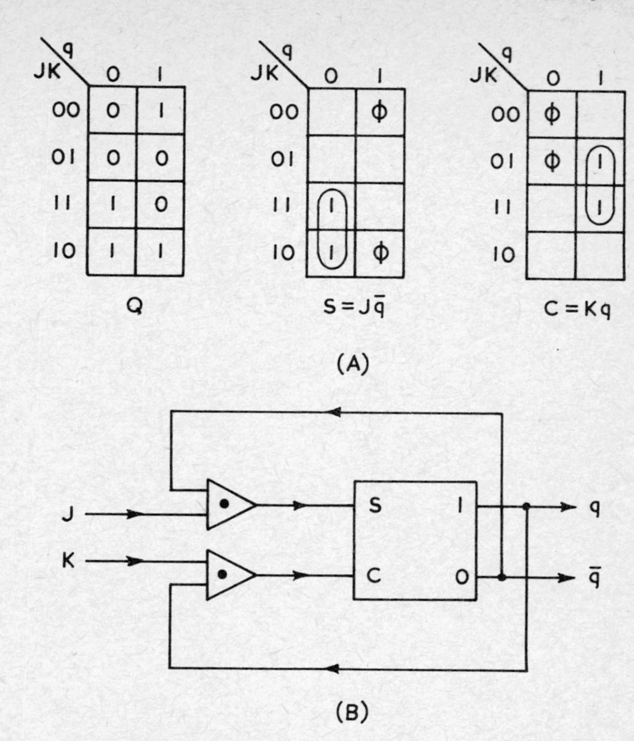

Figure 8-30 (A) Derivation of the S and C functions. (B) A J–K *FF* from an S–C *FF*.

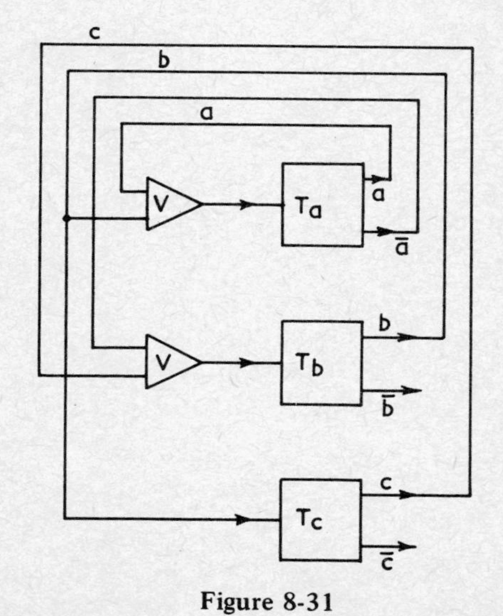

Figure 8-31

Expanding in the canonical form, (Fig.8-32(A))

$a_+(abc) = \vee(2, 3)$

$b_+(abc) = \vee(0, 1, 5, 6)$

$c_+(abc) = \vee(1, 2, 5, 6)$.

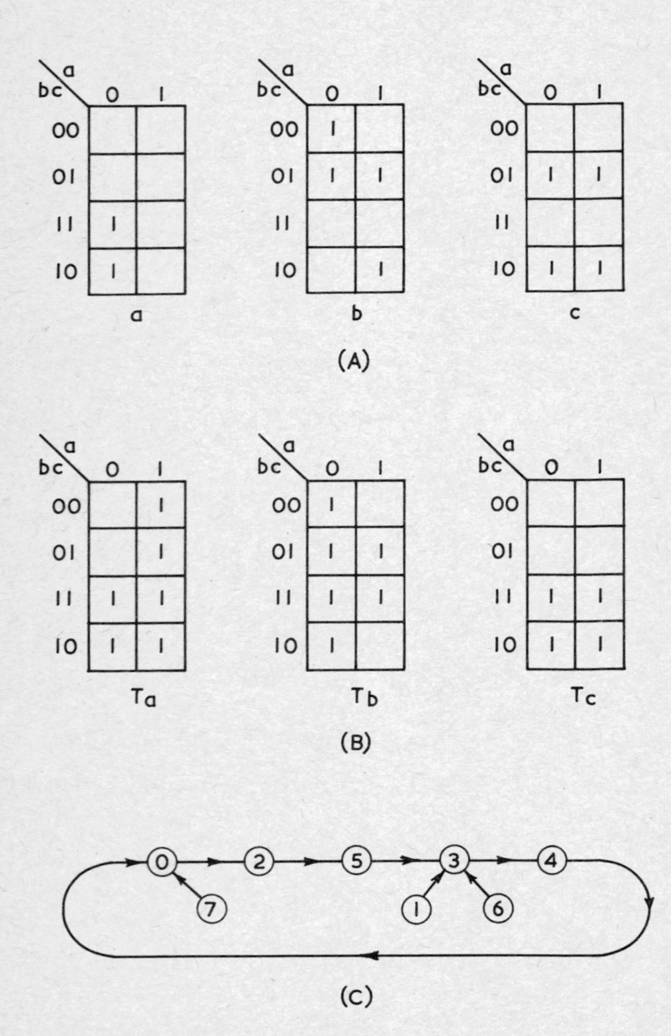

Figure 8-32

This leads to the flow table as given in Fig.8-33.

Alternatively, the flow table can be found out as follows: Expanding the input equations in the canonical form (Fig.8-32(B)).

$T_a = \vee\ (2, 3, 4, 5, 6, 7)$

$T_b = \vee\ (0, 1, 2, 3, 5, 7)$

$T_c = \vee\ (2, 3, 6, 7)$.

These are now plotted in a table (Fig.8-33). From the properties of Trigger *FF*, it is known that when the T line is 1 the *FF* changes state. Thus knowing values of T_a column, the column a_+ can be determined, and similarly the columns b_+ and c_+ can be written out from T_b and T_c respectively. Thus the flow table takes the form as given in Fig.8-33.

	a	b	c	T_a	T_b	T_c	a_+	b_+	c_+
0	0	0	0	0	1	0	0	1	0
1	0	0	1	0	1	0	0	1	1
2	0	1	0	1	1	1	1	0	1
3	0	1	1	1	1	1	1	0	0
4	1	0	0	1	0	0	0	0	0
5	1	0	1	1	1	0	0	1	1
6	1	1	0	1	0	1	0	1	1
7	1	1	1	1	1	1	0	0	0

Figure 8-33

From the flow table, we see the transition of the counter from a certain present state to the next state. Designating each state by its decimal number, the flow table leads to the flow diagram of Fig.8-32(C). The flow diagram brings out that the states 1, 6 and 7 are redundant. It also shows the displaying sequence. The corresponding flow table is shown in Fig.8-34.

a	b	c	Decimal cell	a_+	b_+	c_+
0	0	0	0	0	1	0
0	1	0	2	1	0	1
1	0	1	5	0	1	1
0	1	1	3	1	0	0
1	0	0	4	0	0	0
0	0	1	1	ϕ	ϕ	ϕ
1	1	0	6	ϕ	ϕ	ϕ
1	1	1	7	ϕ	ϕ	ϕ

Figure 8-34

8.10 TABULAR METHOD FOR FLIP-FLOP INPUT EQUATIONS

The algorithm method of deriving input equations is a very convenient one. But, since the derivations are carried out on maps, the method tends to become inconvenient for large sequential circuits where the transition functions of the flip-flops are functions of more than four variables. If the number of variables is more than six, then the map method virtually breaks down. In this section a tabular method is presented, wherein it is shown how the underlying principle of the map method can easily be extended to cover all types of sequential circuits, large as well as small. A computer program capable of producing the excitation functions has also been reported (Biswas, 1970). The method can be best described with the help of an illustration. Let us consider a decade counter which counts the clock pulses. The counter needs four flip-flops, say, y_1, y_2, y_3 and y_4. The flow table of such a counter is given in Fig.8-35. The transition functions of the flip-flops are five-variable Boolean functions. Their disjunctive canonical forms are (from the flow table) as follows:

$$Y_1(c_1y_2y_3y_4) = \vee\,(8, 9, 23, 24)$$
$$\vee\,\phi(10\text{–}15, 26\text{–}31)$$
$$Y_2(c_1y_2y_3y_4) = \vee\,(4, 5, 6, 7, 19, 20, 21, 22)$$
$$\vee\,\phi(10\text{–}15, 26\text{–}31)$$
$$Y_3(c_1y_2y_3y_4) = \vee\,(2, 3, 6, 7, 17, 18, 21, 22)$$
$$\vee\,\phi(10\text{–}15, 26\text{–}31)$$
$$Y_4(c_1y_2y_3y_4) = \vee\,(1, 3, 5, 7, 9, 16, 18, 20, 22, 24)$$
$$\vee\,\phi(10\text{–}15, 26\text{–}31).$$

Let it also be decided that the counter is to be implemented by $S\text{–}C$ flip-flops. It can be seen that in order to be able to apply this algorithm for any flip-flop, it is necessary to recognize the 'not half' and the 'true half' for the particular flip-flop. It is comparatively easy to recognize these two halves for any variable on the map. If the map method is utilized for the derivation of the excitation functions of the various flip-flops, then a five variable map as shown in Fig.8-36 has to be used. The map shows the decimal designation of each cell, and also the not and the true halves of the variables y_1, y_2, y_3, and y_4. It can be seen that the cells 0 through 7 and 16 through 23 belong to the $\bar{y}_1$ half, whereas the cells 8 through 15 and 24 through 31 belong to the y_1 half. It is now interesting to note that this information is also available in the flow table as given in the table of Fig.8-35, inasmuch as the cells constituting the $\bar{y}_1$ half (0–7, 16–23) have 0's below the column of y_1, and the cells constituting the y_1 half (8–15, 24–31) have 1's below the y_1 column. Similarly the not and the true halves of

	2	3	4	5	6	7	8	9	10	11	12	13
Cell No.	c	y_1	y_2	y_3	y_4	Y_1	Y_2	Y_3	Y_4		S_1	C_1
0	0	0	0	0	0	0	0	0	0	0	0	ϕ
1	0	0	0	0	1	0	0	0	1	0	0	ϕ
2	0	0	0	1	0	0	0	1	0	0	0	ϕ
3	0	0	0	1	1	0	0	1	1	0	0	ϕ
4	0	0	1	0	0	0	1	0	0	0	0	ϕ
5	0	0	1	0	1	0	1	0	1	0	0	ϕ
6	0	0	1	1	0	0	1	1	0	0	0	ϕ
7	0	0	1	1	1	0	1	1	1	0	0	ϕ
8	0	1	0	0	0	1	0	0	0	3	ϕ	0
9	0	1	0	0	1	1	0	0	1	3	ϕ	0
0	0	1	0	1	0	ϕ	ϕ	ϕ	ϕ	ϕ	ϕ	ϕ
1	0	1	0	1	1	ϕ	ϕ	ϕ	ϕ	ϕ	ϕ	ϕ
2	0	1	1	0	0	ϕ	ϕ	ϕ	ϕ	ϕ	ϕ	ϕ
3	0	1	1	0	1	ϕ	ϕ	ϕ	ϕ	ϕ	ϕ	ϕ
4	0	1	1	1	0	ϕ	ϕ	ϕ	ϕ	ϕ	ϕ	ϕ
5	0	1	1	1	1	ϕ	ϕ	ϕ	ϕ	ϕ	ϕ	ϕ
6	1	0	0	0	0	0	0	0	1	0	0	ϕ
7	1	0	0	0	1	0	0	1	0	0	0	ϕ
8	1	0	0	1	0	0	0	1	1	0	0	ϕ
9	1	0	0	1	1	0	1	0	0	0	0	ϕ
0	1	0	1	0	0	0	1	0	1	0	0	ϕ
1	1	0	1	0	1	0	1	1	0	0	0	ϕ
2	1	0	1	1	0	0	1	1	1	0	0	ϕ
3	1	0	1	1	1	1	0	0	0	1	1	0
4	1	1	0	0	0	1	0	0	1	3	ϕ	0
5	1	1	0	0	1	0	0	0	0	2	0	1
6	1	1	0	1	0	ϕ	ϕ	ϕ	ϕ	ϕ	ϕ	ϕ
7	1	1	0	1	1	ϕ	ϕ	ϕ	ϕ	ϕ	ϕ	ϕ
8	1	1	1	0	0	ϕ	ϕ	ϕ	ϕ	ϕ	ϕ	ϕ
9	1	1	1	0	1	ϕ	ϕ	ϕ	ϕ	ϕ	ϕ	ϕ
0	1	1	1	1	0	ϕ	ϕ	ϕ	ϕ	ϕ	ϕ	ϕ
1	1	1	1	1	1	ϕ	ϕ	ϕ	ϕ	ϕ	ϕ	ϕ

Figure 8-35 Flow table for a decade counter.

any other variable can be recognized by noting the 0's or 1's in the column under the particular variable.

The excitation functions of the S and C lines of the y_1 flip-flop which is an $S-C$ type flip-flop can now be derived following the steps as given below:

Step 1: Multiply (arithmetically) the column under y_1 (column #3) by 2 and add (arithmetically) to the column under Y_1 (column #7). Any number added to ϕ will yield ϕ only.

This results in the new column #11. It is easy to see that the 0's and 1's of

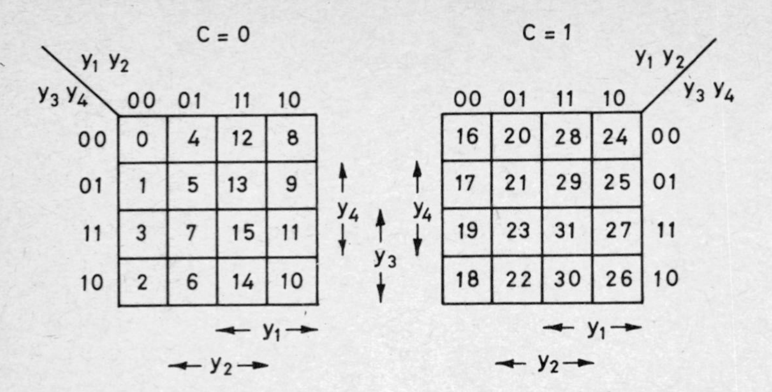

Figure 8-36 Five variable map showing true- and not-halves of each of the variables y_1 through y_4

the $\bar{y}_1$ half and the y_1 half of the Y_1 function are given by the numbers 0, 1 and 2, 3 respectively. The don't care terms remain ϕ's. The plotting algorithm of the S–C flip-flop must now be modified to suit the new designations. Accordingly, the new algorithm should be as given in the Table of Fig.8.37. Following these rules the columns for S_1 and C_1 can be computed as given in columns 12 and 13 of Fig.8-35. Thus the next step is:

Step 2: Produce two new columns from column 11, following the rules for plotting as given in Fig.8.37. These columns (#12 and 13) are the truth table representations of the excitation functions of the S–C type y_1 flip-flop. The excitation functions of the other flip-flops can be determined in a similar manner. The algorithms for the T, and the J–K flip-flops have also been included in the table of Fig.8-37.

		For				
FF	in column	0	1	2	3	ϕ
S–C	*S*	0	1	0	ϕ	ϕ
	C	ϕ	0	1	0	ϕ
J–K	*J*	0	1	ϕ	ϕ	ϕ
	K	ϕ	ϕ	1	0	ϕ
T	*T*	0	1	1	0	ϕ

Figure 8-37 Modified rules for the flip-flops.

REFERENCES

Bartee, T. C. and D. J. Chapman: "Design of an accumulator for a general purpose computer", *IEEE Trans. on Electronic Computers,* **EC-14**, pp.570–574, August, 1965. (This paper describes an application of $T-G$ flip-flop.)

Biswas, N. N.: "The logic and input equations of flip-flops", *Electronic Engineering (London),* pp.107–111, Feb.1966.

—— : "Computer derivation of flip-flop excitation functions", *Proc. National Electronics Conference,* 1970, (Chicago, U.S.A.), Vol.26, pp.75–77.

—— : "$T-G$ flip-flop input equations", *International J. of Control (London),* **15**, pp.191–196, Jan.1972.

Grabbe, E. M., S. Ramo and D. E. Wooldridge: *Handbook of Automation, Computation, and Control,* Vol.2, John Wiley and Sons Inc., New York, 1959.

Graham, P. J. and R. J. Distler: *RST flip-flop input equations, IEEE Trans on Electronic Computers,* **EC-16**, pp.443–445, August 1967.

Phister, M.: *Logical Design of Digital Computers,* John Wiley and Sons Inc., New York, 1959.

Pressman, A. I.: *Design of Transistorized Circuits for Digital Computers,* John F. Rider, Publisher, Inc., New York, 1959.

Richards, R. K.: *Digital Computer Components and Circuits,* D. Van Nostrand Co. Inc., Princeton, N.J., 1957.

PROBLEMS

8.1 The transition function of the 'a' *FF* is given by $f(xabc) = \vee\,(0, 2, 4, 6, 8, 10, 11)$. Derive excitation function for the *FF*, if this *FF* is a) $S-C$ type, b) $T-G$ type.

8.2 The transition function of the 'y_2' *FF* is given by $f(xy_1y_2y_3) = \vee\,(2, 3, 5, 9, 15) \vee \phi(0, 1, 8, 10)$. Derive excitation function for the y_2 *FF* if this is a) $J-K$ type, b) T type.

8.3 Using $S-C$ *FF*'s design a decade counter which counts from 0 through 9 upwards, and then recycles. The digits are represented by the binary code. Use AND, OR logics for interconnections.

8.4 Using $J-K$ *FF*'s design a decade counter which utilizes the excess-three code. Use AND, OR logics for interconnection.

8.5 Using T *FF*'s design an octal up counter (counts 0 through 7 upwards and then recycles). The digits are represented by the binary code. Use minimal NOR logic for interconnection. (*Hint:* Express the excitation functions of the *FF*'s as octal functions. Then find the minimal NOR circuit consulting Smith's catalog.)

8.6 Use $J-K$ *FF*'s to design a counter which displays the binary code of the following numbers in a cyclic sequence. 1, 2, 3, 6, 7, 9, 0, 5, 4, 8, 1, 2, . . . etc.

8.7 Use *T–G FF*'s to design a down octal counter using binary code. Use minimal NAND gate circuits for interconnection.

8.8 Use *S–C–T FF*'s to design a counter which is to display in the following sequence. 0, 2, 4, 6, 8, 10, 12, 14, 0, 2, etc.

8.9 Show how delay flip-flops can be used to make a) an *S–C FF*, b) a *J–K FF* and c) a *T–G FF*.

8.10 You have been given an *S–C* flip flop. Show how you can convert it into *T–G FF*.

8.11 Show how a *T–G FF* can be converted into a) an *S–C FF*, and b) a *J–K FF*.

8.12 For an autonomous counter using *J–K FF*'s w, x, y, and z, the excitation functions are as follows.

$$J_w = z \qquad K_w = x \vee y \vee z$$

$$J_x = w \qquad K_x = \bar{w}$$

$$J_y = x \qquad K_y = \bar{x}$$

$$J_z = \bar{w}y \qquad K_z = \bar{w}\bar{y}.$$

Find the displaying sequence of the counter. Assume that binary code has been used.

8.13 Complete the design of the example in Sec.8.10.

8.14 The transition functions of the c and f *FF*'s are given by

$$c_+ = g(abcdef) = \vee\,(0, 2, 5, 9, 13, 18, 22, 31, 45, 46, 52, 54, 56, 59) \vee \phi(60–63)$$

$$f_+ = h(abcdef) = \vee\,(1, 3, 5, 7, 8, 12, 15, 20, 31, 32, 36, 38, 41, 42, 43, 50, 52, 55) \vee \phi(60–63).$$

Find the excitation functions if the *FF*'s are of *T*-type.

8.15 An n-stage shift-register can be designed with n flip-flops connected in cascade through identical interconnecting networks between any two successive stages. Each *FF* stores either a 0 or a 1. When a shift pulse S is applied ($S = 1$), a 0 or a 1 stored in stage m moves to stage $m + 1$. Design the interconnecting gate network of such a shift register, using *J–K FF*'s and NAND gates only.

8.16 Can an *S–C FF* be made of only NOR (NAND) gates? Discuss.

Chapter 9

SYNCHRONOUS SEQUENTIAL MACHINES

9.1 INTRODUCTION

In Chapter 6 we studied combinational circuits where the outputs are determined solely by the inputs. In the last chapter, while studying the behavior of flip-flops we noticed that these circuits produce different outputs at different situations although the input may be the same. There are many more examples of such circuits which are known as sequential circuits. Take for example the response of an elevator when a passenger waiting on a floor and wanting to go up presses the UP button. When the elevator comes to that floor, it does not stop, if it is going down. On the other hand, it stops and the door opens to let the passenger in, if it is going up. Thus the elevator responds differently, although the input is the same, namely, pressing the UP button, depending on whether it is going up or down. Therefore, it may be rightly said that the response or the output depends not only on the input, but also on the situation, or the state in which the responder is in at that time.

Another familiar example of a sequential circuit is the serial binary adder, whose block diagram is shown in Fig.9-1. Let the adder add the two binary

Figure 9-1 Block diagram of a serial binary adder.

numbers, 0110 and 0111. Let the bits of the first number be incident serially on the x_1 line, and those of the second number on the x_2 line. Further let it be assumed that the operation of the circuit is synchronized with a synchronizing pulse often called a *clock pulse* as it has a definite pulse repetition frequency. Under this condition during an interval say t_1, the first bits of the two numbers to be added, namely 0 and 1 arrive at x_1 and x_2 respectively, and they are added by the adder. The output of the adder is 1, the result of the addition. During this interval no other operation takes place. During the next interval t_2, the second bits, 1 and 1 arrive at x_1 and x_2 and are added by the adder, producing an output of 0. However, this time, the circuit has to remember the *carry* of 1. This the circuit does, and consequently, when in the next interval t_3, two 1's are incident at x_1 and x_2, the output of the circuit is not 0 but 1. During this interval also, the circuit has to remember the carry of 1. During the next interva when two 0's are the input to the adder, the adder produces an output of 1, but it is not required to remember the carry, as it is not produced as a result of the addition. It will be obvious by now, that the circuit must be capable of being in two internal states; in one, it remembers that no carry has resulted during the addition operation of the previous interval, in the other it remembers that a carr has resulted. Let these two states of the circuit be denoted by A and B respectively. Then the operation of the circuit can be as shown in the table of Fig.9-2.

t_4	t_3	t_2	t_1	Interval
0	1	1	0	Augend (x_1)
0	1	1	1	Addend (x_2)
1	1	0	1	Sum
B	B	A	A	Present state
A	B	B	A	Next state

Figure 9-2

In this table, the next state at interval t_1, is the present state at interval t_2, an so on. It is interesting to note from Fig.9-2, that although the inputs are the sar at intervals t_2 and t_3, the outputs are not the same. This is because the present states are different. Thus, in this sequential circuit the output is a function of *both* the input and the present state. So also is the case with the next state, that is, it is also a function of the input and the present state. Thus, the behavio of a sequential circuit is completely defined by a table where the values of the next state and output are shown for all combinations of the input and the prese state. Such a table for the serial binary adder is given in Fig.9-3.

	NS, z			
		x_1x_2		
PS	00	01	11	10
A	A, 0	A, 1	B, 0	A, 1
B	A, 1	B, 0	B, 1	B, 0

Figure 9-3 State table of the serial binary adder.

As a sequential circuit has to have different internal states which remember different data, it must have memory. The flip-flops discussed in the last chapter form the memory units of various sequential circuits. A single flip-flop can have two different states, and therefore will be adequate for the serial binary adders which need to have only two states. Let y be the flip-flop, and further let the 0 state of the FF be the A state of the adder, and the 1 state of the FF the B state of the adder. Incorporating this information, we may derive the *transition* and *output tables* from the *state table*. These are shown in Fig.9-4(A) and (B).

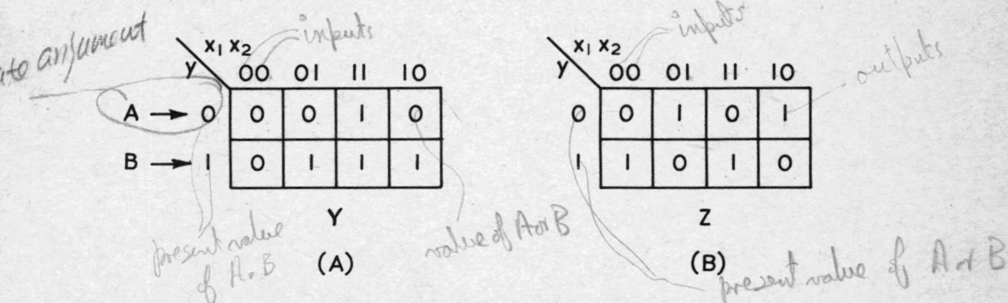

Figure 9-4 (A) Transition, and (B) output tables of the adder.

It will be seen later that the transition and the output tables are the starting point for the design of a sequential circuit. The state table as given in Fig.9-3 is said to define a sequential machine. The term *sequential machine* is a more general term. The circuit which implements a sequential machine is known as a *sequential circuit*. Thus there can be a variety of sequential circuits implementing the same sequential machine.

Definition 9.1.1 The state table of Fig.9-3 describes a sequential machine whose output depends on both the input and the present state. Such a machine is called a *Mealy machine*. In another model, the output depends only on the present state and is independent of the input. Such a machine is called a *Moore Machine*.

PS	NS				z
	x_1x_2				
	00	01	11	10	
1	1	2	3	2	0
2	1	2	3	2	1
3	2	3	4	3	0
4	2	3	4	3	1

Figure 9-5 State table of the Moore model of the serial binary adder.

It is obvious that by incorporating more internal states, it is possible to conver a Mealy machine into a Moore machine. To convert the Mealy model of the serial binary adder into the Moore model, we must have four internal states, corresponding to $A, 0; A, 1; B, 0$, and $B, 1$. Let the four states, 1, 2, 3 and 4 represent these respectively. Then the states 1 and 3 must produce an output of 0, whereas the states 2 and 4 produce an output of 1. The resultant state table is shown in Fig.9-5.

9.2 BASIC DEFINITIONS

From what we have discussed so far, we may conclude that a sequential machine must have the following:

1) A finite number, say p, of internal states. Let the set S contain all the internal states so that $S = \{ S_1, S_2, \ldots, S_p \}$

2) A finite number, say n, of input sequences, which is called the *input alphabet.* Each member of the alphabet is called an input symbol, or simply an input. Thus, if there are two input lines, x_1 and x_2, then there are four input symbols, namely when $x_1 x_2$ are 00, 01, 11 and 10 respectively. The input alphabet then comprises of these four input symbols. Let X be the set representing the input alphabet. Then $X = \{ X_1, X_2, \ldots, X_n \}$.

3) A finite number, say m, of output sequences known as the *output alphabet.* Each member of the output alphabet is called an output symbol, or simply an output. Let Z denote the set of output symbols, so that $Z = \{ Z_1, Z_2, \ldots, Z_m \}$

4) A characterizing function f, which uniquely defines the next state S^{t+1} as a function of the present state S^t and the present input X^t, so that $S^{t+1} = f(S^t, X^t)$.

5) A characterizing function g, which uniquely defines the output as a functio of the input and the internal state, for a Mealy machine, so that $Z^t = g(S^t, X^t)$. For a Moore machine, the output is a function of the internal state only, so that $Z^t = g(S^t)$.

A formal definition of a sequential machine can now be given as follows:

Definition 9.2.1 A sequential machine M is a quintuple, $M = (X, Z, S, f, g)$ where X, Z and S are the finite and non-empty sets of inputs, outputs and states respectively. f is the next-state function, such that

$$S^{t+1} = f(S^t, X^t), \tag{9.1}$$

and g is the output function, such that

$$Z^t = g(S^t, X^t) \text{ for a Mealy machine} \tag{9.2}$$

and

$Z^t = g(S^t)$ for a Moore machine. (9.3)

It is apparent that a state table gives all these informations about a sequential machine, and therefore is frequently used to describe a machine. Hence, the first step in the synthesis of a sequential machine is to produce a state table from the word statements or specifications. For example, from the specification known about the serial binary adder, we produced the state table of Fig.9-3. A more systematic procedure for this will be discussed in a later section. However, it must be mentioned here that the state table that we may get from the word statement may not always contain the minimum number of states that may be adequate to implement the sequential machine. Consequently, there must be ways to reduce a given state table to a form where it has the minimum number of states. This technique is known as the state minimization.

9.3 STATE MINIMIZATION

Let us consider the state table as given in Fig.9-6. The sequential machine M_1 and others described in this chapter may not always describe practical machines, but are good enough for the academic exercise for which they are introduced. Machine M_1 has eight states, A through H, two input symbols,

M_1		
	NS, z	
	x	
PS	0	1
A	A, 0	D, 1
B	C, 1	D, 0
C	B, 0	E, 1
D	D, 1	A, 1
E	E, 0	G, 1
F	G, 0	E, 0
G	D, 1	A, 1
H	D, 1	C, 1

Table 9-6 State table of machine M_1

0 and 1, and two output symbols, 0 and 1. We must investigate if some of the states turn out to be equivalent to some other. For this we must be quite clear about what is meant by equivalent states. So let us first determine the features which distinguish one state from another. Confining our attention to a Mealy machine, we can see from Eqns (9.1) and (9.2), that two states P and Q are uniquely determined by

$$P^{t+1} = f(P^t, X^t), \text{ and } Q^{t+1} = f(Q^t, X^t)$$

and

$$Z_P^t = g(P^t, X^t), \text{ and } Z_Q^t = g(Q^t, X^t).$$

Hence P will be indistinguishable from Q, that is, P and Q are *equivalent*, if

$$P^{t+1} = Q^{t+1}$$

for all $X^t \in X$ and

$$Z_P^t = Z_Q^t$$

for all $X^t \in X$. Now, the state table shows both the next states and outputs for all input symbols. Hence an inspection of the state table can tell us if two states are equivalent. For example, taking states A and B of Machine M_1, we see that in order that A and B be equivalent, their next states for $x = 0$ input must be the same or equivalent. In this case, therefore, A and C must turn out to be equivalent. The next states for $x = 1$ input are, however, the same. Also, their outputs must also be identical. While it may not be possible to determine if $A = C$ at this stage, it is easy to find out, that the output sequence for state A, 01, is not equal to the sequence for state B, 10. Hence if the output sequences are not the same, we can at once conclude that the two states cannot be equivalent.

Theorem 9.3.1 Two states of a sequential machine are equivalent if in the state table a) their output sequences are identical, and b) the correspondings next states in their next state sequences are equivalent.

Thus, for Machine M_1, A and B can never be equivalent, as they differ in the outputs. But the states A and C may be equivalent (as both have the same output sequence) if the state pairs, (A, B) and (D, E) are equivalent. Thus the equivalence of the pair (A, C) implies the equivalence of the two pairs (A, B) and (D, E). Following this argument, we can prepare an implication chart from the state table as suggested by Paull and Unger (1959). The implication chart is then used to establish the equivalence of various state pairs. The implication chart for Machine M_1 is as shown in Fig.9-7. Since A and B cannot be equivalent because of their difference in the output a x is placed at the square of column A and row B. Again, A and C have similar outputs. So they may be equivalent if the pairs AB and DE are equivalent. This fact is recorded by writing AB and DE at the square at column A and row C. Thus the various results of comparing A with the rest of the states are written at the various squares of column A. Similarly, the results of comparing state B with states C through H are recorded at the squares of column B. At the end of this initial comparison, the implication chart appears as shown in Fig.9-7(a). Now the entry at the square of column A and row C is AB and DE. So we must check if A and B are equivalent according to the chart. We find that they are not. So, we place a x in the square. This

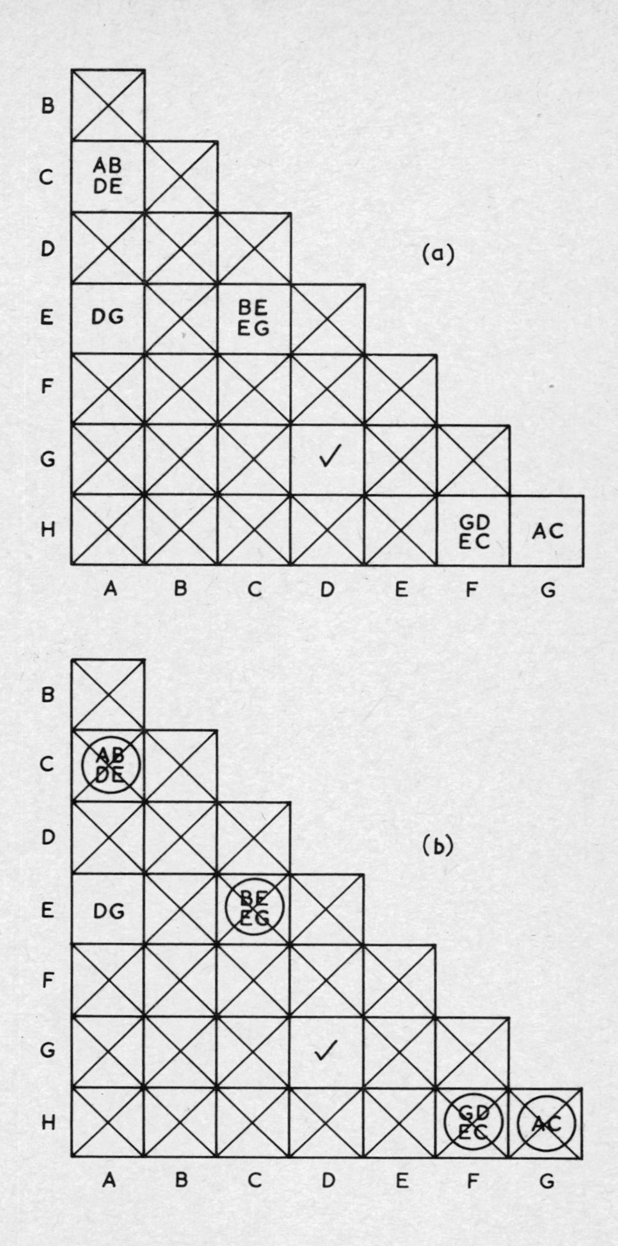

Figure 9-7 Implication chart for machine M_1. (a) Initial form, (b) final form.

gives rise to a new fact that the pair AC is not equivalent. Hence if AC appears in any square, a x is also placed in that square. After all squares with AC have been searched and crossed out, a circle is placed in the square of column A and row C. This procedure is repeated for all squares with entries of probable equivalent pairs. At the end of the operation, the implication chart appears as shown in Fig.9-7(b). The squares which have not been crossed out indicate pairs which are equivalent. For Machine M_1, AE and DG are such pairs.

However, we may accomplish this result by arguing from another direction following Huffman (1954) and Moore (1956). The requirement a) of Th.9.3.1 fixes the lower bound of the number of states needed for a sequential machine. For machine M_1, this number is 4. Then why not assume that the machine can be implemented by only four states. Obviousl then we must merge or combine all the states having similar output into a single state. This will result in partitioning the entire set of states into 4 partitions. Then we must find out the next states of each state in terms of the partition. Thus for state A, the next states are A and D. According to the partition P_0 of Fig.9-8, A belongs to the partition 1 and D to partition 3. Hence under the column *NS*, (next states) we write the sequence 13 of the *PS* (present state) A. Similarly, the *NS* sequences of all other present states are written in terms of the partition P_0. If all the states of a partition generate the same sequence of next states, then the partition is retained. But if they generate more than one sequenc then the partition is further partitioned into the number of sequences generated. Following this procedure, we observe that partition 1 of P_0 must be further subdivided into two. Relabeling the partitions we get the new partition P_1 as shown in Table II of Fig.9-8. The *NS* sequences are determined afresh, and the procedu

Table I

P_0	*PS*	*NS*
1	A	13
	C	21
	E	13
2	B	
3	D	31
	G	31
	H	31
4	F	31
	H	31

Table II

P_1	*PS*	*NS*
1	A	14
	E	14
2	C	
3	B	
4	D	41
	G	41
	H	42
5	F	41

Table III

P_2	*PS*	*NS*
1	A	14
	E	14
2	C	
3	B	
4	D	41
	G	41
5	H	
6	F	

Figure 9-8 Minimization of machine M_1.

of subdividing an existing partition is repeated, until we reach a Table, where no further subdivision is needed. The tables of Fig.9-8 depict the situations encountered in minimizing machine M_1. States (A, E) and (D, G) turn out to be equivalent. The reduced state table, where E has been replaced by A, and G by D, is shown in Fig.9-9. In this table states A and E (and also D and G) are said to have *merged* or *combined.*

	NS, z	
	x	
PS	0	1
A	A, 0	D, 1
B	C, 1	D, 0
C	B, 0	A, 1
D	D, 1	A, 1
F	D, 0	A, 0
H	D, 1	C, 1

Figure 9-9 Reduced state table of Machine M_1.

9.4 INCOMPLETELY SPECIFIED MACHINE

The sequential machine discussed above has definite next states and output entries in the state table. However, there are machines for which some of the next states or outputs may not be known, or may be ignored.

Definition 9.4.1 A sequential machine is called an *incompletely specified sequential machine (ISSM)* if its state table has one or more unspecified next state and/or output entries.

M_2

	NS, z	
	x	
PS	0	1
A	C, –	B, 0
B	A, 0	F, 0
C	A, 1	C, 0
D	E, –	D, 0
E	D, 1	D, 0
F	E, 0	B, 0

Figure 9-10 State table of machine M_2.

The state minimization of an incompletely specified machine poses special problems, and cannot be carried out by the methods of the last section. Let this point be elaborated by considering the incompletely specified machine M_2 of Fig.9-10. In this machine the output of the present state A for a 0 input is unspecified. If we are free to choose this output to be either 0 or 1, then we see that state A is a candidate to combine with state B, as well as C. But it can be seen that the states B and C cannot combine as they differ in the output sequences. This is the most important difference between a completely specified and an incompletely specified machine. In a *CSSM*, if state P is equivalent to Q and Q to R, then P is also equivalent to R. Hence the transitive relation holds good. On the other hand, as has been pointed out above this transitive relation does not hold good in an *ISSM*. For this reason the term equivalence of states is not used in an *ISSM*. Instead, two states are called compatible, when they satisfy the condition set forth by the following definition.

Definition 9.4.2 Two states of an incompletely specified sequential machine are *compatible,* if their output sequences are not in conflict, and their corresponding next states in the next state sequences are also compatible.

Let the procedure for minimizing an *ISSM* be explained with the example of Machine M_3, whose state table is shown in Fig.9-11. In this machine, states A and B are incompatible as they differ in output. But A and C may be compatible, as an unspecified output may be considered to be a 0 or a 1 output. However, the

M_3

PS	NS, z			
	x_1x_2			
	00	01	11	10
A	E, 1	E, –	–, –	–, 0
B	B, 0	–, –	F, –	C, 1
C	F, –	D, 1	C, –	–, –
D	–, –	C, 0	–, 1	E, 0
E	B, 0	–, –	G, 1	C, –
F	C, 0	–, –	C, 0	–, –
G	–, –	D, 1	–, 0	A, 1

Figure 9-11 State table of machine M_3.

compatibility of A and C implies the compatibility of the pairs EF and DE. Thus an implication chart for the machine is also determined (Fig.9-12(a)). However, as has been pointed out earlier the transitive relation does not hold good in case of incompletely specified machine. Hence we must define what is called a maximal compatible in the following manner.

Definition 9.4.3 If n number of states are so mutually compatible that every

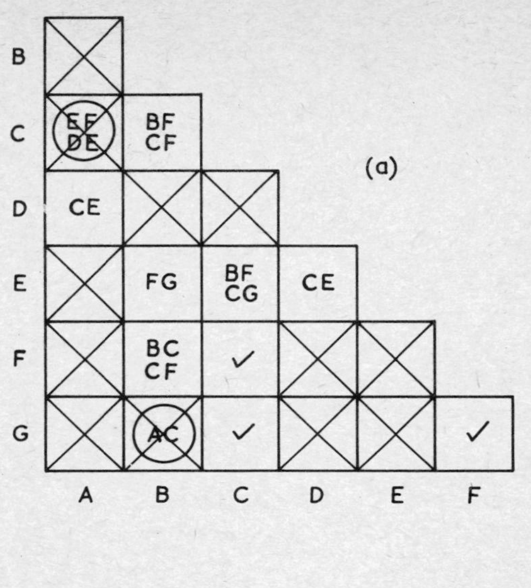

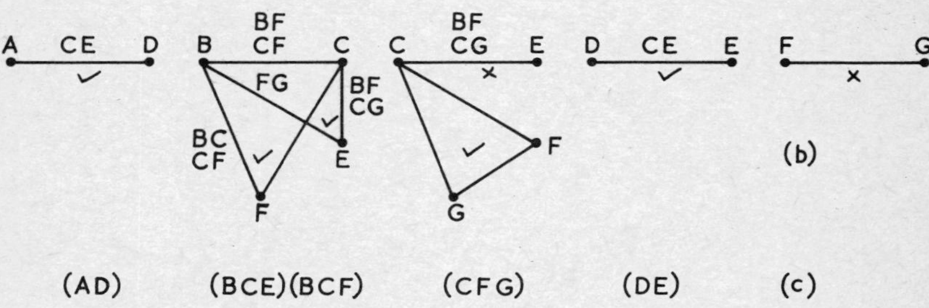

(AD) (BCE)(BCF) (CFG) (DE) (c)

Figure 9-12 (a) Implication chart, (b) compatibility graph, and (c) set of *MC*'s; for machine M_3.

pair of them is compatible, and n is maximum, then the n states form a *maximal compatible (MC)*.

It follows that all states belonging to an *MC* can be merged into a single state. Also, any particular state may belong to more than one *MC*. After the implication chart has been formed, the next step is to determine the set of *MC*'s.

9.5 THE COMPATIBILITY GRAPH

The compatibility graph provides a very convenient way to derive the set of *MC*'s for an *ISSM*. Fig.9-12(b) shows such a graph for Machine M_3. The compatibility graph is constructed in the following way. From column A of the implication chart we find that the state A is compatible with the state D. So, we draw a straight line *AD* to depict this fact. We also write *CE* on this straight line to indicate that the compatibility of *AD* implies the compatibility of the pair *CE*. Next we inspect column *B* of the implication chart. We find that the state *B* is compatible with the states *C, E* and *F*. So, we draw three straight lines *BC, BE* and *BF* to depict this fact. On the straight line *BC*, we write *BF* and *CF* to indicate that the compatibility of *BC* implies the compatibility of *BF* and *CF*. Similarly we write *FG* near the straight line *BE* and *BC* and *CF* near the straight line *BF*. Now we inspect column *C* of the implication chart, and construct the straight lines *CE, CF* and *CG*. We also write *BF* and *CG* near the straight line *CE*, to indicate the compatibility pairs which are implied. At this stage, we must see if there are points in the previous graphs which correspond to the joined pairs in the *C* subgraph. We find that there are such points and consequently we join *CE* and *CF*. Simultaneously, we write the implied compatibility pairs, if any, near these lines. Thus *BF* and *CG* are written near the line *CE* of the *B* subgraph as well. Now we go to column *D* of the implication chart, and draw the *D* subgraph. Here again we must see if the *D* subgraph will initiate further construction in the *C* and *B* subgraphs. The *D* subgraph does not initiate further constructions in the *C* or *B* subgraphs. As column *E* does not show any compatible pair, no *E* subgraph is drawn. After *F* subgraph is drawn, we find that the points *F* and *G* are to be joined in the subgraph *C*. Whenever a new straight line is drawn, the implied compatible pairs are also written near it. After all constructions have been completed, we obtain the complete *compatibility graph.* The compatibility graph for machine M_3 is as shown in Fig.9-12(b). The maximal compatibles can now be chosen by the following theorem, which however must be preceded by the definition of a complete polygon.

Definition 9.5.1 A *complete polygon* is one whose all pairs of vertices are connected so as to form either a side or a diagonal.

Thus straight lines and triangles are complete polygons. A quadrilateral with 2 diagonals, a pentagon with 5, and a hexagon with 9 diagonals are examples of complete polygons. In general, for a polygon with n sides ($n \geqslant 3$), there must be $(n - 3)\, n/2$ diagonals to make it a complete polygon.

Theorem 9.5.1 (Kohavi, 1970). A complete polygon which is not contained in any higher order complete polygon represents a maximal compatible (*MC*).

Proof: By the definition of a complete polygon and by the property of the compatibility graph, all states covered by a complete polygon are pairwise compatible. Hence they constitute a compatible. Since the polygon is not contained in any higher order complete polygon, they (the states) constitute a maximal compatible. ▲

Let the *MC*s be now picked up starting from the *A* subgraph. Here the straight line *AD* gives the maximal compatible *AD*. This straight line is ticked (√) in the subgraph. In the B subgraph the triangles BCE, and BCF give us the maximal compatibles (BCE) and (BCF). These triangles are also ticked in the subgraph. With the selection of the triangles BCE and BCF, the straight line CE (in C subgraph) cannot constitute a maximal compatible. Hence it is crossed off (x). Next going to subgraph C, and then to D, and then to F, and following the same procedure we find two more MCs (CFG) and (DE). Thus the set of maximal compatibles for machine M_3 is

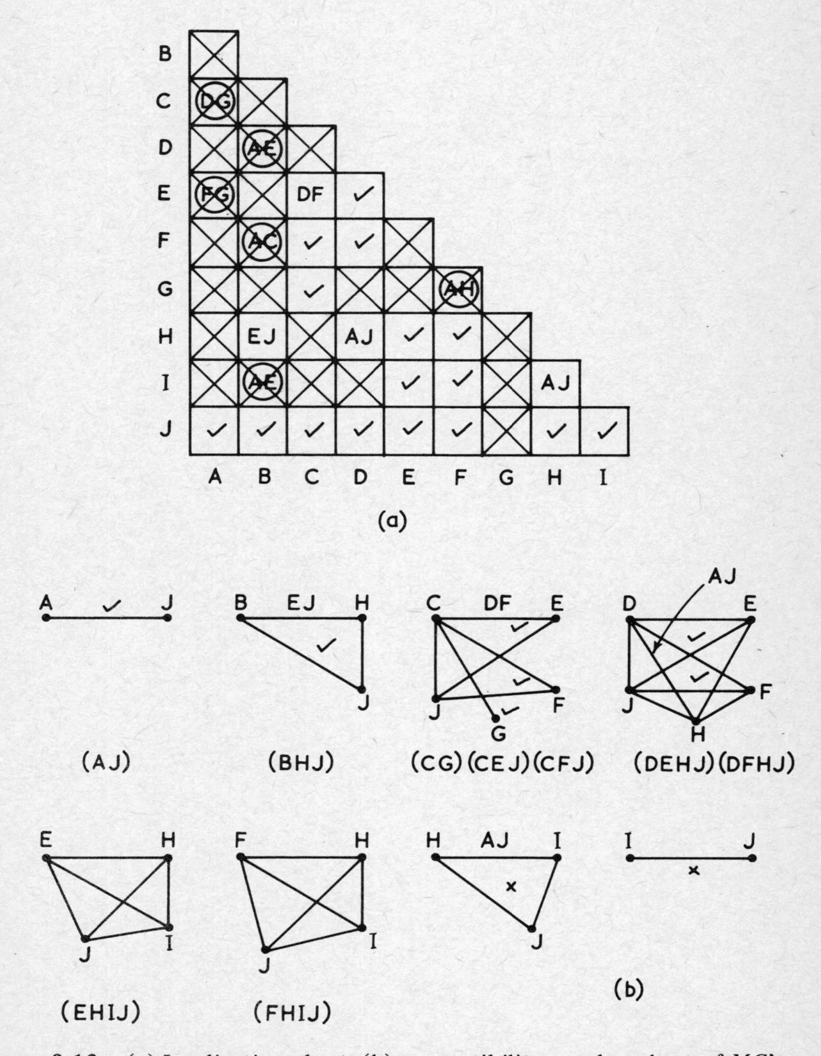

Figure 9-13 (a) Implication chart, (b) compatibility graph and set of MC's, of an ISSM with 10 states.

(AD) (BCE) (BCF) (CFG) (DE).

It should be mentioned here that the compatibility graph as described here is nothing but the *merger graph* of Kohavi (1970), split into parts. It has been the experience of the author that it is easier to detect various complete polygons and reject smaller polygons in the compatibility graph, rather than in the merger graph, specially when the machine has a large number of states. The 'resolving power' of the compatibility graph can be further increased if the straight lines of the various subgraphs which are initially drawn to represent the compatibility pairs of the implication chart are not drawn but kept understood. Figure 9-13 shows the implication chart and the compatibility graph of an *ISSM* with 10 states. Figure 9-14 shows the compatibility graph with high resolving power for the *ISSM* of Fig.9-13. The constructional procedure is modified as follows. To draw the subgraph of column A (Fig.9-13), the point A is drawn above the dotted line (Fig.9-14). Since the state A is compatible with the state J, a point indicating J is drawn below the dotted line. However, no straight line is drawn joining the points A and J. Similarly for the subgraph B, the point B is drawn above the dotted line, and the two points representing the states compatible with the state B, namely H and J are drawn below the line. The points H and J in the

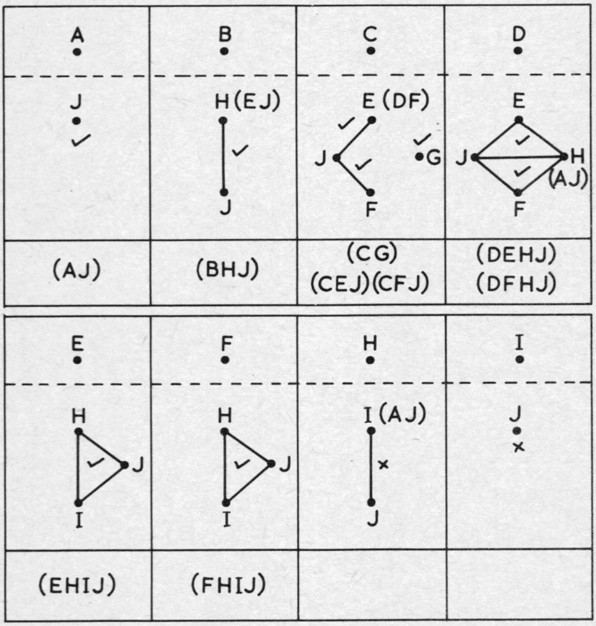

Figure 9-14 Compatibility graph with high resolving power for the ISSM of Fig.9-13.

subgraph *B* get joined by virtue of the fact that the point *J* has appeared below the dotted line in the *H* subgraph. Similarly other subgraphs are drawn. As in the previous case, each time a subgraph is drawn, the constructions wherever called for in the previous subgraphs are also completed. In this way the compatibility graph with higher resolving power is completed. An *MC* is still recognized by a complete polygon of the subgraphs, appearing below the dotted line. However, it should be borne in mind that *a point which has not been included in a bigger polygon also represents an MC.* For example, the point *J* in the *A* subgraph represents the *MC* (*AJ*). The *MC*s of other subgraphs have been written below the subgraphs.

9.6 THE IMPLICATION TREES: BISWAS (1974)

In case the number of members in the set of maximal compatibles is less than the number of original states, then clearly this number of *MC*s gives the upper bound of states which will be adequate to cover the machine. However, very often there may exist a covering machine with a smaller number of states. In order to find this lower bound, we must investigate the implication trees associated with each of the *MC*s. Even when the number of original states is smaller than or equal to the number of *MC*s, the search for the lower bound starts from the implication trees generated by the *MC*s. Let the procedure to construct the implication trees be explained taking the case of machine M_3. The relevant data for the maximal compatible *AD* is available in the subgraph *A* of the compatibility graph (Fig.9-12(b)). The compatibility of *AD* depends on the compatibility of *CE*. So a branch is drawn from *AD* which terminates on *CE*. Again from *C* subgraph we find that the compatibility of *CE* implies the compatibility of *BF* and *CG*. So, two branches start from *CE* and terminate on *BF* and *CG*. From *B* subgraph the compatibility of *BF* implies the compatibility of *BC* and *CF*. So two branches emanate from *BF* and terminate on *BC* and *CF*. The compatibility of *BC* again implies the compatibility of *BF* and *CF*. But both *BF* and *CF* have already appeared as nodes in the tree. Hence, we terminate *BC* on 0. Since the compatibility of *CG* does not depend on anything, we terminate *CG* also on 0. This completes the construction of the implication tree for the maximal compatible *AD*. For the implication tree of the maximal compatible (*BCE*), we see that the compatibility of *BC* implies the compatibility of *BF* and *CF*, that of *CE* implies the compatibilities of *BF* and *CG*, and that of *BE, FG*. Hence four branches start from *BCE* and terminate on *BF, CF, CG* and *FG*. The compatibility of *BF* is again dependent on those of *BC* and *CF*. But *BC* is included in *BCE*, and *CF* has already appeared as a node of the tree. Hence the branch from *BF* terminates on a 0. By following similar reasonings, the branches from *CF, CG* and *FG* all terminate on 0. After all branches terminate on 0's, the implication tree is complete. For the maximal compatible *BCF*, the implied compatibility pairs are *BF, CF* and *BC*. But all these pairs are included in *BCF*. Hence the branch from *BCF* terminates on 0. The implication trees for *CFG* and *DE* are similarly completed. Let these implication trees from the *MC*s be called *primary trees*. After all the primary trees have been drawn, we must also draw what may be called the *secondary* trees. Secondary trees are generated from the compatibles comprising all the subsets of the states constituting an *MC*. An *MC* consisting of n states may generate

$$\sum_{r=1}^{n-1} \binom{n}{r}$$

secondary trees. The generating compatible of a secondary tree is included either in an *MC* or a higher order secondary tree. Hence if a secondary tree shows more nodes than any of the primary trees or higher order trees, then the secondary tree does not serve any purpose. For this reason, the *MC*s which terminate on node 0, such as (*BCF*) and (*CFG*) of Fig.9-15, need not generate their secondary trees. Fig.9-15(b) shows the secondary trees of machine M_3 whose constructional procedure is the same as those of the primary trees. The number of compatibles is further reduced in individual trees by what we shall call a) substitution and b) bunching.

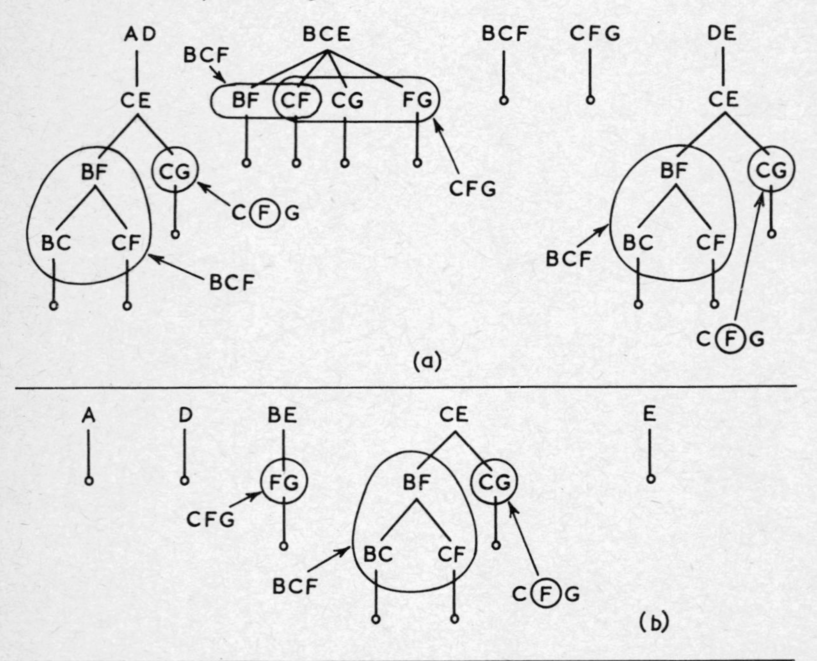

Figure 9-15 (a) Primary trees, and (b) secondary trees, of machine M_3.

Definition 9.6.1 If in the family of implication trees, there exists a tree generated by a compatible C_o with only one branch terminating on node 0, then any node C_i (in any other tree) along with its entire subtree can be *substituted* by C_o terminating on node 0, provided

$$C_i \subset C_o.$$

In Fig.9-15, in the trees of *AD, DE* and *CE*, the node *BF* and its subtree can be substituted by *BCF*.

Definition 9.6.2 In an implication tree if the compatibles $C_1, C_2, \ldots, C_n$ form parallel branches each terminating on node 0, and if there exists another tree generated by the compatible C_0 with only one branch terminating on node 0, such that

$$C_o \supseteq C_1 \cup C_2 \cup \ldots \cup C_n$$

then the parallel branch compatibles can be *bunched* into a single compatible C_o.

It should be mentioned here that a particular compatible can form part of more than one bunch. In Fig.9-15, the four nodes *BF, CF, CG* and *FG* of the *BCE* tree are bunched into two nodes, *BCF* and *CFG*.

While counting the nodes of a tree, the nodes which are bunched or substituted are counted as one node. These two processes, therefore, reduce the number of compatibles constituting an implication tree.

The covering capability of a tree is increased by carrying out *augmentation* as defined below.

Definition 9.6.3 If the tree or the subtree of a compatible C_i which may be either a generating or an intermediate node, has a structure identical with the tree of a generating compatible C_g, then C_i can be *augmented* to C_g, provided

$$C_i \subset C_g.$$

A simple case of augmentation is encountered when C_i does not have any tree but simply terminates on node 0. In Fig.9-15 in the tree of *BE, FG* is augmented to *CFG*. Consequently the tree can now cover the additional state *C*. If the additional state which is introduced into the tree already exists in the tree, then the process of augmentation will be indicated by keeping the additional state circled. In Fig.9-15, the node *CG* in the trees of *AD, DE* and *CE* can be optionally augmented to *C* Ⓕ *G*. After substitution, bunching and augmentation, the primary and secondary trees of Fig.9-15 appear as shown in Fig.9-16. Comparing subtrees or tree of the node *CE* with the tree of *BCE*, it can be seen that *CE* can be optionally augmented to Ⓑ *CE*, if *C* Ⓕ *G* is taken as *CFG*. The minimal state machine which will cover an *ISSM* can now be obtained from the following theorem.

Theorem 9.6.1 The minimal state machine covering an incompletely speci-

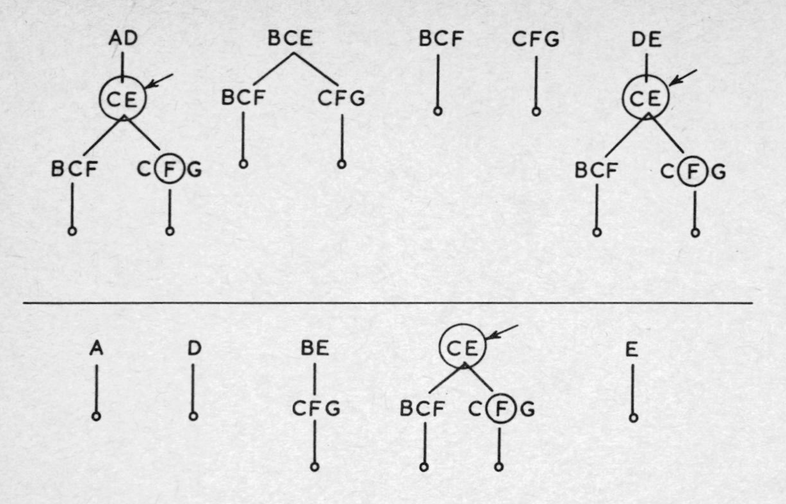

Figure 9-16 Primary and secondary trees of machine M_3 after substitution, bunching, and augmentation. The node *CE* can be optionally augmented to (*B*) *CE* if *C* (*F*) *G* is taken as *CFG*.

fied sequential machine must be a cover of its implication trees satisfying a) consistency, that is, whenever a tree is covered it must be covered in its entirety, and b) closure with minimality, that is, the trees covered will have minimum number of compatibles, and the compatibles must include all states of the original machine.

Proof: The validity of the theorem is obvious from the method of construction of the implication trees. ▲

Applying this theorem, the minimal cover of the machine M_3 is given by the following four solutions. Individual trees have been indicated by parentheses. Because of the optional augmentation, the tree of *AD* yields three solutions:

1) (*AD*, *BCE*, *BCF*, *CFG*)
2) (*AD*, *CE*, *BCF*, *CFG*)
3) (*AD*, *CE*, *BCF*, *CG*)
4) (*BE*, *CFG*), (*A*), (*D*).

Example 9.6.1 Determine a minimal machine for Machine M_2 of Fig.9-10.

Solution: The implication chart for the machine is shown in Fig.9-17. The compatibility graph and the set of *MC*s are shown in Fig.9-18. The set of *MC*s turns out to be

(*ADE*) (*ADF*) (*BDF*) (*CDE*).

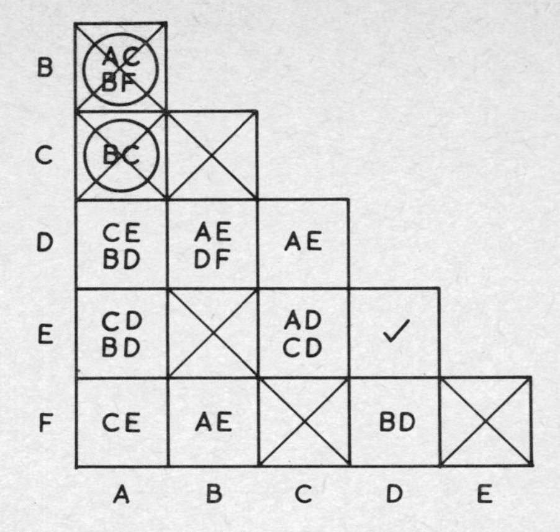

Figure 9-17 Implication chart of machine M_2.

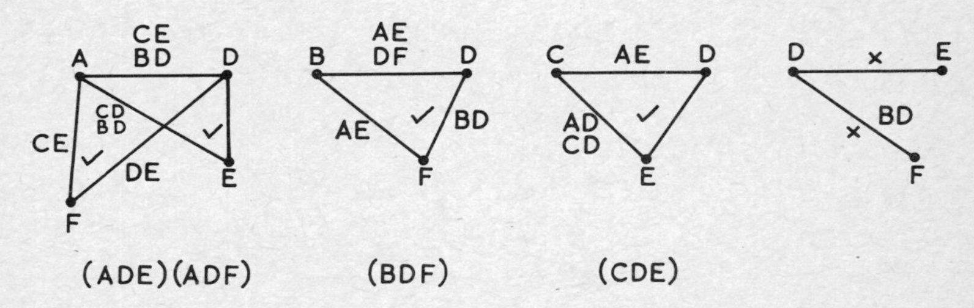

Figure 9-18 Compatibility graph and the set of *MC*'s for machine M_2.

The primary and the secondary implication trees are shown in Fig.9-19. From these the only minimal cover is as given by the tree of the *MC* (*BDF*), namely *BDF, AE, CD.*

Writing 1, 2 and 3 for *BDF, AE* and *CD* respectively, the minimal machine is as shown in Fig.9-20.

It will now be shown that the concept of an essential maximal compatible in the set of *MC*s as advocated by some authors is misleading and must be abandoned. The only way to realize a minimal closed cover from the set of *MC*s is to construct the primary and secondary implication trees for all members of the set, carry out bunching, and then to choose one or more trees satisfying Th.9.6.1.

To demonstrate the non-existence of an essential *MC* we shall give a counter example. Consider machine M_4 of Fig.9-21. Its implication chart, compatibility graph and the set of *MC*s are shown in Fig.9-22 and the associated implication trees are shown in Fig.9-23. If the definition of an essential *MC* is maintained, then the maximal compatible (*ABDF*) is essential as it is the only compatible

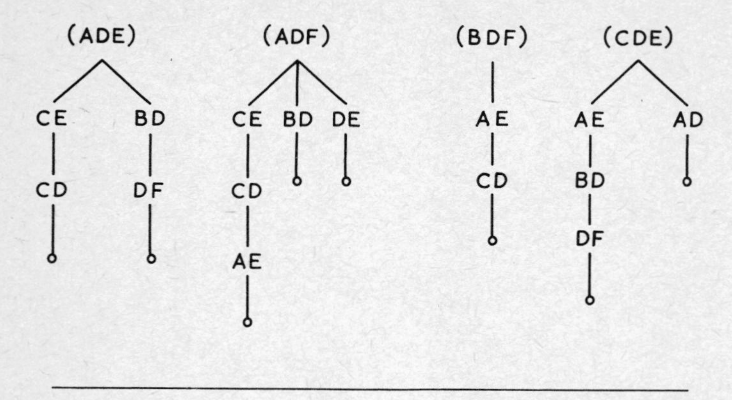

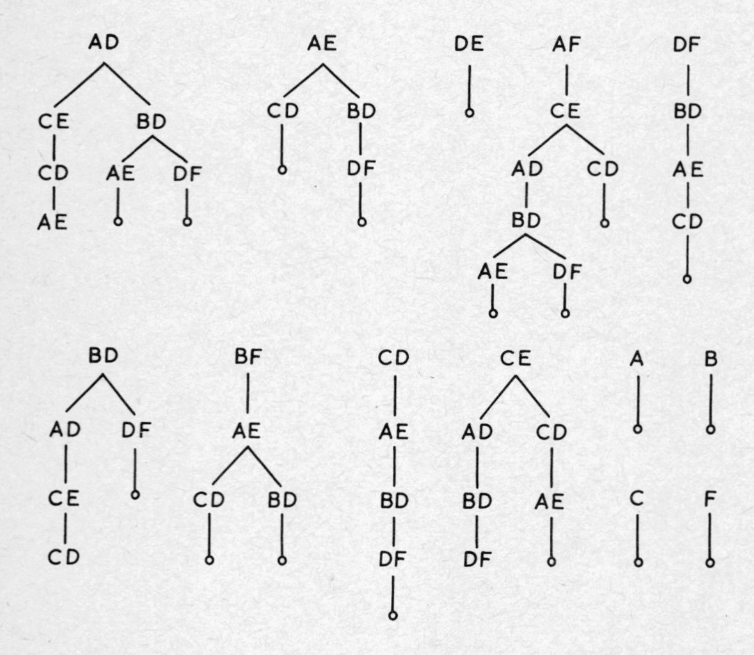

Figure 9-19 Primary and secondary implication trees for machine M_2.

	NS, z	
	x	
PS	0	1
1	2, 0	1, 0
2	3, 1	1, 0
3	2, 1	3, 0

Figure 9-20 A minimal machine covering machine M_2.

which includes state D in it. But it can be easily seen that if ($ABDF$) is retained in the covering machine, the compatibles AE, CF and CE are also to be retained. This will produce a 4 state machine which is also the upper bound. On the other hand the implication tree of CE gives a closed covering with only 3 states. From the implication trees of Fig.9-23 many other minimal covers for machine M_4 can be found. All are listed below. The parentheses identify individual trees:

1) (CE, AB, DF)
2) (ADF, CE, AB)
3) (BDF, AE, CF)
4) (ABD, AE), (CF)
5) (BD, AE), (CF).

There exists, however, an essential implication tree which may be defined as follows:

Definition 9.6.4 If a state of an *ISSM* is covered by one and only one of its implication trees, then the particular tree is an *essential implication tree.*

An essential implication tree must be included in the set of trees covering the minimal state machine.

M_4

	NS, z			
	x_1x_2			
PS	00	01	11	10
A	E, 0	–	–	–
B	–	F, 1	E, 1	A, 1
C	F, 0	–	A, 0	F, 1
D	–	–	A, 1	–
E	–	C, 0	B, 0	D, 1
F	C, 0	C, 1	–	–

Figure 9-21

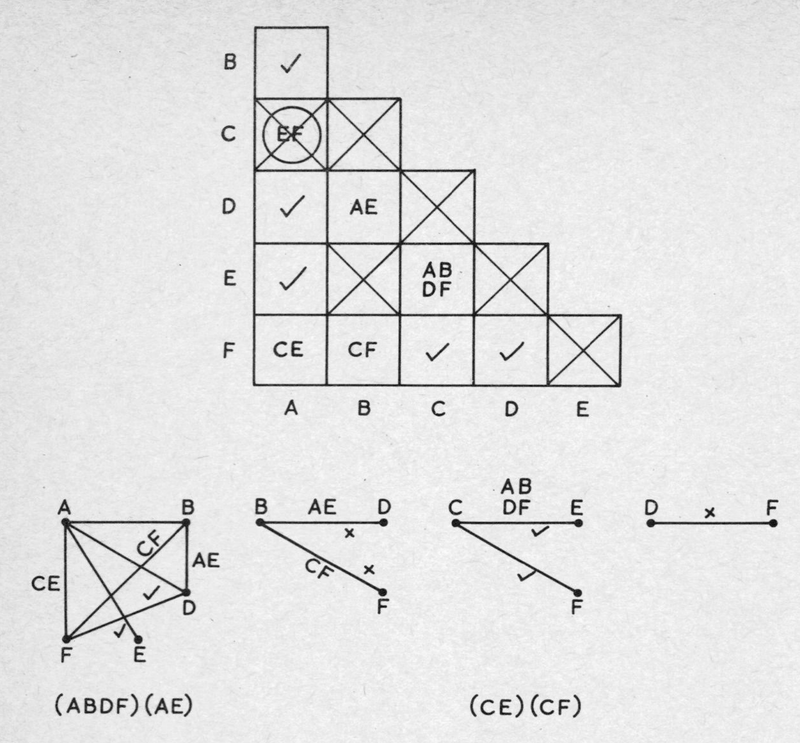

Figure 9-22 Implication chart, compatibility graph, and set of *MC*'s for the machine M_4.

The compatibility graph as has been described in the previous section will be a convenient tool in deriving the set of *MC*s when the number of states of the *ISSM* is not very large, and when the derivation is made by hand computation. For *ISSM*s with excessively large number of states it is advisable to establish the set of *MC*s with the help of a computer program carrying out exhaustive comparison for compatibility among various compatible pairs. As no *MC* can be considered essential, implication trees are constructed from each one of the *MC*s. This fact makes the method quite rigorous and the determination of the minimal machine becomes a systematic rather than a trial and error procedure. However, as the secondary trees are also to be constructed, the method becomes unwieldy if the *MC*s consist of large number of states. Hence, the procedure is recommended for small and medium size *ISSM*s. It is hoped that the method presented here will be as useful in the minimization of incompletely specified sequential machines as the map method is in the minimization of Boolean functions.

For an improved tree method see Appendix D.

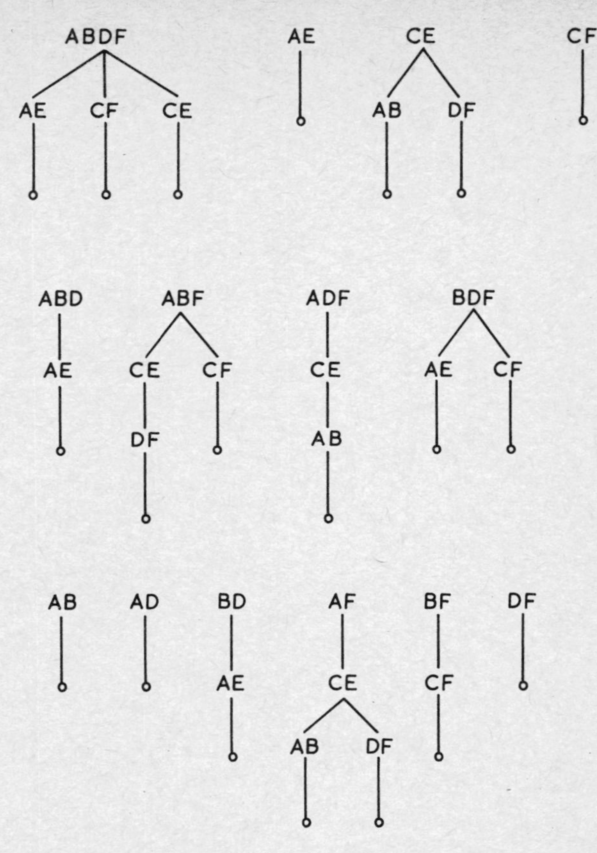

Figure 9-23 Primary and secondary implication trees for machine M_4.

9.7 SYNTHESIS

Let the various steps in the synthesis of synchronous sequential machines be discussed by considering the following machine.

Example 9.7.1 A sequential machine M_5 has one input line where 0's and 1's are being incident. The machine has to produce an output 1 only when exactly two 0's are followed by a 1 or exactly two 1's are followed by a 0. Use any assignment, and *J–K* flip-flops to synthesize the machine.

The first task is to produce a state table from this word statement. For this purpose we shall first write out a primitive state table. Let the initial state (or the 'power-on' state) of the machine be *A*. In this state the machine has not yet received any 0 or 1 at its input. So its memory is blank. This fact will be

PS (Memory)	NS (Memory), z	
	x	
	0	1
A (−)	B (0), 0	C (1), 0
B (0)	D (00), 0	C ~~E~~ (~~0~~1), 0
C (1)	B ~~F~~ (~~1~~0), 0	G (11), 0
D (00)	H (000), 0	C ~~I~~ (~~00~~1), 1
G (11)	B ~~J~~ (~~11~~0), 1	K (111), 0
H (000)	H ~~L~~ (~~0~~000), 0	C ~~M~~ (~~000~~1), 0
K (111)	B ~~N~~ (~~111~~0), 0	K ~~P~~ (~~1~~11), 0

Figure 9-24 Primitive state table for machine M_5.

recorded in the primitive state table by writing (−) after the present state *A* (Fig.9-24). For an input 0, let the machine go from state *A* to state *B*. In this state the machine remembers to have received one 0. This fact is recorded by writing (0) after *B*. Similarly for a 1 input, the machine goes to state *C* and remembers the 1. Accordingly the entry for the *A* row and $x = 1$ column is *C*(1). For both inputs the output is 0. We start the second row with *B*(0) as the present state. For the 0 input the *SM* goes from state *B* to *D* which now remembers to have received two 0's in succession. The output continues to be 0. For 1 input the *SM* goes to state *E* which has 01 in its memory. Now, by studying the requirement of the machine it can be easily concluded that no useful purpose is served by remembering the sequence 01. It is enough if the machine remembers

PS	NS, z	
	x	
	0	1
A	B, 0	C, 0
B	D, 0	C, 0
C	B, 0	G, 0
D	H, 0	C, 1
G	B, 1	K, 0
H	H, 0	C, 0
K	B, 0	K, 0

Figure 9-25 Primitive state table of machine M_5.

only the 1. Hence we strike off the 0 from the sequence 01 of the memory. This makes the state E equivalent to state C, which also remembers a single 1. Hence we replace state E by C. We next start the third row with state $C(1)$ as the present state, and complete the row by new states and memories. Every time the memory of a new state is checked. Whenever possible the bits in the memory are removed and a new state is replaced by an old state. A new row must be started for any new state encountered. In this case after the 7th row with K as the present state is completed, no new state is encountered. Thus the construction of the primitive state table is completed. This is shown in Fig.9-25. The next step is to obtain the final state table by minimizing the primitive state table. This is carried out in the tables of Fig.9-26. No reduction becomes possible.

P_0	PS	NS
1	A	11
	B	21
	C	13
	H	11
	K	11
2	D	
3	G	

P_1	PS	NS
1	A	23
	H	13
	K	21
2	B	
3	C	
4	D	
5	G	

Figure 9-26 Minimization of machine M_5.

	NS, z	
	x	
PS	0	1
1	2, 0	3, 0
2	4, 0	3, 0
3	2, 0	5, 0
4	6, 0	3, 1
5	2, 1	7, 0
6	6, 0	3, 0
7	2, 0	7, 0

Figure 9-27 State table of machine M_5.

Writing 1, 2, 3, 4, 5, 6, and 7, for *A, B, C, D, G, H* and *K* respectively, the final state table is as shown in Fig.9-27. Using the following assignment

1 : 001
2 : 010
3 : 011
4 : 100
5 : 101
6 : 110
7 : 111

we can derive the transition table of Fig.9-28 from the state table. From the transition table, the transition functions of the three flip-flops and the output function are also written in their canonical forms. The transition functions are plotted on maps. From these maps, the maps of the excitation functions are constructed following the rules as given in the table of Fig.8-28. These are shown in Fig.9-29. The excitation functions are then minimized. The *z* map, and the minimized output function are shown in Fig.9-30.

$y_1 y_2 y_3$	$Y_1 Y_2 Y_3$, Z (x = 0)	$Y_1 Y_2 Y_3$, Z (x = 1)	Cell Designation with $x y_1 y_2 y_3$
001	011, 0	010, 0	1 9
010	001, 0	100, 0	2 10
011	101, 0	010, 1	3 11
100	001, 1	110, 0	4 12
101	101, 0	010, 0	5 13
110	001, 0	110, 0	6 14
111	010, 0	111, 0	7 15
000	Not used		0 8

$Y_1(xy_1y_2y_3) = V(2,4,6,11,13,15)\,V\phi(0,8)$

$Y_2(xy_1y_2y_3) = V(1,3\text{-}7,9,10,12-15)\,V\phi(0,8)$

$Y_3(xy_1y_2y_3) = V(9-15)\,V\phi(0,8)$

$z(xy_1y_2y_3) = V(5,12)\,V\phi(0,8)$

Figure 9-28 Transition table for machine M_5.

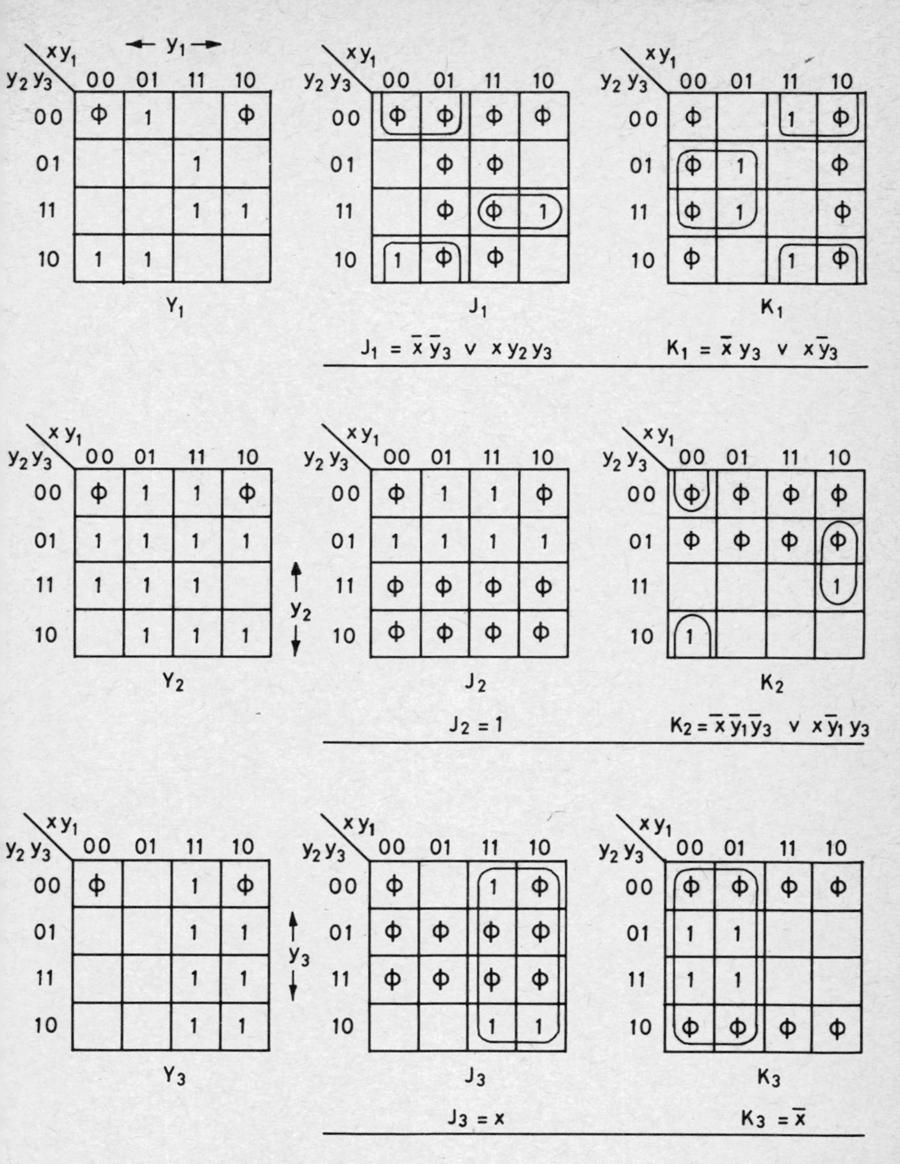

Figure 9-29 Transition and excitation functions for the three $J-K$ flip-flops.

The sequential circuit implementing the machine is now shown in Fig.9-31. The clock pulse c is the synchronizing pulse.

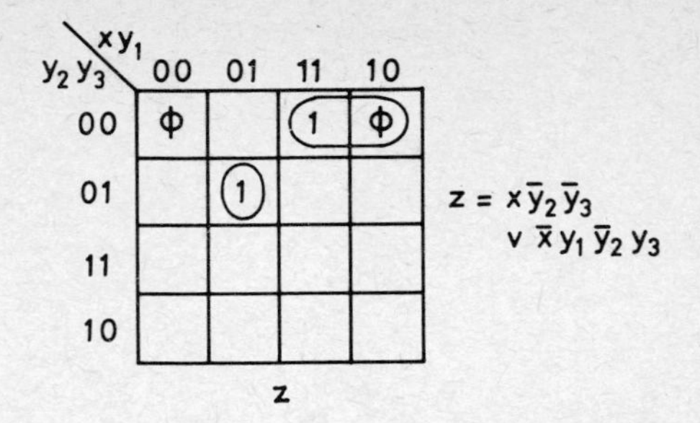

Figure 9-30 The *z*-map for machine M_5.

Example 9.7.2 Determine the state table of a minimal machine M_6 which produces and maintains an output of 1 when it detects the sequence 001. The output returns to zero only when another sequence 010 is detected.

Solution: Since the output 1 is maintained, it must be a level output. Hence the output must depend only on the state. Thus the machine can be designed as a Moore machine. Following the procedure of the previous example, the primitive state table turns out to be as shown in Fig.9-32. The row I of this table deserves special attention. The next state of row I for input 0 is J(0010) The state J, however needs to remember only the last 0. At the same time, J cannot be replaced by state B, as the output of state J must be 1, whereas that of B is 0. Similarly the state K in row I is not replaced by state C, although both of them are required to remember a single 1. The replacements in the subsequent rows will now be obvious. In the rewritten form, the primitive state table appears as shown in Fig.9-33. The next step is to minimize this table. It can be seen on this table itself that state C merges with state A, and state K with state I. After carrying out these mergers, we attempt further minimization in the tables of Fig.9-34. No further reduction becomes possible. Writing C, D, E and F for D, I, J and M respectively, the final table appears as shown in Fig.9-35.

The various steps in the synthesis of a synchronous sequential machine can be summarized as follows:

Step 1: Write a primitive state table from the word specification, after deciding if the machine will be a Mealy or a Moore machine.

Step 2: Minimize the state table as obtained in step 1.

Step 3: Determine a suitable state assignment. This problem has been discussed in Section 9.11.

Step 4: Write out the transition table.

Step 5: Decide on the type of flip-flop to be used, if not already mentioned in the original specification.

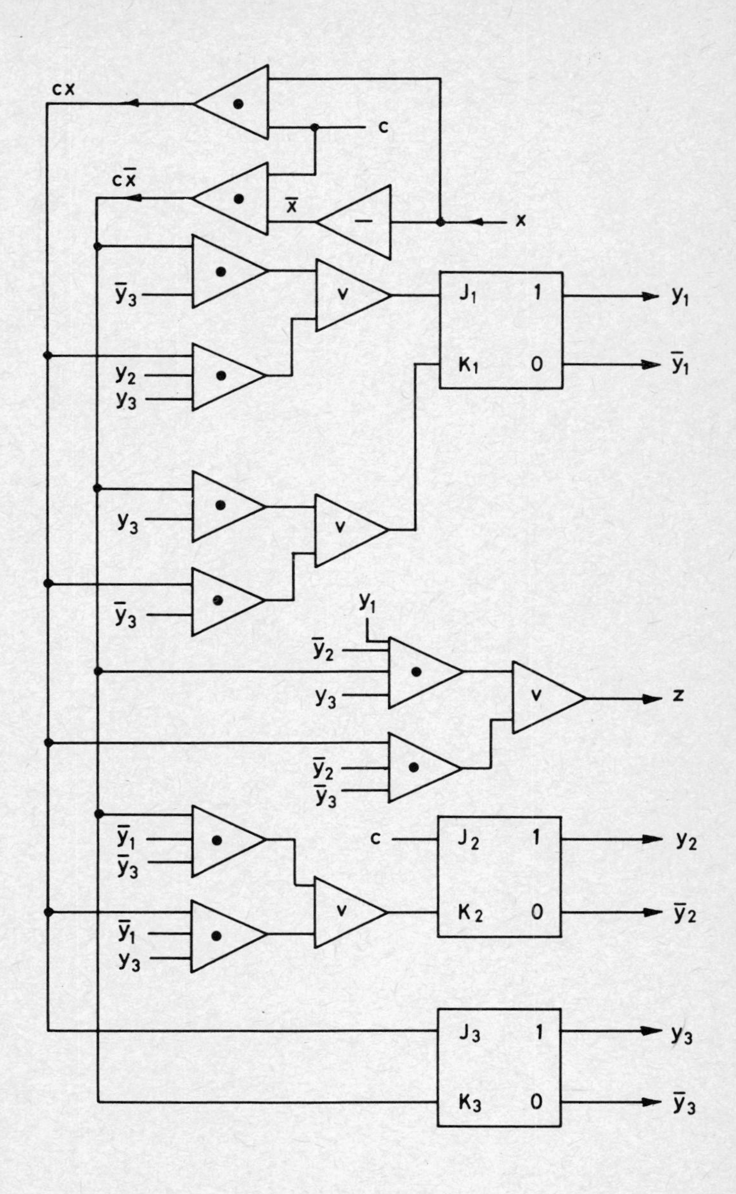

Figure 9-31 Sequential circuit implementing machine M_5.

PS (Memory)	NS (Memory)/z	
	x	
	0	1
A (–)	B (0)/0	C (1)/0
B (0)	D (00)/0	C~~E~~ (Ø1)/0
C (1)	B~~F~~ (~~1~~0)/0	C~~G~~ (~~1~~1)/0
D (00)	D~~H~~ (Ø00)/0	I (001)/1
I (001)	J (ØØ~~1~~0)/1	K (ØØ~~1~~1)/1
J (0)	J~~L~~ (Ø0)/1	M (01)/1
K (1)	J~~N~~ (~~1~~0)/1	K~~P~~ (~~1~~1)/1
M (01)	B~~Q~~ (Ø~~1~~0)/0	K~~R~~ (Ø~~1~~1)/1

Figure 9-32 Primitive state table for machine M_6.

PS	NS		Z
	x		
	0	1	
A	B	A~~C~~	0
B	D	A~~C~~	0
A ~~C~~	~~B~~	~~C~~	0
D	D	I	0
I	J	I~~K~~	1
J	J	M	1
I ~~K~~	~~J~~	~~K~~	1
M	B	I~~K~~	1

C = A
K = I

Figure 9-33 Primitive state table for machine M_6.

P_0	PS	NS
1	A	11
	B	11
	D	12
2	I	22
	J	22
	M	12

P_1	PS	NS
1	A	11
	B	21
2	D	
3	I	33
	J	34
4	M	

Figure 9-34 Minimization of machine M_6.

Step 6: Derive the excitation functions (input equations) of the flip-flops. Also minimize the output function.

Step 7: Implement the excitation and output functions by the type of logic as demanded or desired by the original specification.

9.8 STATE DIAGRAM

The various informations which are contained in a state table can also be depicted by directed graphs. Such a graphical representation of a sequential machine is called a *state diagram.* The state diagrams of the sequential machines M_5 and M_6 are shown in Fig.9-36 and 9-37 respectively. The following conventions are followed in drawing a state diagram.

1) A state is represented by a circle. For a Mealy machine only the name of the state is written within the circle. For a Moore machine the state along with its associated output is written within the circle.

	NS		
	x		
PS	0	1	z
A	B	A	0
B	C	A	0
C	C	D	0
D	E	D	1
E	E	F	1
F	B	D	1

Figure 9-35 State table of machine M_6.

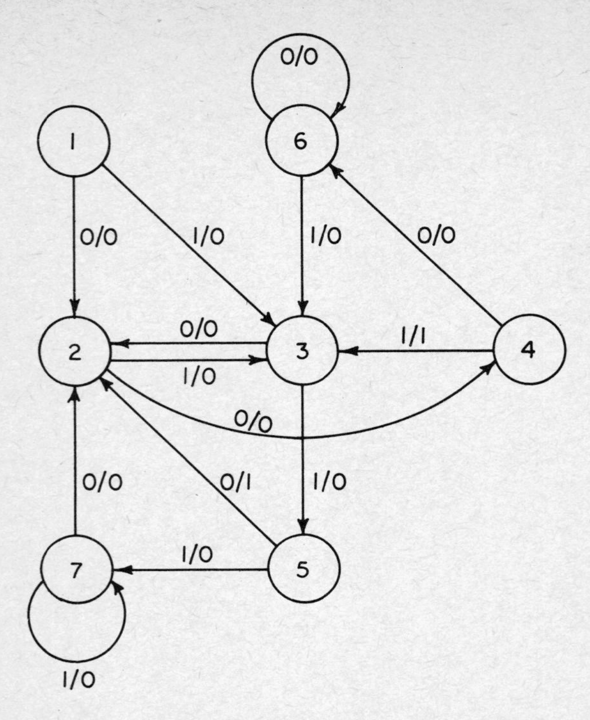

Figure 9-36 State diagram of machine M_5.

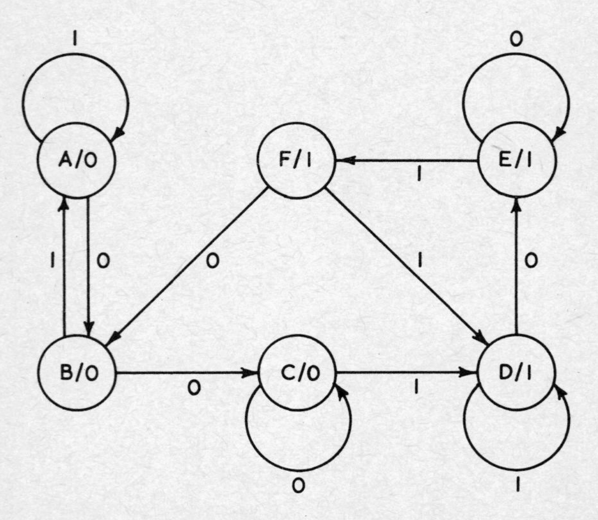

Figure 9-37 State diagram of machine M_6.

2) A directed line connecting two states shows that transition from one state to another happens. The direction of the arrow indicates the direction of the transition. The input symbol which is responsible for the transition is written on the directed line. In a Mealy machine the output is also written along with the input.

The state diagram has the advantage of pictorially depicting the behavior of a machine. It is particularly useful to recognize certain features of a sequential machine. Some of these are defined below.

Definition 9.8.1 A state of a sequential machine which cannot be entered into from any other state is called a *source* state.

Definition 9.8,2 A state of a sequential machine which does not have access to any other state is called a *sink* state.

Both the source and the sink states are known as *terminal* states. The initial state or the 'power-on' state of the *SM* of Example 9.7.1 (Machine M_5) is a source state. (See Fig.9.36.) For an example of a sink state see problem 10.16.

Definition 9.8.3 If a sequential machine can transit from every state all other states either directly or via other states, then the machine is called a *strongly connected machine.*

It is obvious that the state diagram of such a machine will reveal the input sequence that takes the machine from any one state to any other.

9.9 INITIAL STATES

An important point that must be settled in the design of a sequential machine is that of the status of the initial state. If a sequential machine must have an initial state, then whenever the machine is switched on, it must start from this particular state. Take the case of machine M_6 (Fig.9-37). In order that the machine functions properly, right from the beginning, it must go to state A, when it is turned on. If the specification desires such a function, then an *initializing circuit* must be incorporated so that every time the machine is switched on, it goes to state A. Let us consider the situation where no such initializing circuit is built in, and the machine can go arbitrarily to any state when it is turned on. Suppose it goes to state E. Then it can be easily verified that the machine gives wrong output at the beginning and continues to do so until the sequence 10 is received. But a long string of 0's and 1's can be incident at the machine before the sequence 10 is received. Obviously such a situation is not desirable, and the provision of an initializing circuit becomes imperative.

A study of the state diagram of machine M_5 (Fig.9-36) reveals a different picture. Here, if we are prepared to ignore the first three outputs, then the machine can go to any state when it is turned on. In such a case the provision of the source state 1 becomes redundant, and the machine can be designed with only 6 internal states instead of 7.

Another interesting class of sequential machines are the flip-flops. Considering the $J-K$ flip-flop as a sequential Moore machine with two input and one output lines, its state diagram can be drawn as shown in Fig.9-38. The diagram reveals that the *SM* can be turned on in any state and there is no need to ignore even a single output.

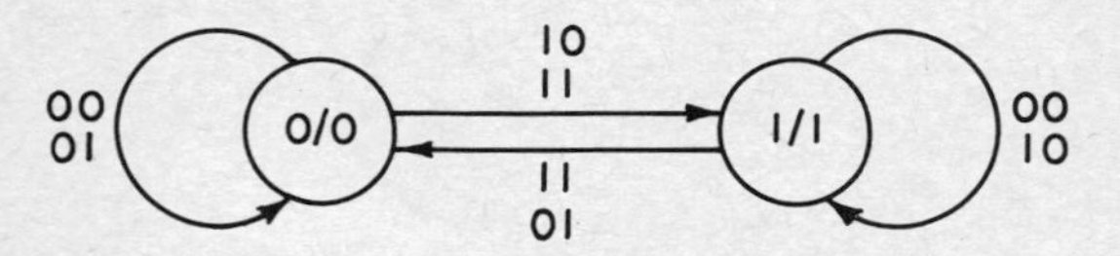

Figure 9-38 State diagram of a $J-K$ flip-flop.

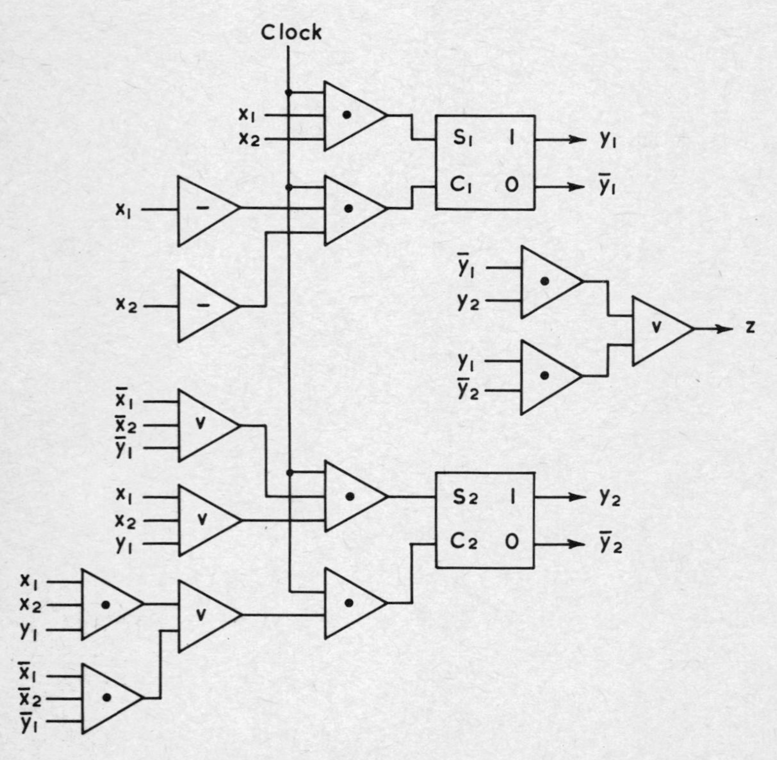

Figure 9-39 Circuit diagram of a sequential machine M_7.

9.10 ANALYSIS

Given a sequential circuit, the procedure to determine its specification in word statement is called analysis. In analyzing a circuit, we, therefore, reverse the steps of synthesis. Up to the derivation of the state table, the procedure is quite straightforward. However, to express in words what the circuit does is often not so simple. The state diagram is more helpful than the state table in this regard. Let the various steps be illustrated by working out the following example.

Example 9.10.1 Analyze the circuit of Fig.9-39, that is, form the state table, draw the state diagram, and describe in words the performance of the circuit.

Solution: The clock pulse is for synchronization, and need not be taken into account, while computing the various equations. Now, from the circuit,

$$S_1 = x_1 x_2$$

$$C_1 = \bar{x}_1 \bar{x}_2$$

$$S_2 = (\bar{x}_1 \vee \bar{x}_2 \vee \bar{y}_1)(x_1 \vee x_2 \vee y_1)$$

$$\therefore \bar{S}_2 = x_1 x_2 y_1 \vee \bar{x}_1 \bar{x}_2 \bar{y}_1$$

$$C_2 = x_1 x_2 y_1 \vee \bar{x}_1 \bar{x}_2 \bar{y}_1 = \bar{S}_2$$

$$z = \bar{y}_1 y_2 \vee y_1 \bar{y}_2.$$

All these variables are functions of the two inputs x_1 and x_2 and the two flip-flops y_1 and y_2. Hence, these are plotted on four variable maps as shown in Fig.9-40. In the excitation map for flip-flop y_1, first the excitation functions S_1 and C_1 are plotted. Depending on the values of S_1 and C_1 and knowing the property of the *SC* flip-flop, the next state function Y_1 can be plotted. This gives the $S_1 C_1 Y_1$ map. Similarly the $S_2 C_2 Y_2$ map is also obtained. These two maps correspond to the excitation tables for the two flip-flops. In the third map, the values of Y_1 and Y_2 as well as the output z are plotted. This map (the $Y_1 Y_2 z$ map) therefore gives the transition table of the circuit. From this table and also from the circuit, it can be seen that the output is independent of the input and is a function only of the present states. Thus the circuit implements a Moore machine. Naming the four internal states of the circuit as shown, the state table of the circuit can be derived. From the state table, the state diagram is drawn (Fig.9-40). Up to this point the method of analysis is quite systematic and straightforward. To obtain a word statement from the state table or state diagram is not so simple. The reader may try his skill on this machine. If he is successful it is a very good performance. If not, he may have a look at the state table of Fig.9-5 for possible help.

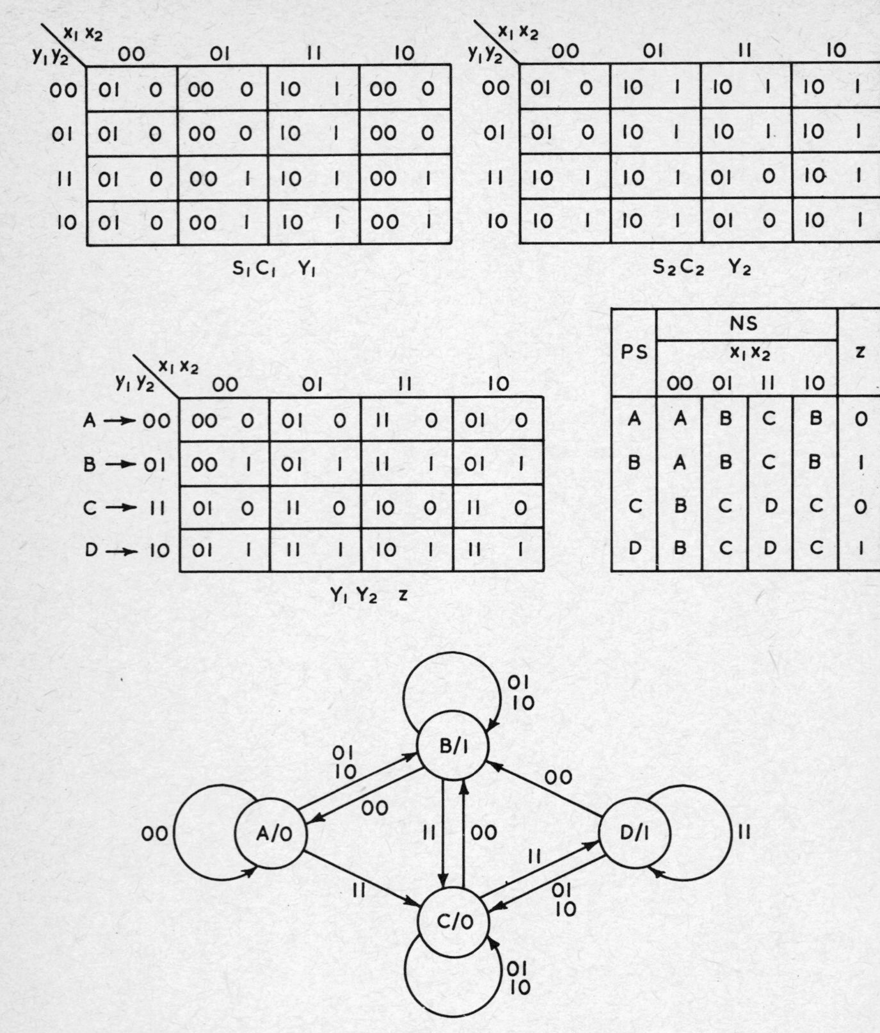

y_1y_2 \ x_1x_2	00	01	11	10
00	01 0	00 0	10 1	00 0
01	01 0	00 0	10 1	00 0
11	01 0	00 1	10 1	00 1
10	01 0	00 1	10 1	00 1

S_1C_1 Y_1

y_1y_2 \ x_1x_2	00	01	11	10
00	01 0	10 1	10 1	10 1
01	01 0	10 1	10 1	10 1
11	10 1	10 1	01 0	10 1
10	10 1	10 1	01 0	10 1

S_2C_2 Y_2

y_1y_2 \ x_1x_2	00	01	11	10
A → 00	00 0	01 0	11 0	01 0
B → 01	00 1	01 1	11 1	01 1
C → 11	01 0	11 0	10 0	11 0
D → 10	01 1	11 1	10 1	11 1

Y_1 Y_2 z

PS	NS x_1x_2 00	01	11	10	z
A	A	B	C	B	0
B	A	B	C	B	1
C	B	C	D	C	0
D	B	C	D	C	1

Figure 9-40 Excitation maps, transition table, state table, and state diagram of machine M_7.

9.11 STATE ASSIGNMENT

After the minimum row state table for a sequential machine is determined, the number of states or rows of the table decides the number of flip-flops to be used. If n is the required number of FF's to implement an SM whose number of states is S, then n is the smallest integer satisfying the following inequality,

$2^n \geqslant S.$

Now, a particular state of a machine is represented by a combination of the n flip-flops. Hence it can be seen that there are many different combinations which may be assigned to represent the states of a machine. Each assignment will yield a set of excitation functions which are to be implemented by logic gates. In order to keep the cost of the design minimum, the aim of the designer is to find the assignment which produces the most simplified excitation functions so that the cost of the implementing hardware is kept minimum. Obviously there are only a few (and more often only one) such assignments. We shall call this the *best* assignment. It will be seen that determination of this best assignment is a very difficult problem. However, there are methods by which a very nearly the best (if not the best) assignment can be found.

Consider the implementation of the sequential machine M_8 whose state table is given in Fig.9-41. As the machine has four internal states, it will require two flip-flops, say y_1 and y_2 to represent these four states. Now the manner in which the four different combinations of y_1 and y_2 are assigned to represent the four states A, B, C and D is called a particular *state assignment* (SA) for the machine.

	NS, z	
	x_1	
PS	0	1
A	A, 0	B, 0
B	C, 0	B, 0
C	D, 0	B, 0
D	A, 0	B, 1

Figure 9-41 State table of machine M_8.

Example 9.11.1 Determine the excitation functions of the delay elements implementing machine M_8 for the five different assignments in Fig.9-42.

Solution: Using Assignment I – Replacing A, B, C, D of the state table by the values of y_1 and y_2, the excitation table for the machine M_8 is as shown in Fig.9-43. In the same way Y_1 and Y_2 for other assignments can be calcu-

	I	II	III	IV	V
A	00	00	00	00	01
B	01	01	11	10	00
C	11	10	01	11	10
D	10	11	10	01	11

Figure 9-42 Five different state assignments.

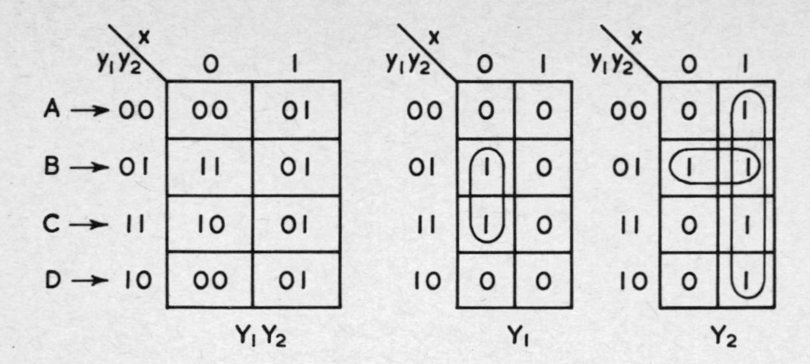

Figure 9-43 Excitation table for the machine M_8.

lated, and these are given in the tables of Fig.9-44 along with the number of diodes required if the combinational logic is implemented by diode logic. Among the first three assignments, *SA I* has the least cost function. Hence to implement the *SM* by delay elements, *SA I* must be chosen. The assignments *IV* and *V* also have the same cost function as *SA I*. This is because, these two assignments are really not distinct from *SA I*, inasmuch as *SA IV* can be obtained from *SA I* by permuting the variables y_1 and y_2 and *SA V* is obtained from *SA I* by complementing the y_2 column. Hence the interconnecting combinational logic for these three *SAs* are identical, only the inputs are different. We can now define distinct assignment.

Definition 9.11.1 Two state assignments are called *distinct* if one cannot be obtained from the other by complementing and/or permuting one or more variables.

It can easily be verified that in case of 4 internal states, there are only three distinct state assignments. These are the *SA I, II* and *III* of the previous example. The table of Fig.9-45 gives the number of distinct *SA*s for machines of up to 9 internal states.

Assg.	Y_1 (No. of diodes)	Y_2 (No. of diodes)	Total No. of diodes
I	$\bar{x}y_2$ (2)	$x \vee \bar{y}_1y_2$ (4)	6
II	$\bar{x}(\bar{y}_1y_2 \vee y_1\bar{y}_2)$ (8)	$x \vee y_1\bar{y}_2$ (4)	12
III	$x \vee \bar{y}_1\bar{y}_2$ (4)	$x \vee y_1y_2$ (4)	8
IV	$x \vee y_1\bar{y}_2$ (4)	$\bar{x}y_1$ (2)	6
V	$\bar{x}\bar{y}_2$ (2)	$\bar{x}(y_1 \vee y_2)$ (4)	6

Figure 9-44 Excitation functions and the number of diodes required to implement these functions for the five state assignments.

No. of internal states	No. of FF's required	No. of distinct SAs
2	1	1
3	2	3
4	2	3
5	3	140
6	3	420
7	3	840
8	3	840
9	4	10,810,800

Figure 9-45 Number of distinct state assignments for machines of up to 9 internal states.

It is evident from the table of Fig.9-45 that for machines of up to 4 internal states, it may be possible to derive the excitation functions for all the three distinct *SA*s and then to choose the most economical one. But as the number of states increases, the labor involved to work out each assignment will be enormous. Hence other methods must be found to arrive at the best assignment. Unfortunately up til now, there is no method which can lead to the *best* assignment. However, number of workers have expounded many methods which produce very nearly the best assignment. It should be mentioned here that the best *SA* also depends on the type of flip-flop chosen to design the sequential circuit. Let this be illustrated by the following example:

Example 9.11.2 Determine which of the three assignments has the least cost function if the machine of the previous example is implemented by trigger flip-flops.

Solution: The $Y_1 Y_2$ map and the corresponding T_1 and T_2 maps for the three assignments are shown in Fig.9-46. Accordingly, the excitation function and their costs are given in the table of Fig.9-47.

Hence assignment II turns out to be the best in this case. However, the criterion which is very often fixed to determine the best assignment is the one which produces the most economical next state function. The works of Hartmanis (1961), Stern and Hartmanis (1961), and Dolotta and McCluskey (1964) are particularly useful in obtaining a very good assignment. However, even these methods do not always produce the best assignment, and are very involved. Hence these methods are not being discussed in this text. However, we may mention the following two rules which although very simple may produce results which may not be very inferior to those obtained by more sophisticated and involved methods.

Rule 9.11.1: Two states which are previous states of a single state should be given adjacent assignments.

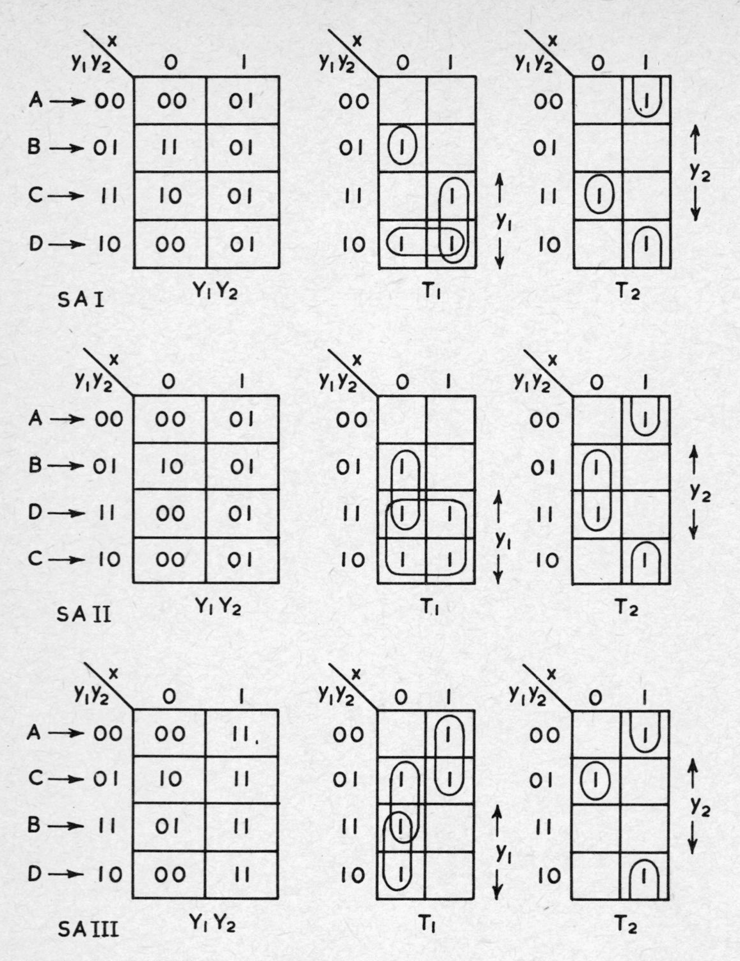

Figure 9-46 The $Y_1 Y_2$ and the corresponding T_1 and T_2 maps for the three assignments.

By *adjacent assignments* is meant that the two assignments differ in one variable only. Thus the assignment 010 is adjacent to the assignments 110, 000 and 011. It can be easily verified that when two states having the same next state are given adjacent assignments, the next state function is so mapped that a sub-cube of 1's or 0's become available in the column. This tends to minimize the next state function in such a way that the resultant *MSP* form may contain a smaller number of literals.

Rule 9.11.2: Two states which are the next states of a single state should be given adjacent assignments.

Assg.	T_1 (No. of diodes)	T_2 (No. of diodes)	Total No. of diodes
I	$x\bar{y}_1y_2 \vee xy_1 \vee y_1y_2$ (10)	$\bar{x}y_1y_2 \vee x\bar{y}_2$ (7)	17
II	$y_1 \vee \bar{x}y_1$ (4)	$\bar{x}y_2 \vee x\bar{y}_2$ (6)	10
III	$x\bar{y}_1 \vee \bar{x}y_2 \vee \bar{x}y_1$ (9)	$xy_2 \vee \bar{x}\bar{y}_1y_2$ (7)	16

Figure 9-47 Excitation functions and cost functions of trigger flip-flops for the three state assignments.

It can be easily verified that this rule produces such a mapping that there is the possibility of having subcubes of 1's or 0's in the rows of the next-state function map.

It must be mentioned that while trying to apply rules 1 and 2 to a particular state table, it may generate conflicting demands. In such cases, the best compromise is considered to be the best assignment.

Example 9.11.3 Determining a good *SA* for the machine M_8.

Solution: The state table of the machine is shown in Fig.9-41. The table of Fig.9-48 is derived from the state table.

Previous states	Present states	Next states
AD	A	AB
ABCD	B	CB
B	C	DB
C	D	AB

Figure 9-48

According to rule 1, the state pairs in column 1 should be given adjacent assignments, and according to rule 2, those in column 3 are to be given adjacent assignments. An assignment can be worked out where the adjacent states are as follows:

ABCDA.

From the map of Fig.9-49, the assignment which satisfies the required adjacencies is as follows:

A : 00
B : 01
C : 11
D : 10.

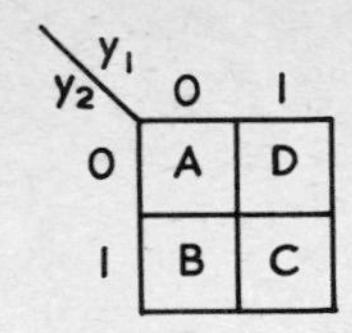

Figure 9-49 The adjacency map of the states.

This is the same as *SA I* which turned out to be the best assignment for delay elements. But this assignment is certainly not the best if the machine is to be implemented by trigger *FF*'s. It would, therefore, appear that the various methods to produce a good state assignment will be inadequate if they only produce the best next state function. They should also somehow take the type of flip-flop into consideration.

REFERENCES

Armstrong, D. B.: "A programmed algorithm for assigning internal codes to sequential machine", *IEEE Trans. Electron. Computers,* **EC-11**, No.5, pp.466–72, August 1962.

—— : "On the efficient assignment of internal codes to sequential machines", *IEEE Trans. Electron Computers,* **EC-11**, No.5, pp.611–622, Oct.1962.

Biswas, N. N.: "State minimization of incompletely specified sequential machines" *IEEE Trans. Electron. Computers,* **C-23**, No.1, pp.80–84, Jan.1974.

Dolotta, T. A. and E. J. McCluskey: "The coding of internal states of sequential circuits", *IEEE Trans. Electron Computers,* **EC-13**, No.5, pp.549–562, Oct.1964.

Grasselli, A. and F. Luccio: "A method for minimizing the number of internal states in incompletely specified sequential networks", *Trans. Electron. Computers,* **EC-14**, pp.350–359, June 1965.

Hartmanis, J.: "On the state assignment problem for sequential machines I", *IRE Trans. Electron Computers,* **EC-10**, pp.157–165, June 1961.

Huffman, D. A.: "The synthesis of sequential switching circuits", *J. Franklin Institute,* **257**, pp.275–303, March–April 1954.

Karp, R. M.: "Some techniques of state assignment for synchronous sequential machines", *IEEE Trans. Electron Computers,* **EC-13**, No.5, pp.507–518, Oct.1964.

Kella, J.: "State minimization of incompletely specified sequential machines", *IEEE Trans. Computers,* **C-19**, pp.342–348, April 1970.

Kohavi, Z.: *Switching and Finite Automata Theory,* McGraw-Hill, New York, 1970.

McCluskey, E. J.: *Introduction to the Theory of Switching Circuits,* McGraw-Hill, New York, 1965.

McCluskey, E. J. and S. H. Unger: "A note on the number of internal assignments for sequential switching circuits", *IRE Trans. Electron. Computers,* **EC-8**, No.4, pp.439–440, Dec.1959.

Mealy, G. H.: "A method for synthesizing sequential circuits", *BSTJ,* **34**, pp.1045–1079, Sept.1955.

Meisel, W. S.: "A note on internal state minimization in incompletely specified sequential networks", *IEEE Trans. Electron. Computers,* **EC-14**, pp.508–509, August 1967.
Moore, E. F.: "Gadanken-experiments on sequential machines", pp.129–153, *Automata Studies,* Princeton University Press, Princeton, N.J. 1956.
Paull, M. C. and S. H. Unger: "Minimizing the number of states in incompletely specified sequential switching functions", *IRE Trans. Electron Computers,* **EC-8**, No.3, pp.356–367, Sept.1959.
Sterns, R. E. and J. Hartmanis: "On the state assignment problem for sequential machines, II", *IRE Trans. Electronic Computers,* **EC-10**, No.4, pp.593–603, Dec.1961.

PROBLEMS

9.1 Convert the following Mealy machines into Moore machines:

(a)

	NS, z	
	x	
PS	0	1
A	B, 1	A, 0
B	A, 0	C, 1
C	C, 0	A, 0

(b)

	NS, z	
	x	
PS	0	1
A	A, 0	B, 0
B	A, 0	C, 0
C	A, 0	D, 1
D	A, 0	B, 1

(c)

	NS, z			
	x_1x_2			
PS	00	01	11	10
A	A, 0	B, 1	B, 0	C, 1
B	B, 0	C, 1	D, 0	A, 1
C	C, 0	D, 1	A, 1	B, 1
D	D, 0	B, 0	C, 0	D, 0

9.2 Minimize the following state tables:

(a)

PS	NS, z: x = 0	NS, z: x = 1
A	B, 0	C, 0
B	C, 1	D, 0
C	F, 0	A, 0
D	F, 1	G, 1
E	B, 1	D, 1
F	A, 1	G, 0
G	B, 1	E, 1

(b)

PS	NS, z: x_1x_2 = 00	01	11	10
A	A, 0	C, 1	F, 0	G, 1
B	B, 0	D, 1	F, 0	H, 1
C	C, 0	G, 0	H, 0	F, 1
D	D, 0	A, 0	B, 0	F, 1
E	E, 0	B, 0	H, 0	F, 1
F	F, 0	E, 1	A, 1	H, 0
G	G, 0	E, 1	F, 0	B, 1
H	H, 0	D, 1	F, 0	A, 1

9.3 Minimize the following state tables:

(a)

PS	NS, z: x = 0	NS, z: x = 1
A	C, –	E, 1
B	D, 0	–
C	E, 1	F, –
D	F, 1	E, 0
E	A, –	C, 1
F	B, 0	D, –

(b)

PS	NS, z: x = 0	NS, z: x = 1
A	B, 0	C, –
B	C, –	D, 0
C	F, 0	A, 0
D	F, –	–
E	–, 1	D, –
F	A, –	G, 0
G	B, –	–, 1

(c)

PS	NS, z: x_1x_2 = 00	01	11	10
A	A, 0	–	E, –	B, 1
B	E, –	C, 1	B, –	–
C	–	B, 0	–, 1	D, 0
D	A, 0	–	F, 1	B, –
E	B, 0	–	B, 0	–
F	–	C, 1	–, 0	C, 1

(d)

PS	NS, z: x_1x_2 = 0	1	3	2
A	E, 0	–	–	–
B	–	F, 1	E, 1	A, 1
C	F, 0	–	A, 0	F, 1
D	–	–	A, 1	–
E	–	C, 0	B, 0	D, 1
F	C, 0	C, 1	–	–
G	E, 0	–	–	A, 1

(e)

PS	NS, z				
	$x_1x_2x_3$				
	0	1	2	4	7
A	A, –	B, 0	G, 0	–, 1	–, 1
B	A, –	–	–, 1	E, 0	C, 0
C	A, –	–	D, 0	–, 1	–
D	A, 0	–	–	A, 0	–
E	A, –	–	F, 0	–	–, 1
F	A, 0	–	–	–	A, 0
G	A, –	–, 1	–	–, 1	H, 0
H	A, –	I, 0	–	J, 0	–
I	A, 0	–	–	A, 0	–
J	A, 0	–, 0	–	–	–

9.4 Construct minimum-row state tables for the following synchronous sequential machines with one input and one output lines:

a) Produces an output 1, only when the input sequence has exactly one 0 followed by two 1's, that is, for the input sequence 01000110110, the output sequence will be 00000000010.

b) Produces an output 1, whenever any of the sequences 000, 010, and 111 is detected, e.g. for the input sequence, 101000111, the output sequence is 000101001.

9.5 Construct a minimum row state table for a synchronous sequential machine with one input and one output line, which produces an output 1 whenever any of the sequences 0001, 1000, and 1001 is detected. Once a sequence is detected no part of it can be taken to form part of another sequence.

9.6 Implement the above machine with S–C flip-flops and AND-OR logic gates, using any assignment.

9.7 A synchronous sequential machine has two input lines and one output line. Whenever $x_1 = 1$, the machine starts looking for the sequences 010, 110, and 111 incident on the x_2 line. Whenever any one of these sequences is detected, the machine produces an output 1. The output is maintained at 1 until the next 1 is detected on the x_1 line. Construct the minimum row state table of the machine.

9.8 Implement the above machine with J–K flip-flops and AND-OR logic gates, using a reasonably good assignment.

9.9 A synchronous sequential machine has two input lines, and one output line. The machine produces an output 1, whenever the pair of bits on the x_1 and x_2

lines is identical with the pair immediately preceding it. Implement the machine with $S-C$ FF's and AND-OR logic gates. Show the clock pulse on the circuit. Should the machine have an initial state? Can the initial state be ignored under any circumstances?

9.10 Implement the machine of the above problem with $S-C$ FF's and only NAND gates. Show the connection of the clock line on the circuit.

9.11 A synchronous sequential machine has one input and one output line. The digits 0 through 9 in 8-4-2-1 binary code along with a parity bit (even) come serially at the input, the least significant bit coming first, and the parity bit last. The output z is 1 whenever there is an error in the incoming digit. Derive a minimum row state table for the machine.

9.12 Draw the state diagram of a synchronous sequential machine to detect error in a 2-out-of-5 code data system operating in the serial mode. Is the machine strongly connected?

9.13 Find a reasonably good state assignment for the reduced state table of problem 9.2(a). Implement the machine with this SA, first with $S-C$ and then with T FF's, and AND OR gates. Which of the two types of FF turns out to be more economical on the basis of gate count only?

9.14 Analyze, derive the state tables, and draw the state diagrams of the sequential circuits of Fig.8-29(B), and 8-30(B). Assume a clock pulse (not shown in the figures) for the purpose of synchronization.

Chapter 10

ASYNCHRONOUS SEQUENTIAL MACHINES

10.1 INTRODUCTION

As the name suggests, the sequential machines which are not synchronous are asynchronous machines. The circuits of these machines, therefore, do not have any clock or synchronizing pulse. Consequently, no interval is specifically earmarked during which the input should change or during which the circuit should respond to the change in input. An asynchronous circuit can be thought of being ever alive and responsive to the situation brought about by the change of input. Keeping this essential nature of difference, an asynchronous sequential machine is operated mainly in two modes, fundamental mode and pulse mode.

10.2 FUNDAMENTAL MODE OPERATION

It is known that whenever a change in input occurs, certain changes follow in the memories and other associated circuits. These changes may again effect further changes, and so on. Thus, if it is so stipulated that an input symbol will not change until the machine undergoes the entire chain of changes so as to reach a state where no further change takes place, then the machine is said to operate in the *fundamental mode.*

This stipulation and the absence of clock pulses give rise to another interesting restriction on the inputs. Let us consider a machine having two inputs x_1 and x_2. The value of 0 of x_1 and x_2 is indicated by zero voltage, and the value of 1 is indicated by a pulse of any duration. Let us now envisage a situation where the values of x_1x_2 changes from 00 to 11, that is, both the inputs have changed simultaneously. This is depicted in the left half of the graph of Fig.10-1. In the right half of the graph the x_2 pulse has been delayed with respect to the x_1 pulse. This is quite possible, and in fact, is very likely to happen as the x_1 and x_2 pulses may have to travel through different paths introducing unequal delays. As has been

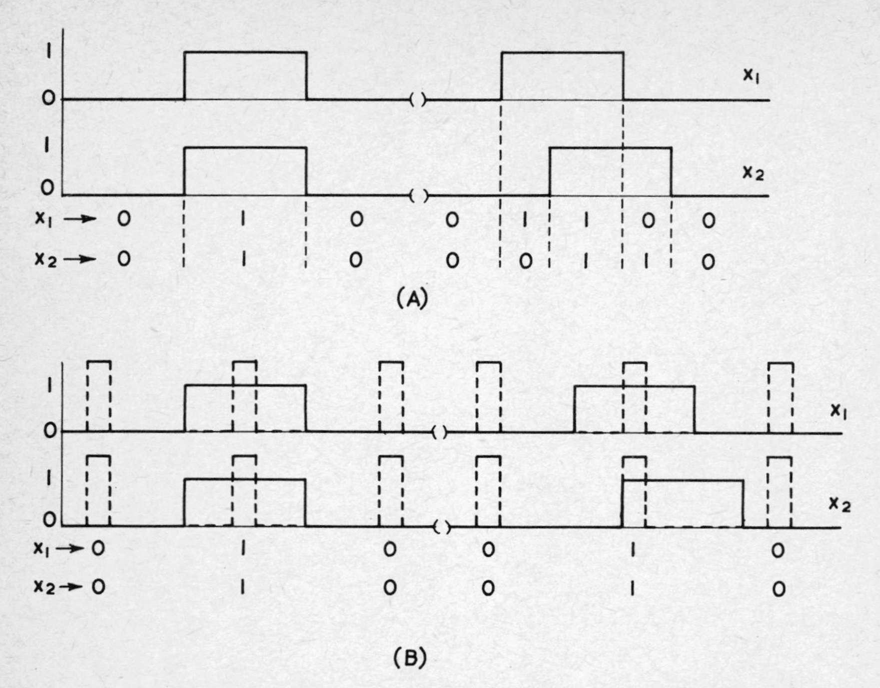

Figure 10-1 Interpretation of input pulses by (A) an asynchronous machine, and (B) a synchronous machine.

shown in Fig.10-1(A), an asynchronous machine will interpret the inputs as changing from 00 to 10, to 11, to 01, and then finally to 00 again. Thus we find that unless the ideal case as depicted by the left half of Fig.10-1 can be guaranteed for all the time, the machine *thinks* that only one change in input occurs at a certain interval. Hence this constraint is deliberately accepted as a precondition for the design of asynchronous machines. Figure 10-1(B) shows how in a similar situation a synchronous machine, due to the sampling operation of the clock pulses, cannot distinguish between the situations depicted in the left and right halves of Fig.10-1.

The fundamental mode operation also gives rise to the concept of stable states. It has been mentioned earlier that in this mode of operation no change in input is permitted unless the entire chain of changes that is initiated by the present input is completed. Let this be discussed in relation to a machine M_1, whose state table is shown in Fig.10-2(A). Let the present state of the machine be A and the input be 00. The machine, therefore, remains in state A. Hence the next state entry A in the row A and column 00 is a stable state, and is indicated by circling the state. If now, the input changes from 00 to 10, the machine goes to state C. The next state C, soon becomes the present state, and in response to the input 10, goes to state D, and then from state D to state B. However, for the

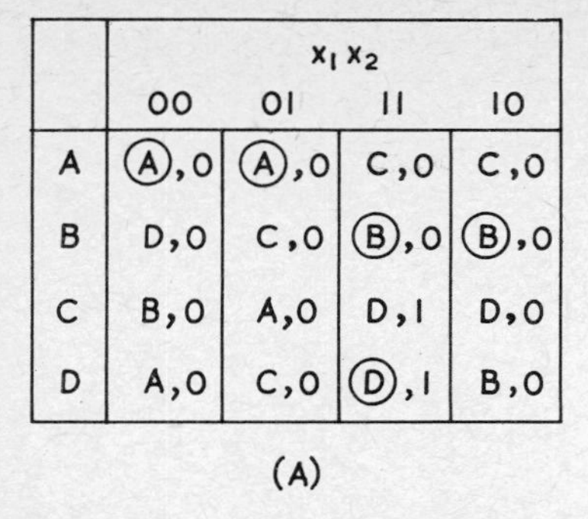

	$x_1 x_2$			
	00	01	11	10
A	(A),0	(A),0	C,0	C,0
B	D,0	C,0	(B),0	(B),0
C	B,0	A,0	D,1	D,0
D	A,0	C,0	(D),1	B,0

(A)

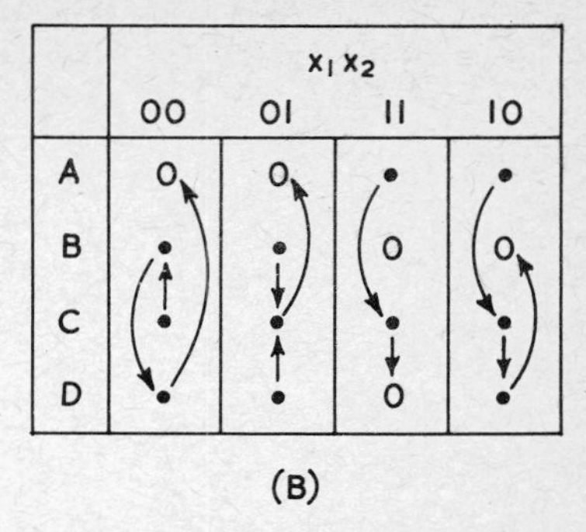

(B)

Figure 10-2 (A) State table of machine M_1, and (B) transition diagram of machine M_1.

present state B and input 10, the next state is also B, and therefore no further state change takes place and the machine stabilizes in this state. The stable states can be defined as follows.

Definition 10.2.1 In an asynchronous machine operating in fundamental mode, a state which the machine does not leave until the input changes is known as a *stable state*. In the state table such states are those next-states which are the same as the present states.

It must be apparent by now that for each input column the machine has definite ways of reaching a stable state from an unstable state.

Definition 10.2.2 A state table where arrows are drawn to indicate the various transitions from unstable to stable states is known as a *transition diagram*.

The transition diagram for machine M_1 is shown in Fig.10-2(B). Here the unstable states are indicated by dots, and the stable states, by circles. Let the various special problems encountered in a fundamental mode asynchronous sequential machine be discussed as they come up while we attempt to synthesize the following sequential machine.

> **Example 10.2.1** An asynchronous sequential machine M_2 operating in fundamental mode has two input lines x_1 and x_2, and one output line z. The output z is 0 whenever x_2 is 0. With the first change in x_1 occurring while x_2 is 1, z becomes 1, and remains 1 until x_2 returns to 0. Design a relay circuit to implement the machine.

The first step is to write a primitive state table. We follow the same procedure as for the synchronous machine, remembering the difference that a 0 and a 1 in the memory indicates the absence and presence respectively of a pulse. The unspecified entries occur in the state table (Fig.10-3) due to the fact that both inputs cannot change simultaneously. While rewriting the primitive state table, the stable states are circled. Another feature of the rewritten primitive state

PS	NS, z			
	x_1 0, x_2 0	x_1 0, x_2 1	x_1 1, x_2 1	x_1 1, x_2 0
A $\binom{0}{0}$	A $\binom{0}{0}$, 0	B $\binom{0}{1}$, 0	—	C $\binom{1}{0}$, 0
B $\binom{0}{1}$	A $\binom{0}{0}$, 0	B $\binom{0}{1}$, 0	D $\binom{01}{11}$, 1	—
C $\binom{1}{0}$	A $\binom{0}{0}$, 0	—	E $\binom{\not{x}1}{\emptyset 1}$, 0	C $\binom{1}{0}$, 0
D $\binom{01}{11}$	—	F $\binom{\emptyset 10}{\not{x}11}$, 1	D $\binom{01}{11}$, 1	C $\binom{1}{0}$, 0
E $\binom{1}{1}$	—	F$\not{\emptyset}$ $\binom{10}{11}$, 1	E $\binom{1}{1}$, 0	C $\binom{1}{0}$, 0
F $\binom{10}{11}$	A $\binom{0}{0}$, 0	F $\binom{10}{11}$, 1	D$\not{H}$ $\binom{\not{x}01}{\not{x}11}$, 1	—

Figure 10-3 Primitive state-cum-memory table of machine M_2.

table (Fig.10-4) is that no output has been specified for the unstable states. This is because the machine remains in an unstable state only during transition and its output is of no real consequence. The primitive state table shows that the machine is an incompletely specified sequential machine. It is now minimized. Figure 10-5 shows the implication chart, compatibility graph, the set of *MC*'s and the implication trees. The trees of *AB, CE* and *DF* are essential and also form a closed cover. Hence the state table can be minimized to a three row table as shown in Fig.10-6. Two relays Y_1 and Y_2 are required to implement the three row state table. Figure 10-7 shows the transition table for the assignments as shown.

PS	NS, z (x_1x_2) 00	01	11	10
A	(A), 0	B	–	C
B	A	(B), 0	D	–
C	A	–	E	(C), 0
D	–	F	(D), 1	C
E	–	F	(E), 0	C
F	A	(F), 1	D	–

Figure 10-4 Primitive state table of machine M_2.

Before we proceed to implement the table of Fig.10-7, let us examine the table more critically.

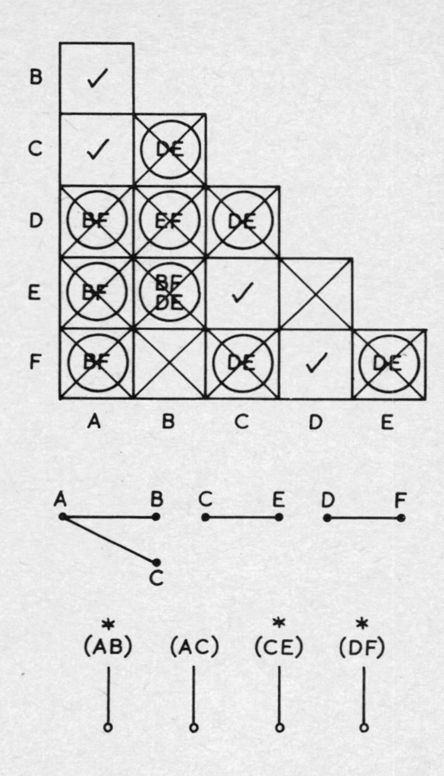

Figure 10-5 Implication chart, compatibility graph, set of *MC*'s, and the implication trees for machine M_2.

PS	NS, z			
	x_1x_2			
	00	01	11	10
AB → 1	(1), 0	(1), 0	3, –	2, –
CE → 2	1, –	3, –	(2), 0	(2), 0
DF → 3	1, –	(3), 1	(3), 1	2, –

Figure 10-6 State table of machine M_2.

y_1y_2 \ x_1x_2	00	01	11	10
1 → 00	(00), 0	(00), 0	11*, –	01, –
2 → 01	00, –	11, –	(01), 0	(01), 0
3 → 11	00, –	(11), 1	(11), 1	01, –
10	—	—	—	—

Y_1Y_2, Z

Figure 10-7 Transition table for machine M_2.

10.3 RACES

Let the circuit be in the stable state (00) with the inputs x_1x_2 as 01. Now let the inputs change to 11. According to the transition table the circuit is to enter the unstable state 11, and then become stable in the same state (11) with inputs as 11. However, in (00) state, both relays are in the released condition, and in the (11) state, both relays are in the operated condition. Hence the transition from (00) to (11) means change in state for both the relays. This means a 'race' condition between the two relays. If both relays operate at the same time, the race ends in a draw, and there will be no problem. But it is very rare that the two relays will have *exactly the same operating time*. Consequently it is more likely that one of the relays will win the race, that is, it will operate earlier than the other. If Y_1 wins the race, then as soon as Y_1 has operated, the circuit goes to state 10. This is however not a stable state. Hence after sometime relay Y_2 also operates and the circuit goes to the stable state (11). But in case Y_2 wins the race, the circuit goes to the state 01. As this is a stable state, the circuit will continue to remain in this state until the input changes. Hence the circuit fails to achieve the desired transition to the stable state (11). Such a situation is called a critical race condition and must be avoided to insure the proper functioning of the circuit.

Definition 10.3.1 If in the transition table of a fundamental mode asynchronous sequential machine, the next state entry differs from the present state by more than one variable, then the column of the next state contains a *race* track. If the column has only one stable state, then the race is *non-critical.* If, on the other hand, the column has two or more stable states then the race is *critical.*

One way to overcome the problem of critical race is to 'fix' the race, so that one particular relay will *always* win the race. In the above example the circuit will function properly, if the operating time of the two relays are so designed that relay Y_1 operates earlier than the relay Y_2.

10.4 VALID ASSIGNMENTS

The approach of fixing the race creates additional design problems to those who manufacture the components. It is always preferable to use as many identical components as possible. Hence, a better way to solve the problem of critical races is by avoiding it altogether by proper choice of secondary assignments. Moreover, this can be handled by the logic designer himself.

Definition 10.4.1 An assignment of secondary variables which eliminates critical races in the state table itself, is known as a *valid assignment.*

It is obvious that in a valid assignment the transition from one state to another is to be effected by a change in *single* variable only. This can be achieved by making two transiting states adjacent. So, first of all, an adjacency diagram is drawn noting the various transitions between the states. For the state table of Fig.10-6 row 1 shows that 1 and 3 must be adjacent as there are two stable states in the column of 11. Hence in the adjacency diagram (Fig.10-8) 1 and 3 are joined by a solid line. Row 2 shows that state 2 should be adjacent to 1 to avoid a non-critical race, and must be adjacent to 3 to avoid a critical race. These are indicated by the dotted line between 2 and 1, and solid line between 2 and 3. Row 3 shows that state 3 should be adjacent to states 1 and 2 to avoid non-critical races. However, these requirements have already been incorporated in

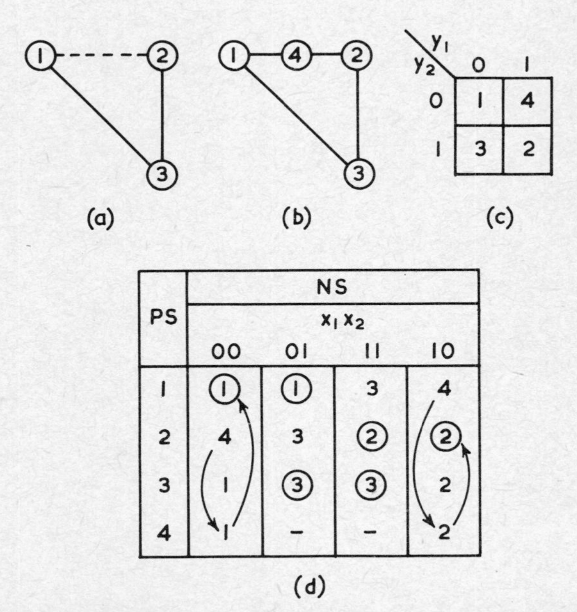

PS	NS x_1x_2 00	01	11	10
1	(1)	(1)	3	4
2	4	3	(2)	(2)
3	1	(3)	(3)	2
4	1	–	–	2

Figure 10-8 The adjacency diagram and the modified state table of machine M_2 after the introduction of a 4th state.

y_1y_2 \ x_1x_2	00	01	11	10
1 → 00	00 , 0	00 , 0	01 , –	10 , –
3 → 01	00 , –	01 , 1	01 , 1	11 , –
2 → 11	10 , –	01 , –	11 , 0	11 , 0
4 → 10	00 , –	– , –	– , –	11 , –

Y_1Y_2 , z

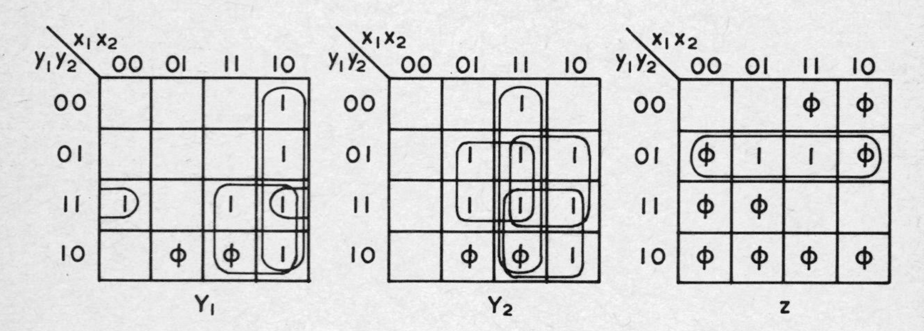

Figure 10-9 Transition table, excitation and output functions of machine M_2.

the adjacency diagram. The adjacency diagram shows that in order to avoid only critical races states 1 and 2 are to be adjacent to 3, whereas to avoid the non-critical race, states 1 and 2 must also be adjacent to one another. But it is not possible to have a two variable assignment where three states can be mutually adjacent to each other. However, in a two variable assignment there are four states. Hence the spare 4th state can be utilized to make two states 'virtually adjacent' *via* this extra state. Introducing this 4th state the adjacency diagram gets modified as shown in Fig.10-8(b). The assignment satisfying the adjacency is shown in Fig.10-8(c), and the modified state table in Fig.10-8(d). The modified state table takes into account the fact that the transition between states 1 and 2 takes place through an intermediate transition via the unstable state 4. We are now ready to proceed to complete the design. Replacing the states of Fig.10-8 by their assignments we get the transition table as shown in Fig.10-9, which also shows the individual maps for Y_1, Y_2 and z. The excitation functions for Y_1 and Y_2 and the output function z turn out to be

$$Y_1 = x_1\bar{x}_2 \vee x_1y_1 \vee \bar{x}_2y_1y_2$$

$$= x_1(\bar{x}_2 \vee y_1) \vee \bar{x}_2y_1y_2$$

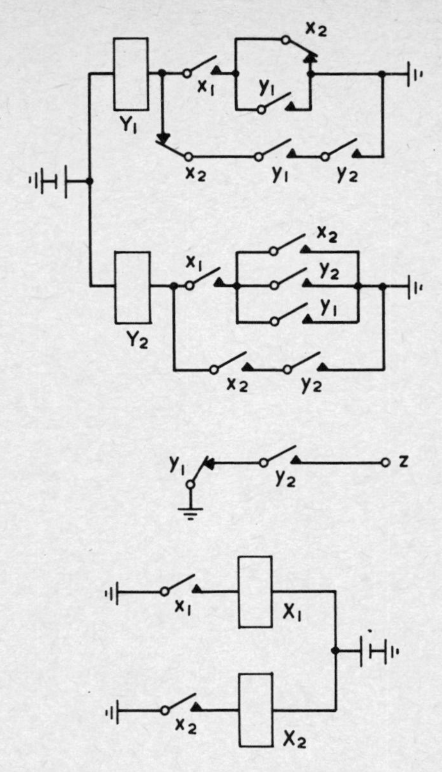

Figure 10-10 The relay circuit implementing machine M_2.

$$Y_2 = x_1x_2 \vee x_1y_2 \vee x_2y_2 \vee x_1y_1$$
$$= x_1(x_2 \vee y_2 \vee y_1) \vee x_2y_2$$
$$z = \bar{y}_1y_2.$$

The relay circuit according to these equations is shown in Fig.10-10. The reader may verify if the circuit functions correctly as per the word statement of the problem.

Although a 4-row state table needs only 2 secondary variables to have an assignment, two variables may not always be adequate to produce a *valid assignment.* Only if the adjacency diagram of the four states turns out to be as given in Fig.10-11, then a valid assignment is possible with two secondary variables. For a machine M_3 with the state table as given in Fig.10-12, the adjacency diagram is as shown in Fig.10-13(a). Here at least two additional states need to be introduced in the adjacency diagram. Since there are six states, three secondary variables are needed. A valid assignment is shown in the map of Fig.10-13(c). The modified state table is shown in Fig.10-14. Here states E and F are unstable states.

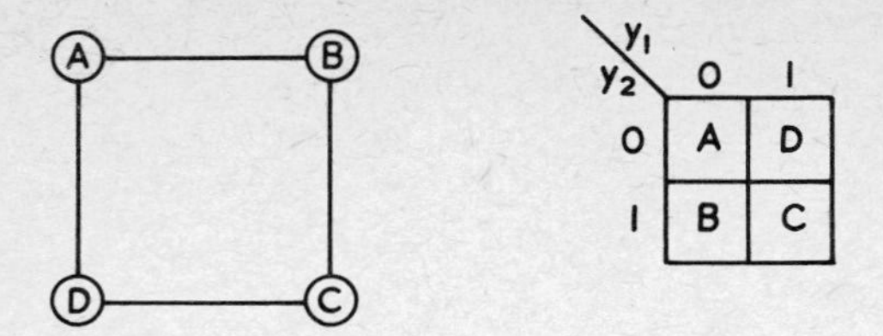

Figure 10-11 The adjacency diagram, and a valid assignment of a 4-row state table.

PS	NS x_1x_2 00	01	11	10
A	(A)	B	C	D
B	A	(B)	C	(B)
C	B	A	(C)	(C)
D	(D)	(D)	A	C

Figure 10-12 State table of machine M_3 with critical races.

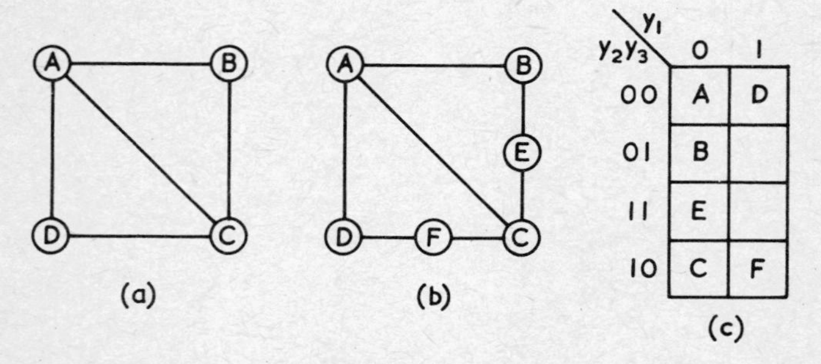

Figure 10-13 (a) Adjacency diagram of machine M_3. (b) Adjacency diagram after the introduction of two new states. (c) A valid assignment with 6 states.

PS	NS			
	x_1x_2			
	00	01	11	10
000 → A	(A)	B	C	D
001 → B	A	(B)	E	(B)
010 → C	E	A	(C)	(C)
100 → D	(D)	(D)	A	F
011 → E	B	–	C	–
110 → F	–	–	–	C

Figure 10-14 Modified state table of machine M_3 with no race.

The transition between states B and C in column 11 is *via* the unstable state E, and that between states D and C in column 10 is *via* the unstable state F. The worst case of adjacency of a 4-row state table is shown in Fig.10-15(a), wherein all states are mutually adjacent to one another. Three additional states need to be introduced (Fig.10-15(b)) and a valid assignment with three secondary variables is determined in the map of Fig.10-15(c).

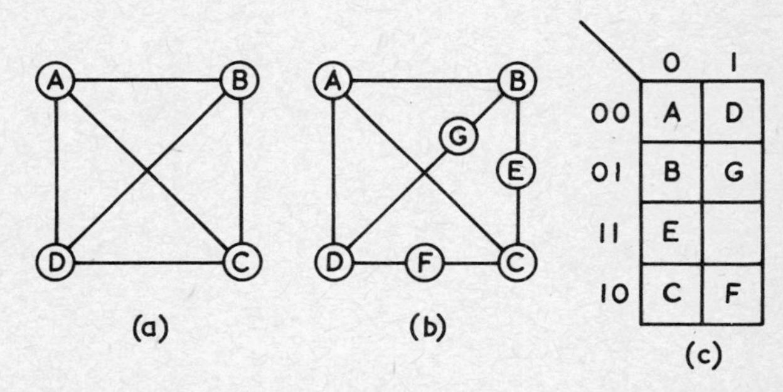

Figure 10-15 (a) Worst case adjacency of a 4-row state table. (b) Adjacency after the introduction of 3 new states to avoid races. (c) A valid assignment with 7 states.

10.5 OUTPUT SPECIFICATION

So far we have maintained that the outputs of the unstable states of a state table are of no consequence. They, therefore, remain unspecified and appear as don't-cares in the z-map. However, if one desires a fast or a slow change of outputs, then the outputs of the unstable states must be specified. For example, if the machine goes from a stable state (A) with output 0, to another stable state (D) with output 1 via the unstable states B and C, then the outputs of B and C

must be specified as 1's if a fast change is desired, and as 0's if a slow change is stipulated. It is evident that before the outputs of the various unstable states can be specified, the transition paths between the stable states must be determined. This is easily done by superimposing the transition arrows in the state table. In Fig.10-16, this has been done for our Example 10.2.1, and the unstable state outputs have been specified for a fast change of output. The resulting z-map is also shown in Fig.10-16. Another desirable feature of the output is that it should be flicker-free, that is, when the output is required to change from one value to another, the change must take place in a single step. For example, if the circuit goes from state (A) with 0 output to state (E) also with 0 output *via* states B and C, then a flicker occurs if either B or C or both have output 1 So both B and C outputs must be specified as 0's, if flicker is to be avoided. If on

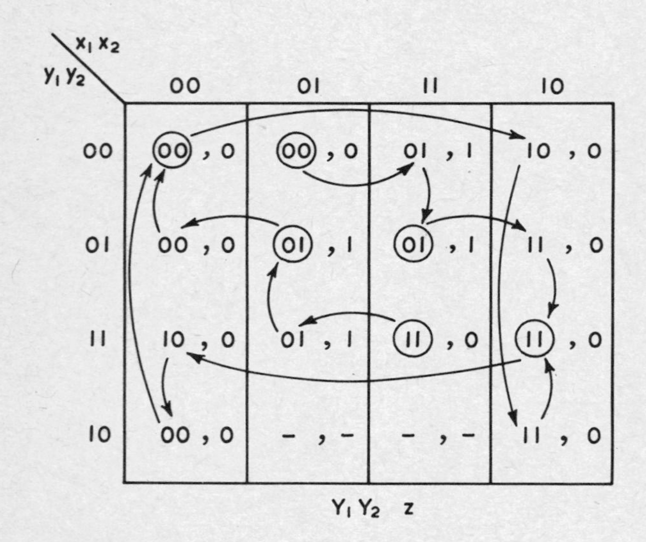

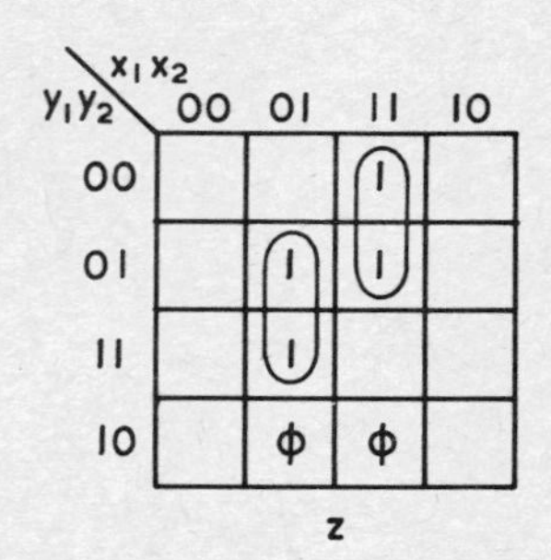

Figure 10-16 The state table with transition arrows and with the specification o unstable state outputs for a fast change of output. The resulting z map is also shown.

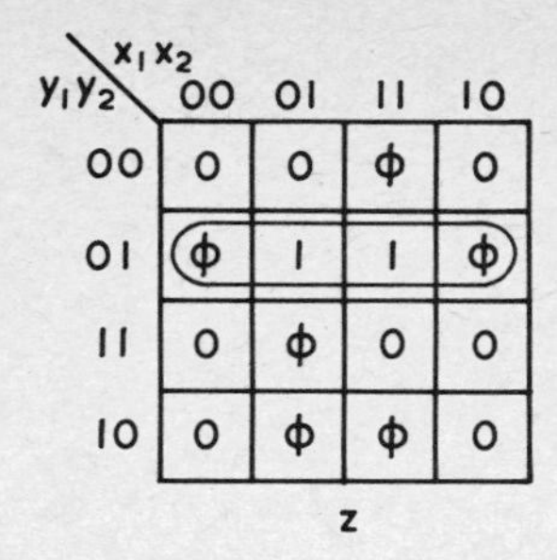

Figure 10-17 The z map for the limited objective of flicker-free output only.

the other hand, the circuit goes from state (A) with a 0 output to state (F) with 1 output *via* the unstable state E, then keeping the output of E unspecified does not prevent the objective of achieving flicker-free output. The specification of outputs for the transition table of Fig.10-16 to achieve the limited objective of flicker-free output only is shown in the z-map of Fig.10-17. It is a matter of chance that the solution for z from the map of Fig.10-17 turns out to be the same as that given by the map of Fig.10-9.

10.6 HAZARDS

A sequential circuit is made up of a number of gates and memory elements which readily respond to a change of inputs. It is, therefore essential that no spurious 0 or 1 is generated by any part of the circuit. This is specially important in a fundamental mode asynchronous circuit which is ever responsive to a change of input. A combinational network which feeds a flip-flop may produce such a spurious output which may prove hazardous to the functioning of the entire circuit, since the circuit may enter into a wrong stable state.

The minimal combinational circuits to implement the S_1 function as plotted in the map of Fig.10-18(A) have been shown both with contact network and gate network in Fig.10-18(A). Here when $y_1y_2 = 01$, the value of S_1 is 1, for both inputs $x_1x_2 = 11$ and $x_1x_2 = 10$. It can be easily verified that the circuits also have an output of 1, for both these inputs with $y_1y_2 = 01$. However, during the interval of transition when x_2 changes its state from 1 to 0, the transfer contact x_2 changes over, and momentarily S_1 becomes 0. This is definitely a hazard condition. The situation can be remedied by using a continuity transfer (make-before break) type of contact for x_2. In the electronic gate circuit, a spurious 0 will appear if the delay Δt_2 introduced by the gate #2 is greater than the delay Δt_3 introduced by the gate #3. Here also the hazard can be removed

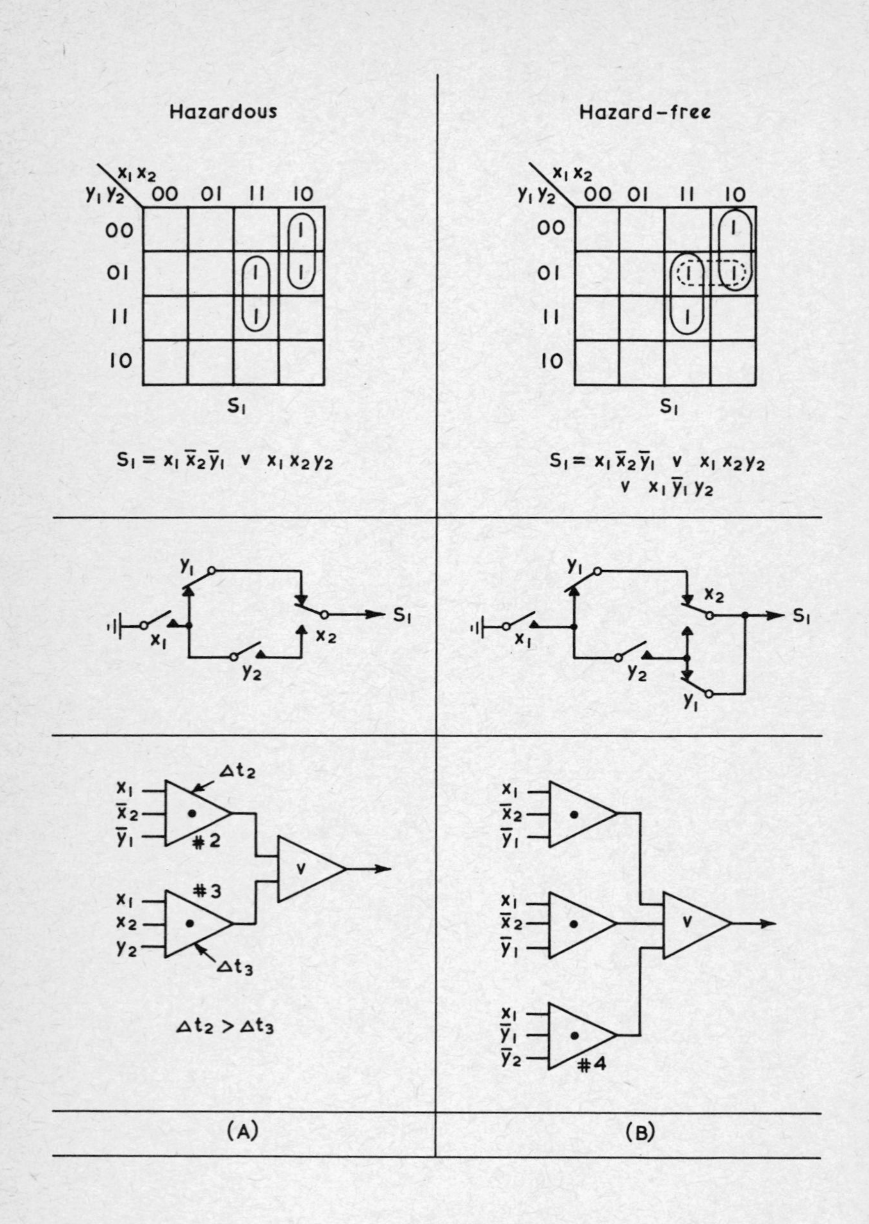

Figure 10-18 The hazardous and hazard-free implementation of the S_1 function.

by making Δt_3 greater than Δt_2. However, in both these types of remedies, the responsibility is passed on to the designer of relays or gates. The logical design itself can remove these hazards easily by introducing additional contacts or gates as shown in Fig.10-18(B). The principle is to incorporate in the Boolean expression for S_1 also the redundant prime implicant shown dotted in Fig.10-18(B). The hazard illustrated in Fig.10-18 is called a static 1 hazard. Similarly one may encounter a static 0 hazard.

Definition 10.6.1 When a pair of adjacent input sequences both produce a 1 output, a *static 1 hazard* exists if it is possible for a momentary 0 output to occur during the transition from one input sequence of the pair to the other.

Definition 10.6.2 When a pair of adjacent input sequences produce a 0 output, a *static 0 hazard* exists if it is possible for a momentary 1 output to occur during the transition from one input sequence of the pair to the other.

Definition 10.6.3 A *dynamic hazard* exists if during the transition between a pair of adjacent input sequences, one of which produces a 0 output, and the other a 1 output, the transition between the 0 and the 1 does not take place in a single step.

It may be observed here that the spurious outputs encountered in the static and dynamic hazards are analogous to the flicker outputs of unstable states as discussed in the previous section.

The static and dynamic hazards originate from the non-uniform delays introduced by the gates and other combinational parts of the circuit. A third type of hazard occurs for the differential delays in the paths through which the signals travel to the flip-flops or delay elements functioning as memory elements. Let us work out the following problem, which is a typical one to encounter such a hazard, known as essential hazard.

PS (Memory)	NS (Memory), z	
	x	
	0	1
A (0)	A (0), 0	B (1), 0
B (1)	C (10), 0	B (1), 0
C (10)	C (10), 0	D (101), 1
D (101)	A ~~E~~ (~~1~~~~0~~~~1~~0), 0	D (101), 1

Figure 10-19 Primitive state-cum-memory table of asynchronous Mod-2 counter.

Example 10.6.1 Design a fundamental mode asynchronous Mod-2 counter using delay elements. The output is to be flicker-free and the combinational circuits are to be free from hazards.

Solution: The primitive state tables are shown in Figs.10-19 and 10-20. The implication chart (Fig.10-21) shows that the state table cannot be minimized further. In Fig.10-22, we draw the adjacency diagram and determine a valid assignment with no races. Figure 10-23 shows transition table with the transition arrows, so that the outputs of unstable states can be specified for a flicker-free output. The figure also shows the excitation maps for Y_1 and Y_2, as well as the z-map. To achieve hazard free combinational networks,

$Y_1 = \bar{x}y_2 \vee xy_1 \vee y_1y_2$

$Y_2 = \bar{x}y_2 \vee x\bar{y}_1 \vee \bar{y}_1y_2$, and

$z = xy_1$.

PS	NS, z (x = 0)	NS, z (x = 1)
A	(A), 0	B, –
B	C, –	(B), 0
C	(C), 0	D, –
D	A, –	(D), 1

Figure 10-20 State table of Mod-2 counter.

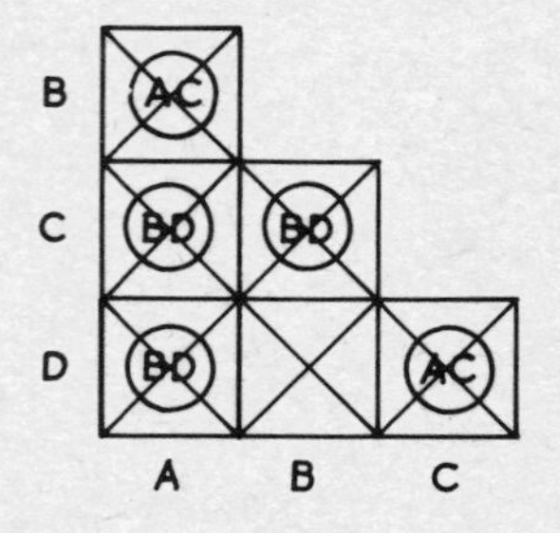

Figure 10-21 The implication chart of the above state table.

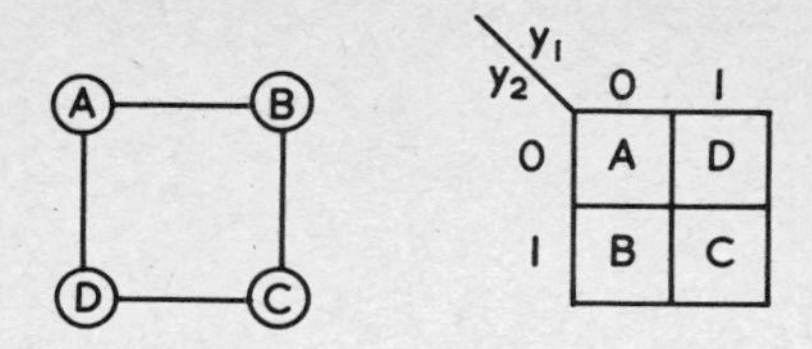

Figure 10-22 The adjacency diagram of the states, and a valid assignment.

The circuit is shown in Fig.10-24. It will now be interesting to investigate the consequences if the inverter gate (gate #8) introduces substantial delay. Let the circuit be in the stable state $y_1y_2 = 00$ with the input $x = 0$. Now, let x change to 1. Gate #6 will now switch D_2 to store a 1, making $y_2 = 1$. This value of y_2 now appears at the input of gate #3. Because of the delay of the inverter gate #8, the other input of gate #3 is still a 1. Hence this gate switches D_1 to store a 1. The circuit thus enters the state $y_1y_2 = 11$. After this has happened, the 1 at

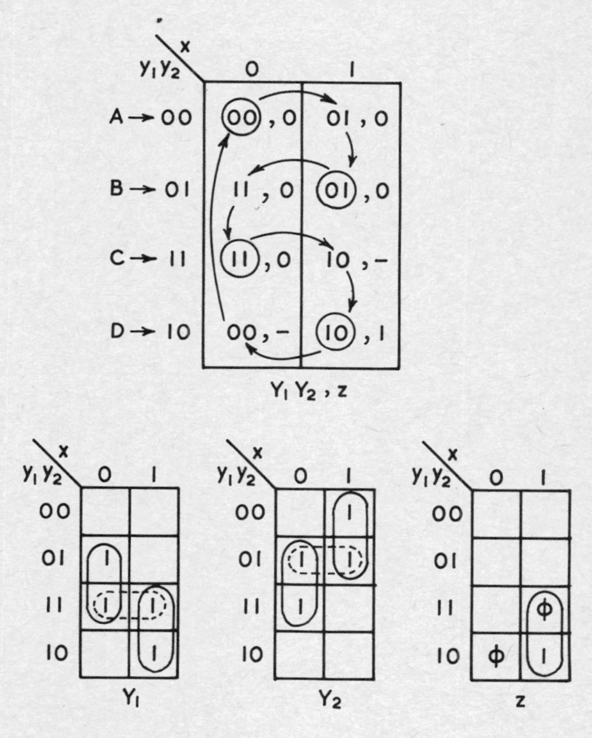

Figure 10-23 The transition table with transition arrows, and with outputs specified for a flicker-free output. The maps for the excitation and output functions are also shown.

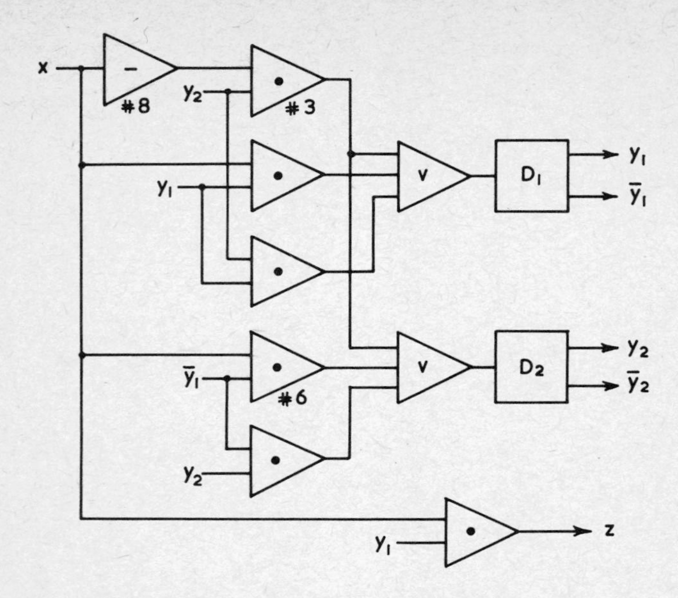

Figure 10-24 The Mod-2 counter circuit.

the input of gate #3 has travelled through the inverter gate. At this stage, the circuit behaves as though it is in the present state of 11, and an input is 1. According to the state table the circuit enters the state 10, and becomes stable at this state. Thus due to the delay of the inverter gate the circuit enters the wrong state 10 instead of the correct state 01. This is the situation of an essential hazard.

Definition 10.6.4 An *essential hazard* exists if due to the differential delays in the transmission paths from input to different memories, a change of input drives the circuit to a wrong stable state.

It may be seen that nothing can be done in the state table to avoid an essential hazard. However, an essential hazard can be detected on the state table by the following theorem due to Unger.

Theorem 10.6.1 (Unger, 1959) An essential hazard exists in a fundamental mode asynchronous sequential machine whenever the state table is such that three consecutive input changes take the circuit to a different stable state than the first change alone.

Proof: The validity of this theorem will be obvious from the above example. ▲

An immediate consequence of Unger's theorem is that the state tables of all counters which progressively move from one state to another new state have essential hazard.

10.7 PULSE-MODE OPERATION

Besides the fundamental mode, an asynchronous circuit can also operate in the pulse mode. Unlike the fundamental mode, in pulse mode operation an input pulse cannot operate for any length of time, but is just broad enough to effect only a single change of the flip-flop. In this respect a pulse mode asynchronous circuit is similar to a synchronous circuit. However, due to the absence of the clock pulses, double change of input is not allowed for reasons as have been explained in Sec.10.1. Moreover, whereas in a fundamental mode, two inputs can both be 1 at certain situation, in a pulse mode no situation is envisaged where more than one input acts on the circuit at the same time. Thus for a circuit with three inputs, x_1, x_2 and x_3, the input sequences which are of consequence are 001, 010 and 100 only. The other sequences with two or more 1's do not occur, and the state table need not have columns for these non-occurring input sequences. The reader may verify that the circuit remains stable between the pulses, even if delay elements are used as memory elements. Since each pulse triggers only a single change, and the circuit remains in this state until another pulse comes, all states are stable. Hence in reality, a pulse mode asynchronous circuit behaves like a synchronous circuit with a more restricted input alphabet.

Example 10.7.1 A pulse mode asynchronous machine, M_4, has two inputs. It produces an output whenever two consecutive pulses occur in one input line only. The output remains 1 until a pulse has occurred in the other input line. Draw the state table for the machine.

PS (Memory)	NS (Memory)/output		
	x_1 0, x_2 0	x_1 0, x_2 1	x_1 1, x_2 0
A $\binom{0}{0}$	A $\binom{0}{0}$/0	B $\binom{0}{1}$/0	C $\binom{1}{0}$/0
B $\binom{0}{1}$	B $\binom{0}{1}$/0	D $\binom{00}{11}$/1	C $\not{E}$ $\binom{\not{0}1}{\not{1}0}$/0
C $\binom{1}{0}$	C $\binom{1}{0}$/0	B $\not{F}$ $\binom{\not{1}0}{\not{0}1}$/0	G $\binom{11}{00}$/1
D $\binom{00}{11}$	D $\binom{00}{11}$/1	D $\not{H}$ $\binom{\not{0}00}{\not{1}11}$/1	C $\not{I}$ $\binom{\not{0}\not{0}1}{\not{1}\not{1}0}$/0
G $\binom{11}{00}$	G $\binom{11}{00}$/1	B $\not{J}$ $\binom{\not{1}\not{1}0}{\not{0}\not{0}1}$/0	G $\not{K}$ $\binom{\not{1}11}{\not{0}00}$/1

Figure 10-25 Primitive state-cum-memory table of machine M_4.

PS	NS x_1x_2 00	01	10	z
A	A	B	C	0
B	B	D	C	0
C	C	B	G	0
D	D	D	C	1
G	G	B	G	1

Figure 10-26 Primitive state table of machine M_4.

Solution: Since the output is maintained between the pulses it must be a Moore machine. The primitive state table can be written by following the same procedure as for synchronous machines. These are shown in Figs.10-25 and 10-26. Figure 10-27 establishes that the stable table of Fig.10-26 is indeed the minimum row state table.

P_0	PS	NS
1	A	111
	B	121
	C	112
2	D	221
	G	212

Figure 10-27 Minimization of M_4.

10.8 CONCLUSION

In the following table the three types of sequential machines are compared. It may be seen that the duration of input pulses in a pulse mode asynchronous sequential machine is very critical. For this reason this mode of operation does not find much favor. Among the remaining two many design requirements are to be met for proper functioning of a fundamental mode asynchronous machine. However the requirements can be met and reliable operation can be achieved. The synchronous sequential machine incorporates a more methodical and reliable approach both in the design and operation. It also has the least restriction on inputs. But its speed of operation is limited by the pulse repetition frequency of the synchronizing pulses. On the other hand an asynchronous machine has no such inhibition and can operate faster.

	Sequential Machines		
		Asynchronous	
Factor	Synchronous	Fundamental Mode	Pulse Mode
1) Input pulses	Minimum duration should be greater than the duration of clock pulses, and must not arrive later than the clock pulses. Maximum duration should be such as to terminate before the arrival of the next clock pulse.	Minimum duration should be such as to effect a single change of the memory element. No limit is imposed on maximum duration. A highly broad pulse is usually interpreted as a level input.	Minimum duration should be such as to effect a single change of the memory element. Maximum duration should be such that it must terminate before changing the state of the memory element twice.
2) Input alphabet	All symbols are permitted. Inputs are free to change in any manner. Inputs however, should not change when the clock pulses are present.	All symbols are permitted. No two inputs are permitted to change simultaneously. An input sequence does not change until the circuit comes to a stable state.	Symbols with single 1's only are permitted.
3) Memory elements	Only bistable flip-flops can be used.	Bistable flip-flops as well as delay elements and relays can be used.	Both bistable flip-flops and delay elements can be used.
4) Stability of states	All next state entries in the state table are stable.	Only the circled next states in the state table are stable.	All next state entries in the state table are stable.
5) Races	No race condition exists.	Critical races may exist. Valid assignments must be chosen to avoid races.	No race condition exists.
6) Hazards	Hazards do not pose special problems.	Static, dynamic and essential hazards should be avoided.	Hazards do not pose special problems.

REFERENCES

Schelberger, E. B.: "Hazard detection in combination and sequential switching circuits", *SCTLD*, pp.111–120, 1964.

Huffman, D. A.: "The synthesis of sequential switching circuits", *J. Franklin Inst.*, **257**, pp.275–303, March–April 1954.

Kohavi, Z.: *Switching and Finite Automata Theory*, McGraw-Hill, New York, 1970.

McCluskey, E. J.: "Fundamental and pulse mode sequential circuits", *IFIP Congress, 1962,* North Holland Publishing Co., Amsterdam, 1963.

Miller, R. E.: *Switching Theory,* Vol.2, John Wiley and Sons, Inc., New York, 1965.

Muller, D. E.: "Treatment of transition signals in electronic switching circuits by algebraic methods", *IRE Trans. Electronic Computers*, **EC-8**, No.3, p.401, Sept.1959.

Unger, S. H.: "Hazards and delays in asynchronous sequential switching circuits", *IRE Trans. Circuit Theory,* **CT-6**, No.12, pp.12–25, March 1959.

—— : *Asynchronous Sequential Machines,* John Wiley and Sons, Inc., 1969.

Yoeli, M. and S. Rino: "Application of ternary algebra to the study of static hazards", *J. ACM,* **11**, No.1, pp.84–97, June 1964.

PROBLEMS

10.1 Determine if there is any race condition in the following transition tables. Which of them are critical?

(a)

	NS, z			
	x_1x_2			
PS	00	01	11	10
00	(00), 0	01, –	01, –	10, –
01	00, –	(01), 1	–	11, –
11	01, –	(11), 0	10, –	(11), 0
10	(10), 0	01, –	(10), 1	–

(b)

PS	NS, z			
	x_1x_2			
	00	01	11	10
000	(000), 0	101, –	001, –	(000), 1
001	010, –	101, –	010, –	010, –
010	100, –	(010), 0	100, –	100, –
100	(100), 1	(100), 1	101, –	111, –
101	100, –	(101), 1	(101), 0	111, –
111	(111), 1	(111), 0	(111), 1	001, –

10.2 Change the assignment of the transition tables of Problem 10.1, so as to avoid (a) all races, (b) only critical races. In both cases use minimum number of states.

10.3 Determine a valid and economical assignment of the following state table and then specify the output for (a) fast and flicker-free output, (b) only flicker-free output.

PS	NS, z			
	x_1x_2			
	00	01	11	10
A	(A), 0	B, –	C, –	D, –
B	C, –	D, –	(B), 1	(B), 0
C	(C), 1	(C), 0	A, –	D, –
D	B, –	(D), 1	(D), 1	C, –

10.4 Implement the machine of Problem 10.3 with the valid assignment and delay elements for (a) fast and flicker-free output, and (b) only flicker-free output.

10.5 The excitation and output functions of an asynchronous sequential circuit are as follows: Investigate them for static and dynamic hazards. Implement them with hazard free (a) gate circuit, (b) relay circuit.

$$Y_1 = x_1x_2 \vee \bar{x}_1y_1y_2$$

$$Y_2 = x_1 \vee \bar{y}_1y_2$$

$$Z = y_2(x_1\bar{y}_1 \vee x_2 y_1) \vee \bar{x}_1 y_1 \bar{y}_2.$$

10.6 Show that the following state table has essential hazard. Implement the circuit with J–K flip-flops. What additional constraints must be placed on the circuit elements to avoid the hazard?

	NS		
	x		
PS	0	1	z
A	(A)	B	0
B	(B)	C	0
C	(C)	D	0
D	(D)	A	1

10.7 Analyse the following circuits for any hazard. Determine the nature of hazard, in case one exists. Suggest ways to render the circuit hazard-free. What are the switching functions realized by the four circuits?

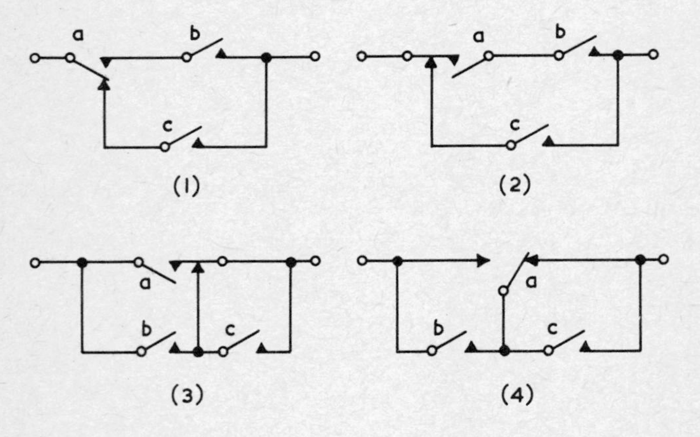

10.8 In a fundamental mode asynchronous sequential circuit the excitation function of relay Y_1 is given by

$$Y_1(x_1 x_2 y_1 y_2) = \vee\,(2, 3, 6\text{–}11, 13, 15).$$

Design a contact network from the *MSP* form of the function. Use only make, break, and transfer contacts. Investigate the circuit for hazard. Suggest means to remove hazards, if any.

10.9 Design the contact network of the above problem from the *MPS* form of the Boolean function. Investigate the circuit for hazards.

10.10 In a fundamental mode asynchronous circuit, the excitation function of a trigger *FF* is given by

$$T_1(x_1x_2y_1y_2) = \vee\,(0\text{–}3, 6, 7, 11, 14, 15).$$

Design a hazard-free gate network with AND OR gates. The flip-flops have bipolar outputs. Are 4 gates adequate?

10.11 For the relay circuit given below, determine (a) the transition table, (b) the state table, and (c) the state diagram.

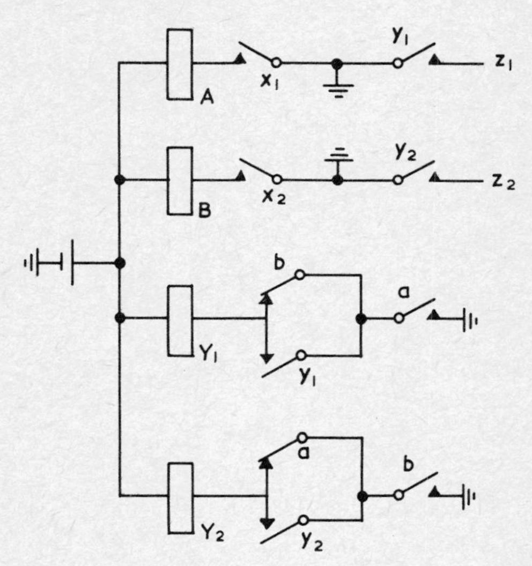

10.12 A fundamental-mode asynchronous sequential machine has two inputs x_1 and x_2 and two outputs z_1 and z_2. When x_1 is 1 with $x_2 = 0$, z_1 becomes 1, and remains 1 until x_1 becomes 0 again. During this period z_1 is not affected by the value of x_2. Similarly z_2 becomes 1 when x_2 becomes 1 with $x_1 = 0$, and remains 1 until x_2 becomes 0 again. Once it has become 1, z_2 too remains unaffected by the value of x_1. Design the circuit with delay elements, and AND OR gates.

10.13 Design a fundamental mode asynchronous mod-2 counter with Trigger *FF*s and AND OR gates.

10.14 Design a pulse mode asynchronous mod-2 counter with Trigger *FF*s and AND OR gates. How does this circuit differ from the circuit of a similar counter working in the synchronous mode?

10.15 A combination lock circuit has two push button switches *a* and *b*. Whenever depressed each switch closes a make contact. The two switches cannot be depressed simultaneously. The circuit has one output *z*. Whenever the pushbuttons are depressed to make the sequence *baa*, z becomes 1 and opens the lock. Design a hazard-free and economical relay circuit for the combination lock.

10.16 A combination lock circuit has two pushbutton switches a and b similar to those of the above problem. This lock circuit, however, has two outputs, z_1 and z_2. Whenever the pushbuttons are depressed to make the sequence *aba*, z_1 becomes 1 and opens the lock. For any other combination z_2 becomes 1 and sounds an alarm. Once either z_1 or z_2 becomes 1, it remains so no matter what happens afterwards. Design a hazard-free and economical relay circuit for this combination lock.

10.17 An elevator serves a building with five floors. Three floors have two push button switches marked UP and DOWN. Once a switch is pressed a corresponding UP or DOWN lamp glows which gets extinguished only when the elevator arrives and stops at the floor. For obvious reasons the first and the fifth floors have only one button each. If the UP button is pressed at a floor, then the elevator stops only if it is going up, and similarly it will stop in response to the DOWN button if it is going down. (The possibility of both the UP and the DOWN buttons being pressed on a floor is not ruled out.) When the elevator is going either up or down it will not respond to a call urging it to move in the opposite direction until it completes the upward or the downward journey, as dictated by the pressed buttons on the floors and inside the elevator. Design a relay circuit to perform as the control circuit of the elevator.

10.18 A traffic light system to control the vehicular traffic at the level crossing of a single-track railway line is to be designed. With no train approaching, the green light will be on. When a train approaches the level crossing from either side, and when it is at a distance of 5000 ft from the level crossing the light must change from green to red. The light will change from red to green when the rear of the train has crossed over and is 100 ft away from the level crossing. Assume that the length of the train does not exceed 4000 ft, and the train does not move backward and forward at the level crossing, that is, the train simply crosses the level crossing by moving in one direction only. Design a relay circuit to control the traffic light system.

Appendix A

SYMBOLS AND CONVENTIONS

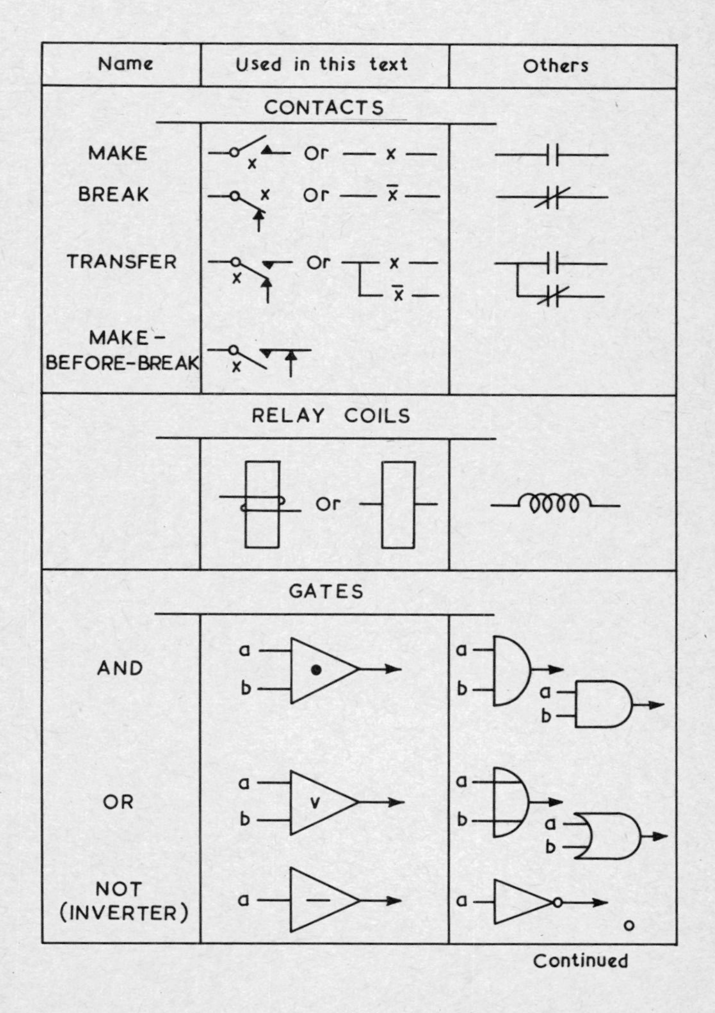

Name	Used in this text	Others
NAND	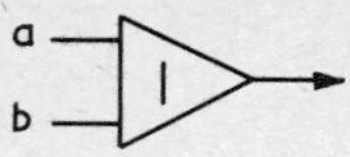 a, b → [|] →	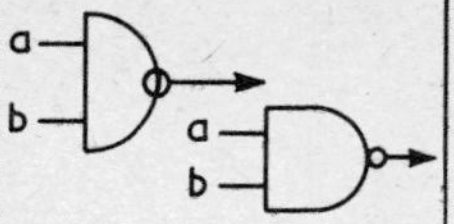 a, b → NAND gate →; a, b → NAND gate →
NOR	a, b → [↓] →	a, b → gate →; a, b → NOR gate →
THRESHOLD	x_1—a_1, x_2—a_2, x_3—a_3, u:l → Or x_1—a_1, x_2—a_2, x_3—a_3, T →	x_1, x_2, x_3 — a_1, a_2, a_3, T →
LOGICAL OPERATIONS		
NOT a	$\bar{a}$	a'
a AND b	ab	$a \bullet b$
a OR b	$a \vee b$	$a + b$
a EXCL-OR b	$a ⓥ b$	$a \oplus b$
a NAND b	$a \mid b$ Or $\overline{ab}$	$(ab)'$
a NOR b	$a \downarrow b$ Or $\overline{a \vee b}$	$(a+b)'$

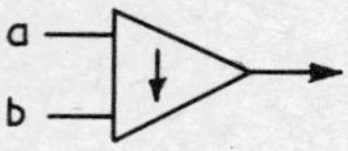

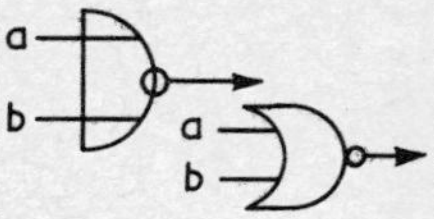

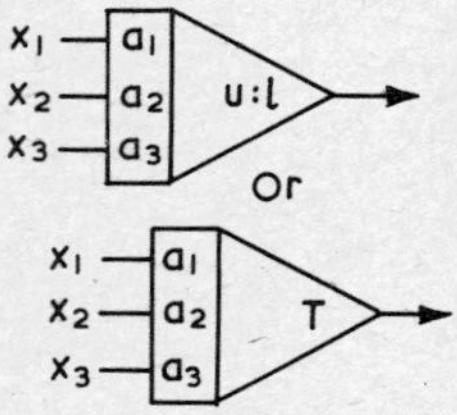

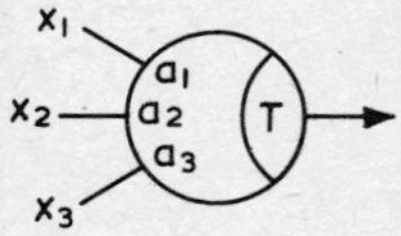

Appendix B

HELLERMAN'S CATALOG OF THREE VARIABLE NOR AND NAND LOGICAL CIRCUITS†

This paper gives a catalog of minimum Or-Invert (NOR) and And-Invert (NAND) circuits for logical functions of three variables. Over the set of circuits, minimum with respect to the logical connective, we catalog all having the least number of connections. A few functions have more than one minimal circuit, but for most of them the minimum is unique.

The maximum fan-in and fan-out for the logic blocks was set equal to the number of variables – three. Although this restriction of the fan limit is somewhat arbitrary, the following supplementary results are partial justification:

There exists just one function of three variables such that relaxing the fan limit from three to four permits a reduction of the number of blocks required for synthesis of the function. We know also that relaxing the fan limit from three to four does not save connection in the syntheses of all three-variable functions – with the possible exception of a few functions requiring seven blocks.

Criteria for minimal syntheses, other than those chosen, are possible. For example, one may associate a distinct cost with each block, depending on number of inputs to the block. A minimal circuit is one of least cost. H. M. McAndrew (in preparation), using programs based on methods of J. Paul Roth and others (1961), has used such a criterion and found a set of minimal NOR circuits. Most of his circuits are minimal with respect to our criterion too, but there are a few differences.

OUTLINE OF METHOD

Our method is exhaustive. That is, we look at every possible combinational network of Or-Invert blocks, and save the desired minimums for each three-variable logic function. It is known that each of these functions can be implemented by seven blocks or fewer (allowing one with a fan of four). It can be shown that

† Reproduced (in abridged form) by permission: Leo Hellerman, "A Catalog of Three-variable Or-Invert and And-Invert Logical Circuits" *IEEE Trans. Electronic Computers,* **12**, pp.198–223, June, 1963.

there are $2^{42} \approx 4 \times 10^{12}$ possible combinational networks with seven or fewer blocks. Using the best program we could devise for the IBM 7090, and taking advantage of all logical short cuts and programming tricks we knew of, this exhaustive procedure was practical – but barely so. Machine time to investigate networks with a given number of blocks increased as follows:

1 to 5 blocks	3 minutes
6 blocks	45 minutes
7 blocks	25 hours.

ELIMINATION OF EQUIVALENT FUNCTIONS CIRCUITS

It is well known (Hellerman, 1961) that the 256 logical functions of three variables can be partitioned into 80 equivalence classes, 68 of which are classes of nondegenerate, three-variable functions. Two functions are equivalent and belong to the same class if, and only if, one can be obtained from the other by a permutation of the input variables. Similarly, networks can be partitioned into equivalence classes with respect to input-line permutations: two networks belong to the same class if, and only if, the connection matrix of one can be obtained from the connection matrix of the other by a permutation of the first three columns.

Permutations of input lines merely juggle the correspondence of lines with variables; permutations of variables do the same. For this reason equivalent networks perform only equivalent functions. Since implementation of any membe of a function class serves (after suitable variable permutations) to implement all members of the class, it is sufficient to evaluate only one network from each equivalence class of networks.

DESCRIPTION AND USE OF THE CATALOG

Each function in our catalog is identified by its truth-table designation number. The assumed form of the truth table is:

a	b	c	f
0	0	0	x_0
0	0	1	x_1
0	1	0	x_2
0	1	1	x_3
1	0	0	x_4
1	0	1	x_5
1	1	0	x_6
1	1	1	x_7

here the x_i's are all zeroes and ones. If we write f horizontally we obtain the
inary number

$$\underbrace{x_7\,x_6}_{O_2} \qquad \underbrace{x_5 x_4 x_3}_{O_1} \qquad \underbrace{x_2 x_1 x_0}_{O_0}$$

rouping the x_i as shown, we obtain the octal number $O_2\ O_1\ O_0$. This octal num-
er designates the function f.

If a logical function is expressed as a sum of elementary products, these ele-
entary products correspond to truth-table rows having the value $x_i = 1$, all
ther rows having the value $x_i = 0$. Thus, for example, $\bar{a}(\bar{b} \vee c)$ has a 1 for $\bar{a}b\bar{c}$,
$\bar{b}\bar{c}$, $\bar{a}bc$, and again $\bar{a}\bar{b}c$. The same function could have been written $\bar{a}\bar{b} \vee \bar{a}bc$
ving the same truth table. There is a unique one-to-one correspondence of
ınctions to designation numbers, whereas there may be many alternative expres-
ons for a single function. This is our main reason for using the function desig-
ation numbers to label a function, rather than some functional expression.

A four-digit decimal number gives certain information about each circuit in
e catalog. The number is denoted by TCL, where T and L are single digits, and
is two digits. The information given is:

– number of transistors (blocks)
– number of connections
– number of levels.

Thus, the circuits for performing the logical function $f = \bar{a}(\bar{b} \vee c)$ are found
y looking up the designation number of f, 13, in the catalog. The entries for
is number tell us the function is performed by the NOR circuit shown as
ircuit 7, using three NOR blocks, five connections, and three levels; and by the
AND circuit 28, using five blocks, seven connections and three levels.

Only 80 of the 256 functions of three variables are listed in our catalog; all
thers are equivalent to a function in the catalog by a permutation of the vari-
bles. Given an arbitrary three-variable function f, we wish to

) find its equivalent, g, in the catalog, and
) permute the variables in the circuits for g, so that these circuits perform
function f.

The procedures for (1) and (2) above are made quite simple by using the
able of Equivalence Classes. We illustrate this with an example:

Suppose the function to be designed has the number 321. The entry for this
umber in the Equivalence Class Table is – 5000 213. This tells us our original
ınction is equivalent to function 213. We have accomplished (1), above. To
ccomplish (2), we note that the high order digit of the entry, 5, tells us which
ermutation to apply to 213 in order to obtain our original function 321.

The permutation numbers and their meanings are:

Number	Permutation	Changes in circuit figures
1	1	none
2	(*abc*)	*a* to *b*, *b* to *c*, *c* to *a*
3	(*acb*)	*a* to *c*, *c* to *b*, *b* to *a*
4	(*bc*)	*b* to *c*, *c* to *b*
5	(*ac*)	*a* to *c*, *c* to *a*
6	(*ab*)	*a* to *b*, *b* to *a*

In the above example the circuits for 213 are shown as Circuits 20 and 31. Applying permutation 5 to these simply interchanges *a* and *c*.

Twelve of the 80 three-variable functions listed in the catalog are degenerate in the sense that they are independent of at least one variable. These twelve include the eight functions of two variables; and the two functions of one variable; and the two functions of no variable, $f = 0$ and $f = 1$. The circuits for the degenerate functions are particularly simple. We saw no need to prove them minimal by our IBM 7090 program. Nevertheless, for completeness, we include these functions in the catalog. The circuit figures for the degenerate functions are marked, "D". They are grouped together at the beginning of the circuit drawings.

Some of the figures for NAND circuits are marked with an X. For example, the NAND circuit for function 30 is shown in Circuit 44X. This means that Circuit 44X differs from Circuit 44 only by some permutation of the variables. This duplication was necessary in some cases to permit a uniform application of the Table of Equivalence Classes, which is shown on the opposite page.

REFERENCES

Hellerman, L., "Equivalence classes of logical functions", IBM Technical Report No.00.819, Nov.1, 1961.

McAndrew, M. H., "A table of minimal circuit realizations" (in preparation).

Roth, J. P., R. M. Karp, F. E. McFarlin and J. R. Wilts, "A computer program for the synthesis of combinational switching circuits", *Proceedings of the Second Annual Symposium on Switching Circuit Theory and Logical Design*, *AIEE*, Sept.1961, pp.182–194.

Equivalence classes of functions of 3 variable

	0	1	2	3	4	5	6	7
0	−1000000	1000001	1000002	−1000003	−3000002	−3000003	1000006	1000007
10	1000010	1000011	−1000012	1000013	−4000012	−4000013	1000016	−1000017
20	−2000002	−2000003	−2000006	−2000007	−3000006	−3000007	1000026	1000027
30	1000030	1000031	1000032	1000033	−4000032	−4000033	1000036	1000037
40	−2000010	−2000011	−6000012	−6000013	−2000030	−2000031	−6000032	−6000033
50	1000050	1000051	1000052	1000053	1000054	1000055	1000056	1000057
60	−2000012	−2000013	−2000016	−2000017	−2000032	−2000033	−2000036	−2000037
70	−6000054	−6000055	−6000056	−6000057	−1000074	1000075	1000076	−1000077
100	−3000010	−3000011	−3000030	−3000031	−3000012	−3000013	−3000032	−3000033
110	−3000050	−3000051	−4000054	−4000055	−3000052	−3000053	−4000056	−4000057
120	−5000012	−5000013	−5000032	−5000033	−3000016	−3000017	−3000036	−3000037
130	−3000054	−3000055	−3000074	−3000075	−3000056	−3000057	−3000076	−3000077
140	−2000050	−2000051	−2000054	−2000055	−5000054	−5000055	−2000074	−2000075
150	1000150	1000151	1000152	1000153	−3000152	−3000153	1000156	1000157
160	−2000052	−2000053	−2000056	−2000057	−5000056	−5000057	−2000076	−2000077
170	−2000152	−2000153	−2000156	−2000157	−3000156	−3000157	1000176	1000177
200	1000200	1000201	1000202	1000203	−3000202	−3000203	1000206	1000207
210	−1000210	1000211	1000212	1000213	−4000212	−4000213	1000216	1000217
220	−2000202	−2000203	−2000206	−2000207	−3000206	−3000207	1000226	1000227
230	1000230	−1000231	1000232	1000233	−4000232	−4000233	1000236	1000237
240	−2000210	−2000211	−6000212	−6000213	−2000230	−2000231	−6000232	−6000233
250	1000250	1000251	−1000252	1000253	1000254	1000255	1000256	−1000257
260	−2000212	−2000213	−2000216	−2000217	−2000232	−2000233	−2000236	−2000237
270	−6000254	−6000255	−6000256	−6000257	1000274	1000275	1000276	1000277
300	−3000210	−3000211	−3000230	−3000231	−3000212	−3000213	−3000232	−3000233
310	−3000250	−3000251	−4000254	−4000255	−3000252	−3000253	−4000256	−4000257
320	−5000212	−5000213	−5000232	−5000233	−3000216	−3000217	−3000236	−3000237
330	−3000254	−3000255	−3000274	−3000275	−3000256	−3000257	−3000276	−3000277
340	−2000250	−2000251	−2000254	−2000255	−5000254	−5000255	−2000274	−2000275
350	1000350	1000351	1000352	1000353	−3000352	−3000353	−1000356	1000357
360	−2000252	−2000253	−2000256	−2000257	−5000256	−5000257	−2000276	−2000277
370	−2000352	−2000353	−2000356	−2000357	−3000356	−3000357	1000376	−1000377

Explanatory example – Class of 321 is given by word at intersection of row 320 and column 1, −5000213. This says 321 is in class of 213 by permutation 5. Negative permutation 1 means the function is degenerate.
Permutation 1 is the identity; permutation 2 is (abc); permutation 3 is (acb); permutation 4 is (bc); permutation 5 is (ac); permutation 6 is (ab)

Catalog of Degenerate Circuits

Function (octal)	Functional expression	Or-Invert				And-Invert			
		Circuit	T	C	L	Circuit	T	C	L
0	$f = 0$	1D	0	00	0	1D	0	00	0
3	$\bar{a}\bar{b}$	5D	1	02	1	10D	4	05	3
12	$\bar{a}c$	7D	2	03	2	8D	3	04	3
17	$\bar{a}$	4D	1	01	1	4D	1	01	1
74	$\bar{a}b \vee a\bar{b}$	12D	5	08	3	11DX	4	08	3
77	$\bar{a} \vee \bar{b}$	10D	4	05	3	5D	1	02	1
210	bc	9D	3	04	2	6D	2	03	2
231	$bc \vee \bar{b}\bar{c}$	11D	4	08	3	12DX	5	08	3
252	c	3D	0	01	0	3D	0	01	0
257	$\bar{a} \vee c$	8D	3	04	3	7D	2	03	2
356	$b \vee c$	6D	2	03	2	9D	3	04	2
377	$f = 1$	2D	0	00	0	2D	0	00	0

Catalog of Nondegenerate Circuits

Function (octal)	Functional expression	Or-Invert				And-Invert			
		Circuit	T	C	L	Circuit	T	C	L
1	$\bar{a}\bar{b}\bar{c}$	1	1	03	1	26	5	07	3
2	$\bar{a}\bar{b}c$	4	2	04	2	12	4	06	3
6	$\bar{a}(\bar{b}c \vee b\bar{c})$	32	5	09	3	48	5	11	3
7	$\bar{a}(\bar{b} \vee \bar{c})$	14	4	06	3	13	4	06	3
10	$\bar{a}bc$	10	3	05	2	6	3	05	3
11	$\bar{a}(\bar{b}\bar{c} \vee bc)$	21	4	09	3	62	6	11	3
						53	6	11	4
13	$\bar{a}(\bar{b} \vee c)$	7	3	05	3	28	5	07	4
16	$\bar{a}(b \vee c)$	3	2	04	2	19	4	07	3
26	$\bar{a}\bar{b}c \vee \bar{a}b\bar{c} \vee a\bar{b}\bar{c}$	58	6	14	3	79	7	15	3
27	$\bar{a}\bar{b} \vee \bar{a}\bar{c} \vee \bar{b}\bar{c}$	27	5	10	3	27	5	10	3
30	$a\bar{b}\bar{c} \vee \bar{a}bc$	59	6	11	4	44X	5	11	3
31	$\bar{a}bc \vee \bar{b}\bar{c}$	63	6	10	4	54X	6	10	3
		52	6	10	4				
32	$\bar{a}c \vee a\bar{b}\bar{c}$	36	5	09	3	43X	5	10	3
33	$\bar{a}c \vee \bar{b}\bar{c}$	29	5	08	4	29X	5	08	4
36	$\bar{a}(b \vee c) \vee a\bar{b}\bar{c}$	37	5	10	3	67	6	14	3
		46	5	10	4				
37	$\bar{a} \vee \bar{b}\bar{c}$	13	4	06	3	14	4	06	3
50	$\bar{a}bc \vee a\bar{b}c$	61	6	10	3	24	4	10	3

(*continued*)

Function (octal)	Functional expression	Or-Invert				And-Invert			
		Circuit	T	C	L	Circuit	T	C	L
51	$\bar{a}\bar{b}\bar{c} \vee \bar{a}bc \vee a\bar{b}c$	76	7	14	4	73	7	15	4
						74	7	15	5
						75	7	15	4
						78	7	15	3
52	$c(\bar{a} \vee \bar{b})$	30	5	07	3	5	3	05	3
53	$c(\bar{a} \vee \bar{b}) \vee \bar{a}\bar{b}$	55	6	10	3	55	6	10	3
54	$\bar{a}b \vee a\bar{b}c$	60	6	10	4	22X	4	09	3
55	$\bar{a}(b \vee \bar{c}) \vee a\bar{b}c$	64	6	11	4	64X	6	11	4
		65	6	11	5	65X	6	11	5
56	$\bar{a}b \vee \bar{b}c$	31	5	08	3	20	4	08	3
57	$\bar{a} \vee \bar{b}c$	28	5	07	4	7	3	05	3
75	$\bar{a}b \vee a\bar{b} \vee \bar{a}\bar{c}$	54	6	10	3	63X	6	10	4
						52X	6	10	4
76	$\bar{a}b \vee a\bar{b} \vee \bar{a}c$	33	5	09	3	49	5	11	3
150	$\bar{a}bc \vee a\bar{b}c \vee ab\bar{c}$	72	7	15	4	51	5	15	3
151	$(\bar{a}b \vee a\bar{b})c \vee (ab \vee \bar{a}\bar{b})\bar{c}$	77	7	16	5	77	7	16	5
152	$(\bar{a} \vee \bar{b})c \vee ab\bar{c}$	68	6	12	4	25	4	10	3
153	$\bar{a}c \vee \bar{a}\bar{b} \vee \bar{b}c \vee ab\bar{c}$	73	7	15	4	76	7	14	4
		74	7	15	5				
		75	7	15	4				
		78	7	15	3				
156	$\bar{a}(b \vee c) \vee \bar{b}c \vee b\bar{c}$	56	6	10	3	23	4	09	3
157	$\bar{a} \vee \bar{b}c \vee b\bar{c}$	53	6	11	4	21	4	09	3
		62	6	11	3				
176	$a\bar{b} \vee b\bar{c} \vee \bar{a}c$	57	6	11	3	50	5	12	3
177	$\bar{a} \vee \bar{b} \vee \bar{c}$	26	5	07	3	1	1	03	1
200	abc	17	4	06	2	2	2	04	2
201	$abc \vee \bar{a}\bar{b}\bar{c}$	50	5	12	3	57	6	11	3
202	$(ab \vee \bar{a}\bar{b})c$	42	5	10	3	35	5	10	3
						38	5	10	4
203	$abc \vee \bar{a}\bar{b}$	49	5	11	3	33	5	09	3
206	$abc \vee \bar{a}\bar{b}c \vee \bar{a}b\bar{c}$	69	6	15	4	66	6	15	3
						70	6	15	4
						71	6	15	4
207	$abc \vee \bar{a}(\bar{b} \vee \bar{c})$	67	6	14	3	37	5	10	3
						46	5	10	4
211	$\bar{a}\bar{b}\bar{c} \vee bc$	23	4	09	3	56	6	10	3
212	$c(\bar{a} \vee b)$	15	4	06	3	11	4	06	4
213	$\bar{a}b \vee bc$	20	4	08	3	31	5	08	3
216	$\bar{a}b \vee \bar{a}c \vee bc$	39	5	10	4	39	5	10	4

(*continued*)

Function (octal)	Functional expression	Or-Invert				And-Invert			
		Circuit	T	C	L	Circuit	T	C	L
		41	5	10	3	41	5	10	3
		45	5	10	4	45	5	10	4
217	$\bar{a} \vee bc$	19	4	07	3	3	2	04	2
226*	$abc \vee a\bar{b}\bar{c} \vee \bar{a}b\bar{c} \vee \bar{a}\bar{b}c$	80	7	20	4	80	7	20	4
227	$\bar{a}\bar{b} \vee \bar{a}\bar{c} \vee \bar{b}\bar{c} \vee abc$	79	7	15	3	58	6	14	3
230	$a\bar{b}\bar{c} \vee bc$	40	5	10	4	34X	5	09	3
232	$c(\bar{a} \vee b) \vee \bar{c}a\bar{b}$	47	5	11	4	47X	5	11	4
233	$bc \vee \bar{b}(\bar{a} \vee \bar{c})$	43	5	10	3	36X	5	09	3
236	$c(\bar{a} \vee b) \vee \bar{a}b \vee a\bar{b}\bar{c}$	66	6	15	3	69	6	15	4
		70	6	15	4				
		71	6	15	4				
237	$bc \vee \bar{a} \vee \bar{b}\bar{c}$	48	5	11	3	32	5	09	3
250	$c(a \vee b)$	8	3	05	2	9	3	06	2
251	$c(a \vee b) \vee \bar{a}\bar{b}\bar{c}$	25	4	10	3	68	6	12	4
253	$c \vee \bar{a}\bar{b}$	5	3	05	3	30	5	07	3
254	$\bar{a}b \vee ac$	16	4	07	3	16X	4	07	3
255	$\bar{a}\bar{c} \vee bc \vee ac$	22	4	09	3	60X	6	10	4
256	$\bar{a}b \vee c$	11	4	06	4	15	4	06	3
274	$a(\bar{b} \vee c) \vee \bar{a}b$	34	5	09	3	40X	5	10	4
275	$a(\bar{b} \vee c) \vee \bar{a}(b \vee \bar{c})$	44	5	11	3	59X	6	11	4
276	$\bar{a}b \vee a\bar{b} \vee c$	35	5	10	3	42	5	10	3
		38	5	10	4				
277	$\bar{a} \vee \bar{b} \vee c$	12	4	06	3	4	2	04	2
350	$ab \vee ac \vee bc$	18	4	09	2	18	4	09	2
351	$ab \vee ac \vee bc \vee \bar{a}\bar{b}\bar{c}$	51	5	15	3	72	7	15	4
352	$ab \vee c$	9	3	06	2	8	3	05	2
353	$ab \vee c \vee \bar{a}\bar{b}$	24	4	10	3	61	6	10	3
357	$\bar{a} \vee b \vee c$	6	3	05	3	10	3	05	2
376	$a \vee b \vee c$	2	2	04	2	17	4	06	2

* There is no circuit with seven blocks or fewer and a fan limit of three for performing this function. The circuit shown, found by Robert Schiller, has a fan-out of four. An eight-block solution is also known, but neither circuit is proved minimal with respect to the number of connections.

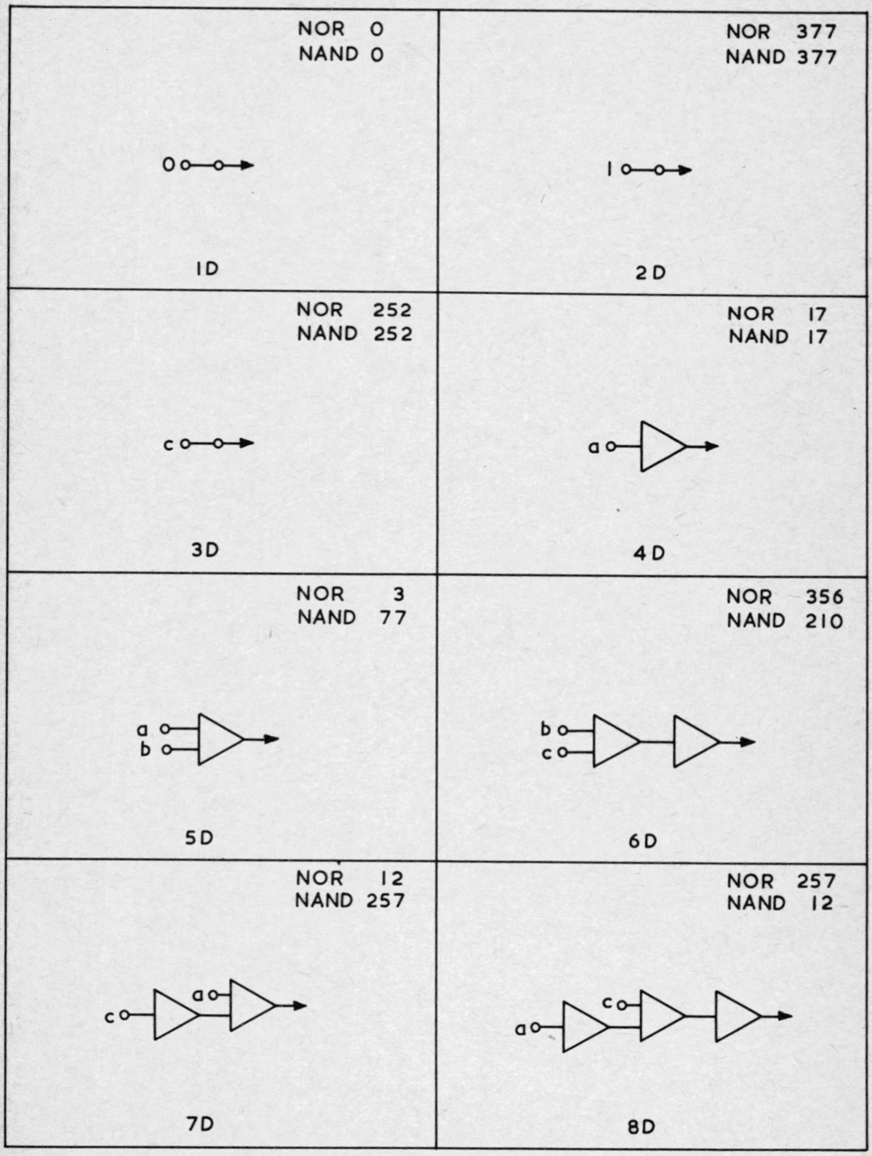

NOR 0
NAND 0

0

1D

NOR 377
NAND 377

1

2D

NOR 252
NAND 252

c

3D

NOR 17
NAND 17

a

4D

NOR 3
NAND 77

a
b

5D

NOR 356
NAND 210

b
c

6D

NOR 12
NAND 257

c
a

7D

NOR 257
NAND 12

a
c

8D

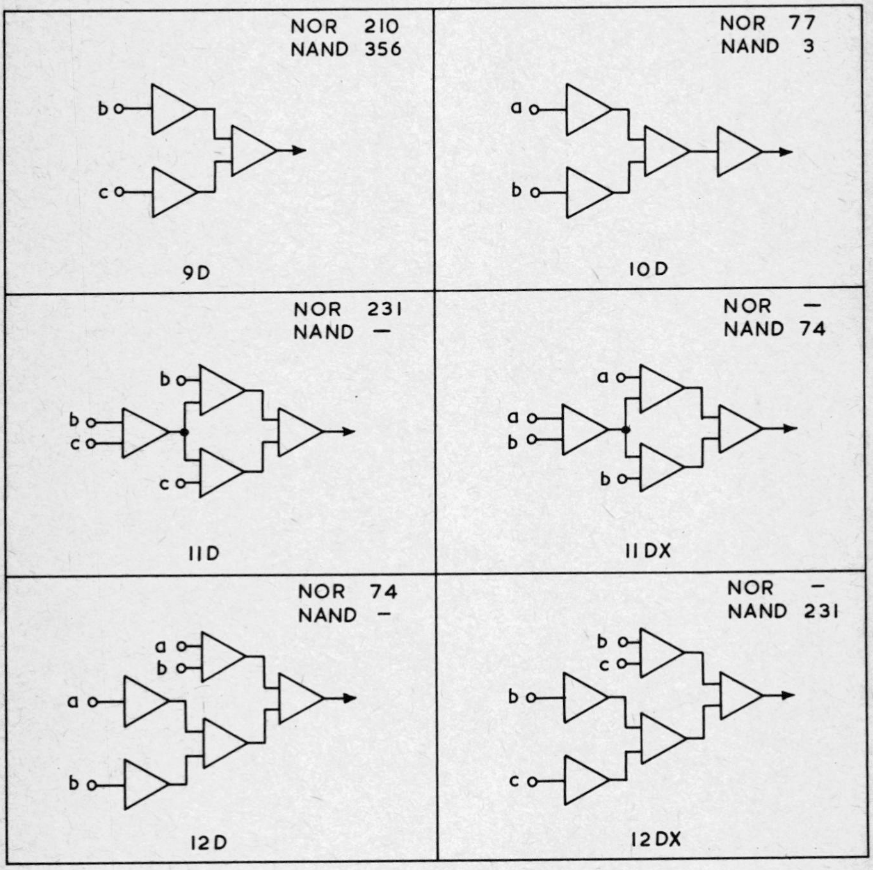

NOR 210
NAND 356
b
c
9D

NOR 77
NAND 3
a
b
10D

NOR 231
NAND —
b
b
c
c
11D

NOR —
NAND 74
a
a
b
b
11DX

NOR 74
NAND —
a
b
a
b
12D

NOR —
NAND 231
b
c
b
c
12DX

NOR 1
NAND 177

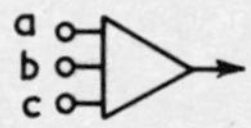

a b c

1

NOR 376
NAND 200

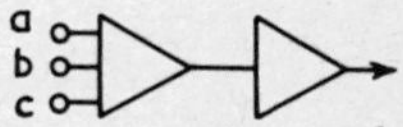

a b c

2

NOR 16
NAND 217

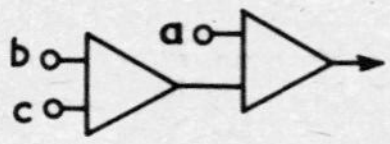

b c a

3

NOR 2
NAND 277

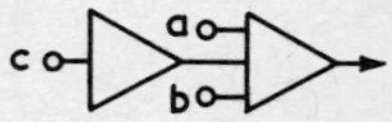

c a b

4

NOR 253
NAND 52

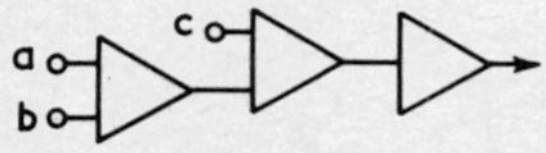

a b c

5

NOR 357
NAND 10

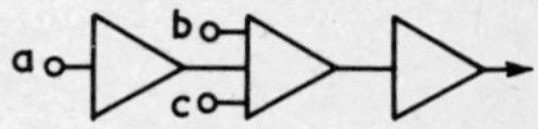

a b c

6

NOR 13
NAND 57

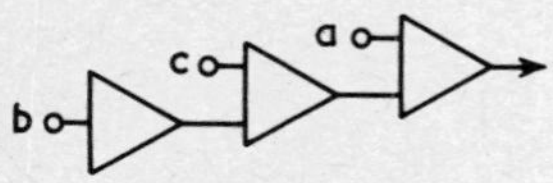

b c a

7

NOR 250
NAND 352

c a b

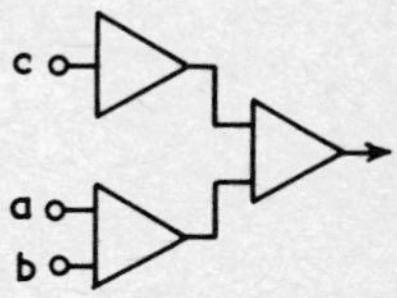

8

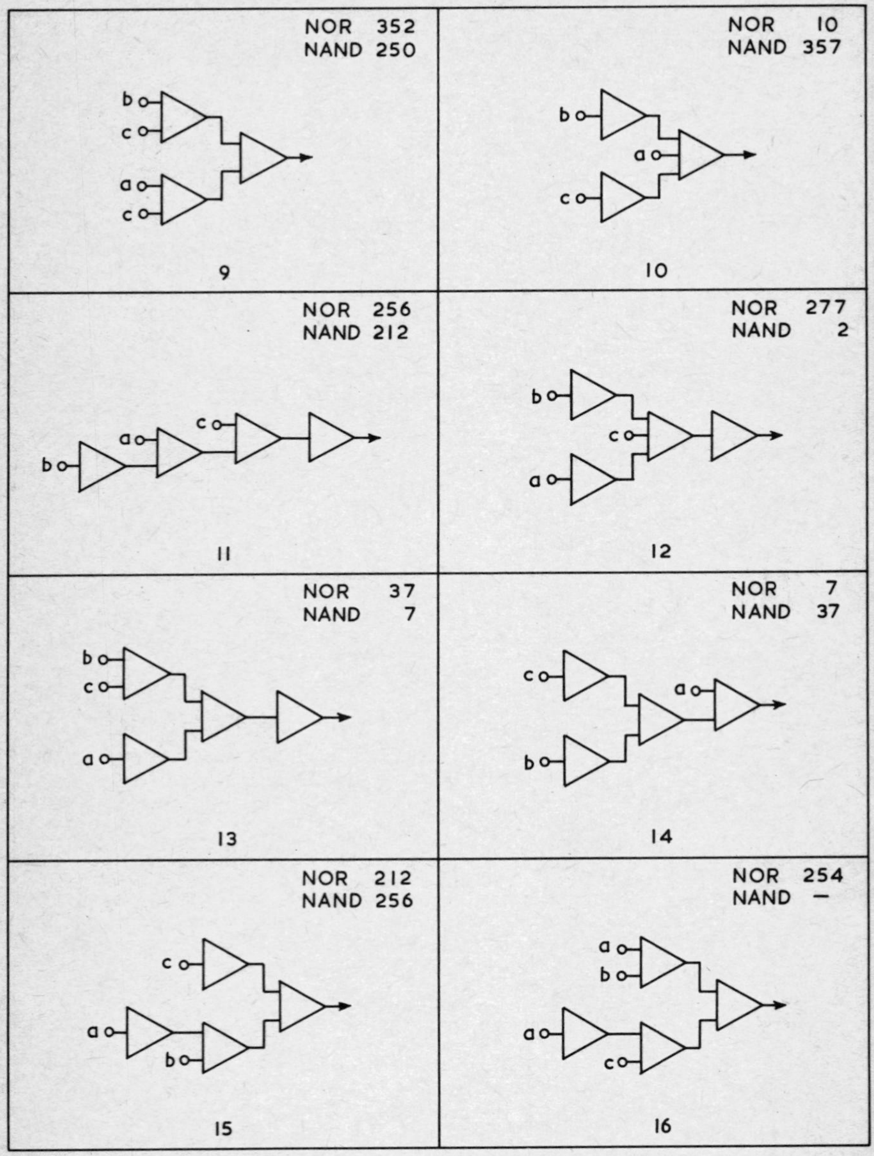
NOR 352
NAND 250
b c a c
9
NOR 10
NAND 357
b a c
10
NOR 256
NAND 212
c a b
11
NOR 277
NAND 2
b c a
12
NOR 37
NAND 7
b c a
13
NOR 7
NAND 37
c a b
14
NOR 212
NAND 256
c a b
15
NOR 254
NAND —
a b a c
16

NOR —
NAND 254

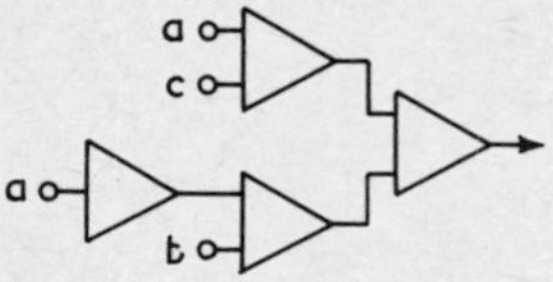

a c a b

16X

NOR 200
NAND 376

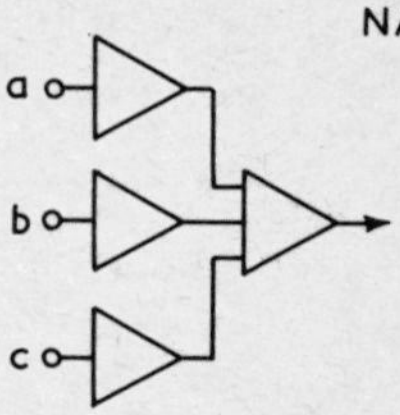

a b c

17

NOR 350
NAND 350

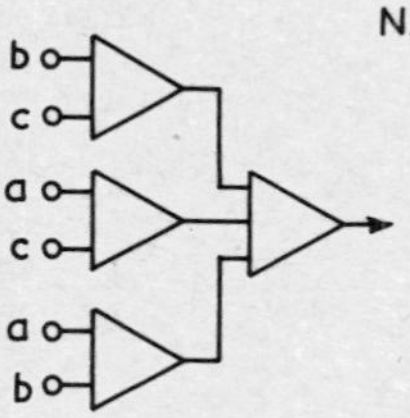

b c a c a b

18

NOR 217
NAND 16

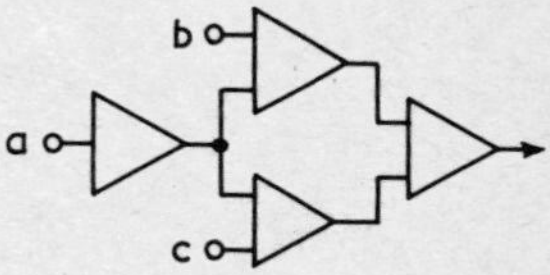

b a c

19

NOR 213
NAND 56

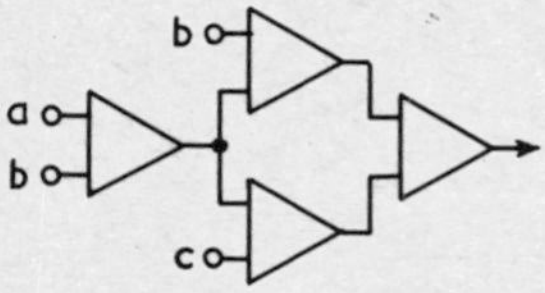

b a b c

20

NOR 11
NAND 157

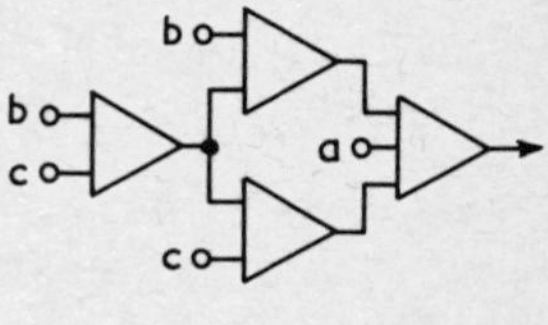

b b c a c

21

NOR 255
NAND —

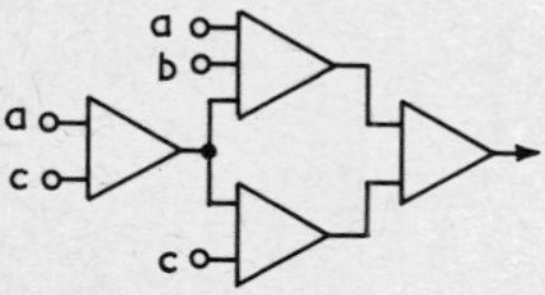

a b a c c

22

NOR —
NAND 54

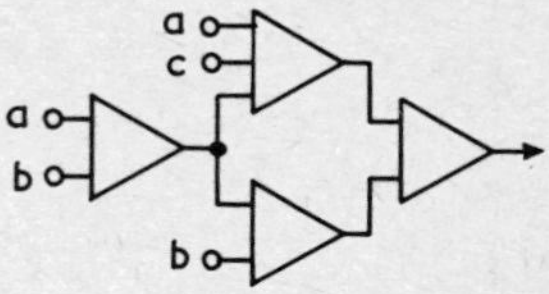

a c a b b

22X

NOR 211
NAND 156

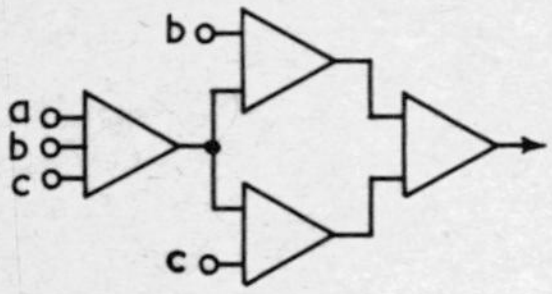

23

NOR 353
NAND 50

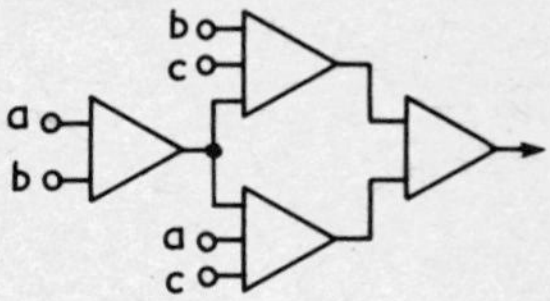

24

NOR 251
NAND 152

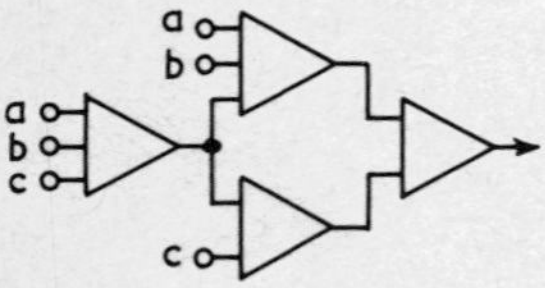

25

NOR 177
NAND 1

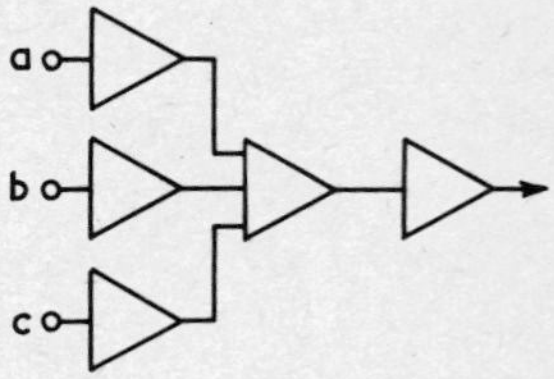

26

NOR 27
NAND 27

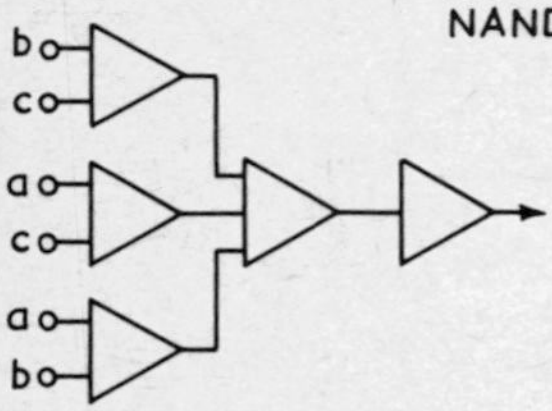

27

NOR 57
NAND 13

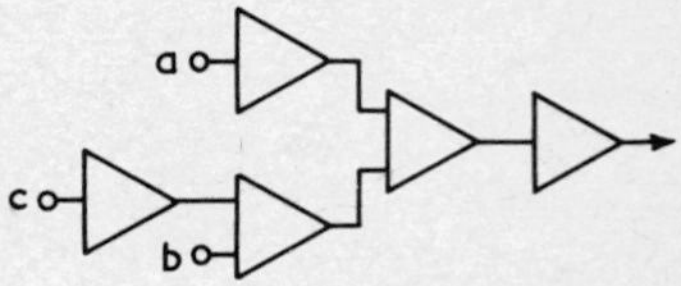

28

NOR 33
NAND —

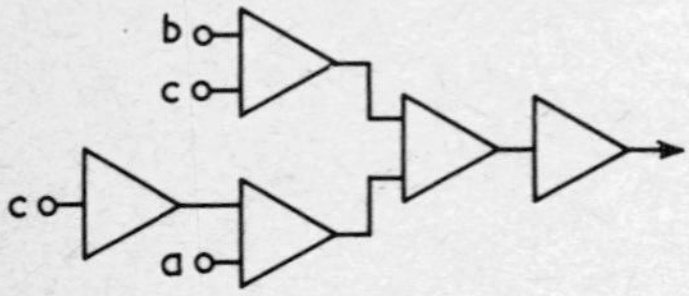

29

NOR —
NAND 33

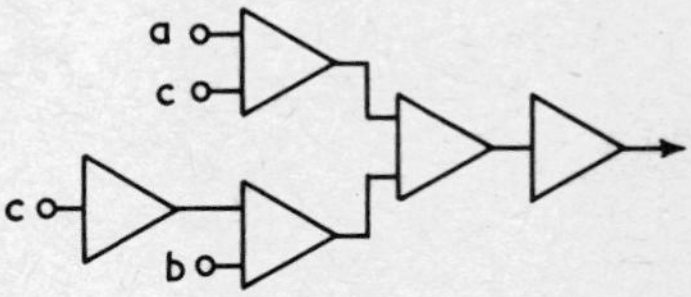

29X

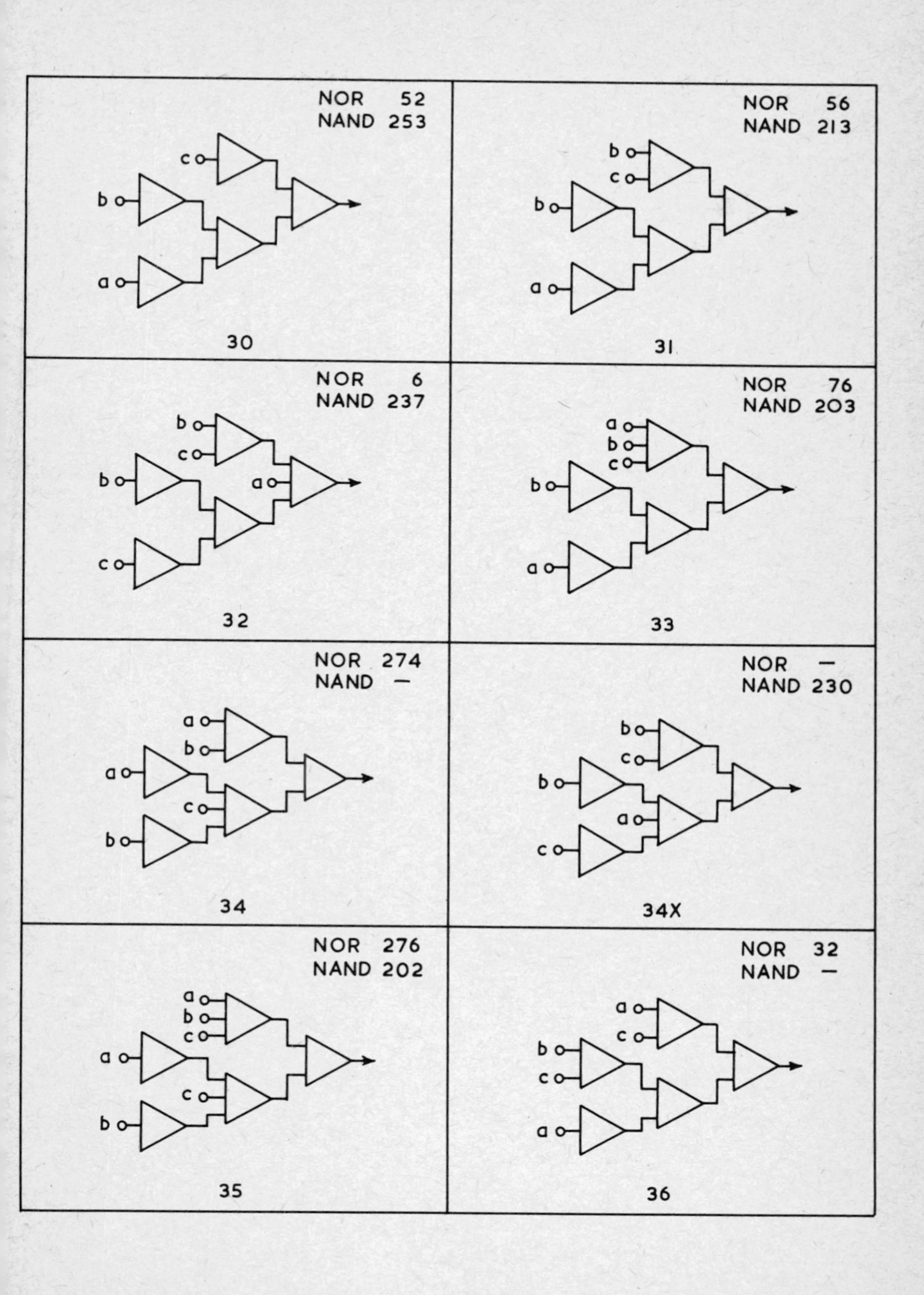
NOR 52
NAND 253
c
b
a
30
NOR 56
NAND 213
b
c
b
a
31
NOR 6
NAND 237
b
c
b
a
c
32
NOR 76
NAND 203
a
b
c
b
a
33
NOR 274
NAND —
a
b
a
c
b
34
NOR —
NAND 230
b
c
b
a
c
34X
NOR 276
NAND 202
a
b
c
a
c
b
35
NOR 32
NAND —
a
c
b
c
a
36

NOR —
NAND 233

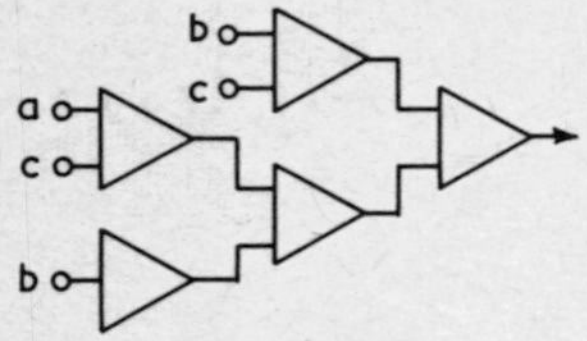

b
c
a
c
b

36X

NOR 36
NAND 207

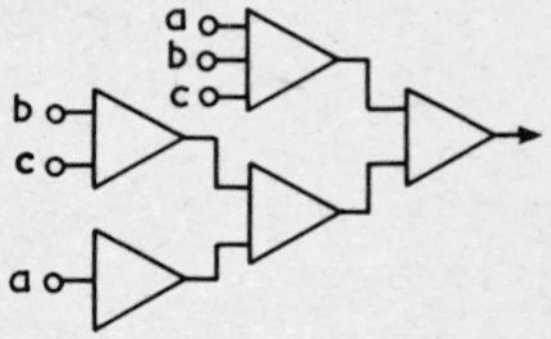

a
b
c
b
c
a

37

NOR 276
NAND 202

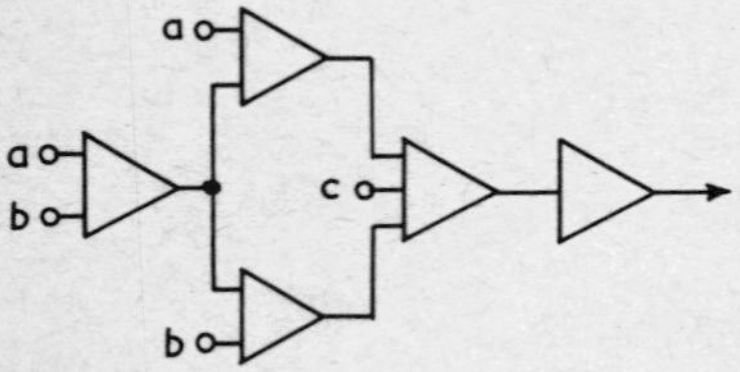

a
a
b
c
b

38

NOR 216
NAND 216

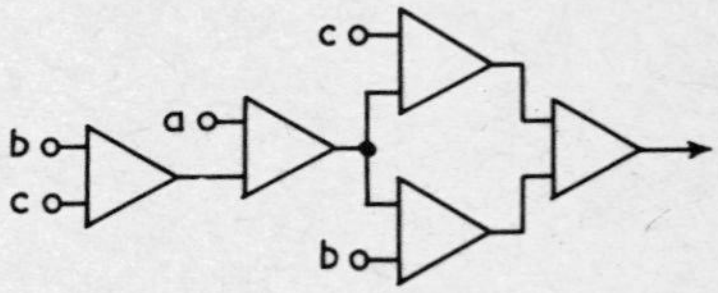

c
a
b
c
b

39

NOR 230
NAND —

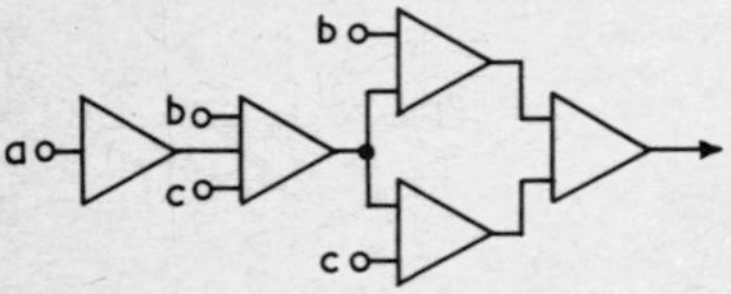

b
a
b
c
c

40

NOR —
NAND 274

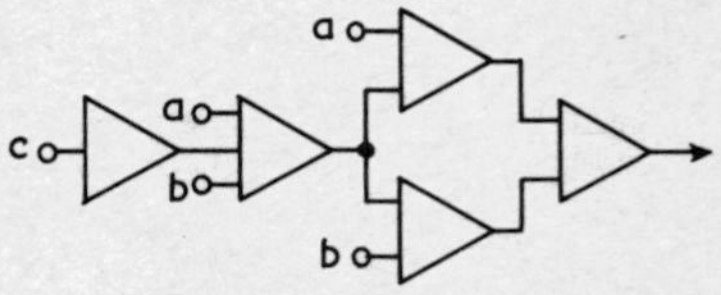

a
a
c
b
b

40X

NOR 216
NAND 216

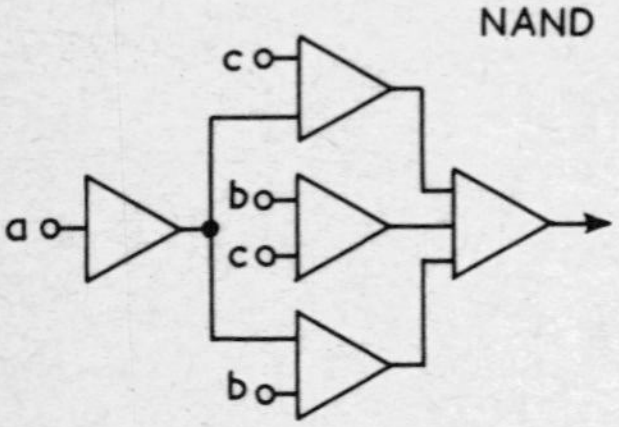

c
b
a
c
b

41

NOR 202
NAND 276

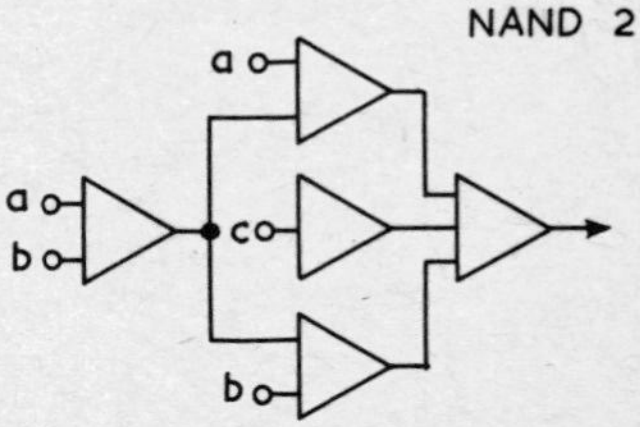

a
a
c
b
b

42

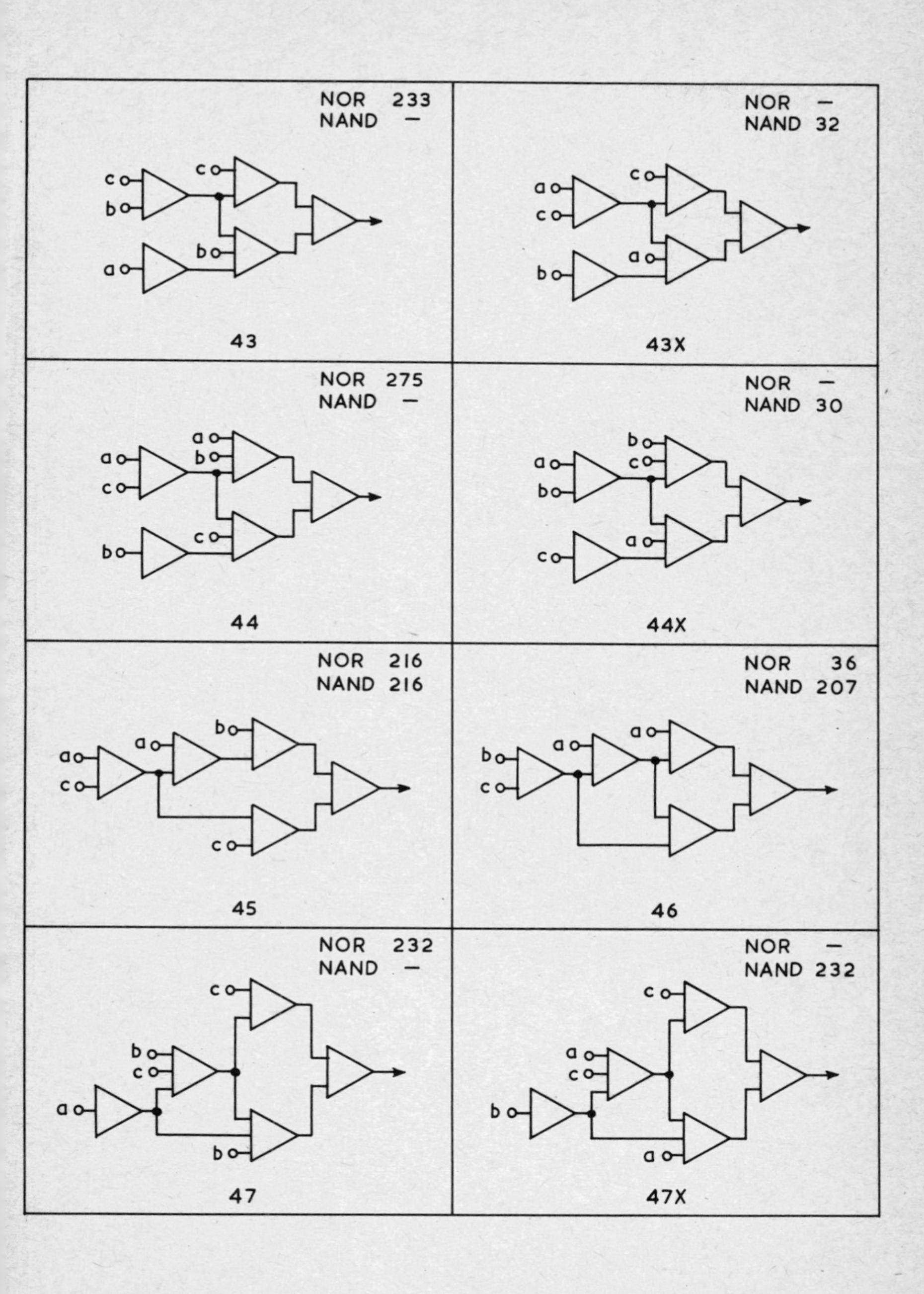
NOR 233
NAND —
43
NOR —
NAND 32
43X
NOR 275
NAND —
44
NOR —
NAND 30
44X
NOR 216
NAND 216
45
NOR 36
NAND 207
46
NOR 232
NAND —
47
NOR —
NAND 232
47X

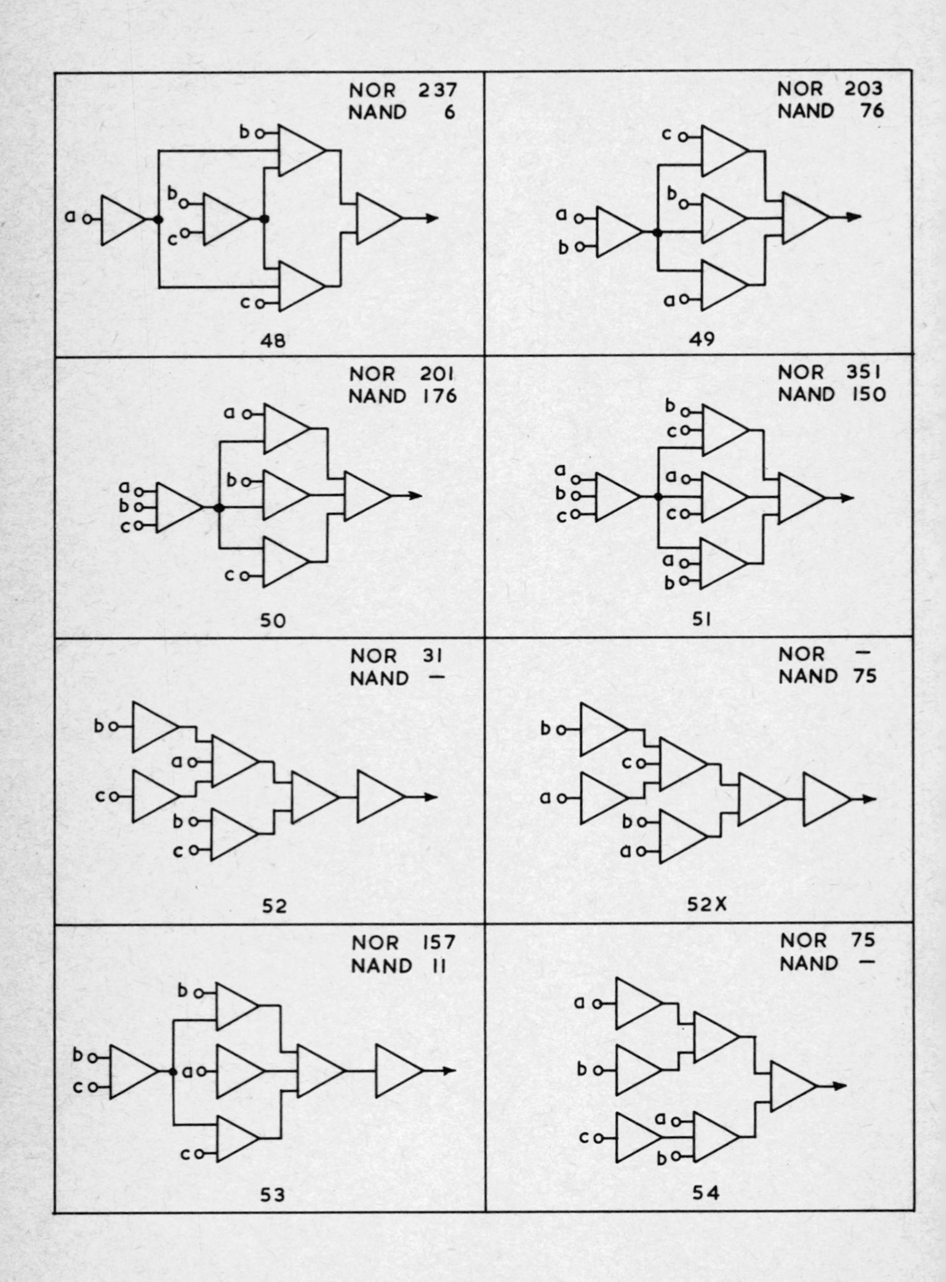
NOR 237
NAND 6
48
NOR 203
NAND 76
49
NOR 201
NAND 176
50
NOR 351
NAND 150
51
NOR 31
NAND —
52
NOR —
NAND 75
52X
NOR 157
NAND 11
53
NOR 75
NAND —
54

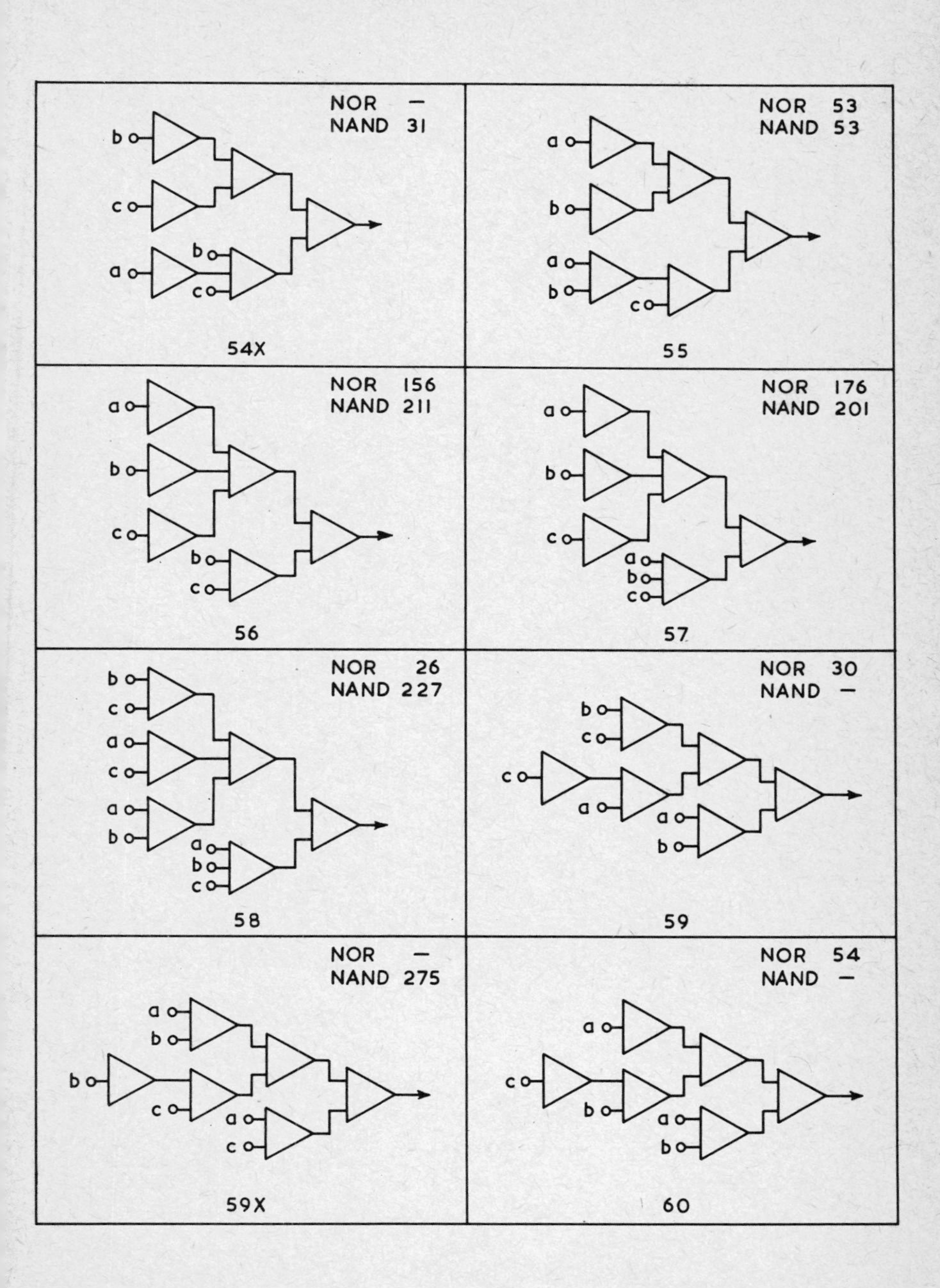
NOR —
NAND 31
b
c
a
b
c
54X
NOR 53
NAND 53
a
b
a
b
c
55
NOR 156
NAND 211
a
b
c
b
c
56
NOR 176
NAND 201
a
b
c
a
b
c
57
NOR 26
NAND 227
b
c
a
c
a
b
a
b
c
58
NOR 30
NAND —
b
c
c
a
a
b
59
NOR —
NAND 275
a
b
b
c
a
c
59X
NOR 54
NAND —
a
c
b
a
b
60

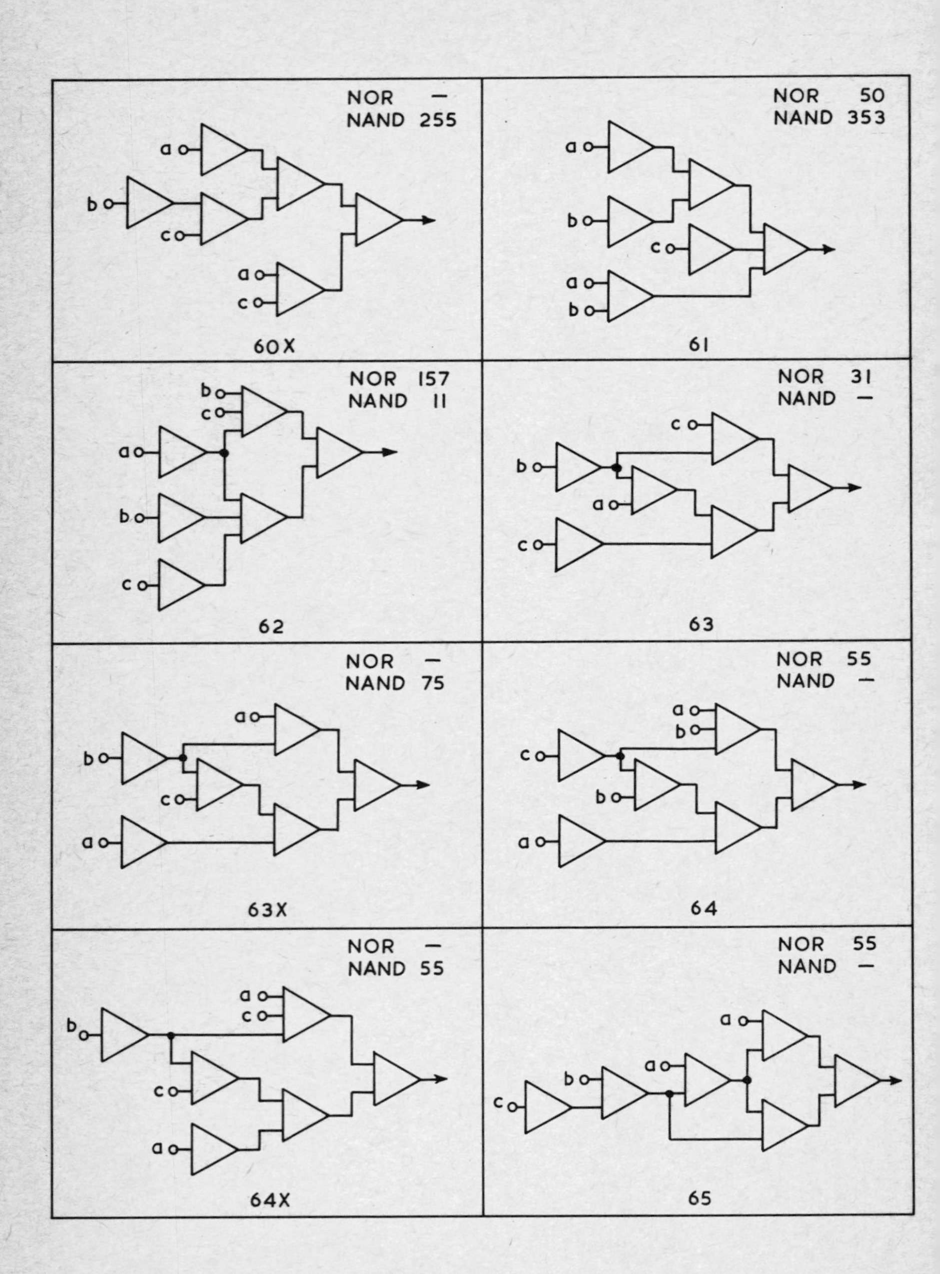
NOR —
NAND 255
a
b
c
a
c
60X
NOR 50
NAND 353
a
b
c
a
b
61
NOR 157
NAND 11
b
c
a
b
c
62
NOR 31
NAND —
c
b
a
c
63
NOR —
NAND 75
a
b
c
a
63X
NOR 55
NAND —
a
b
c
b
a
64
NOR —
NAND 55
a
c
b
c
a
64X
NOR 55
NAND —
a
a
b
c
65

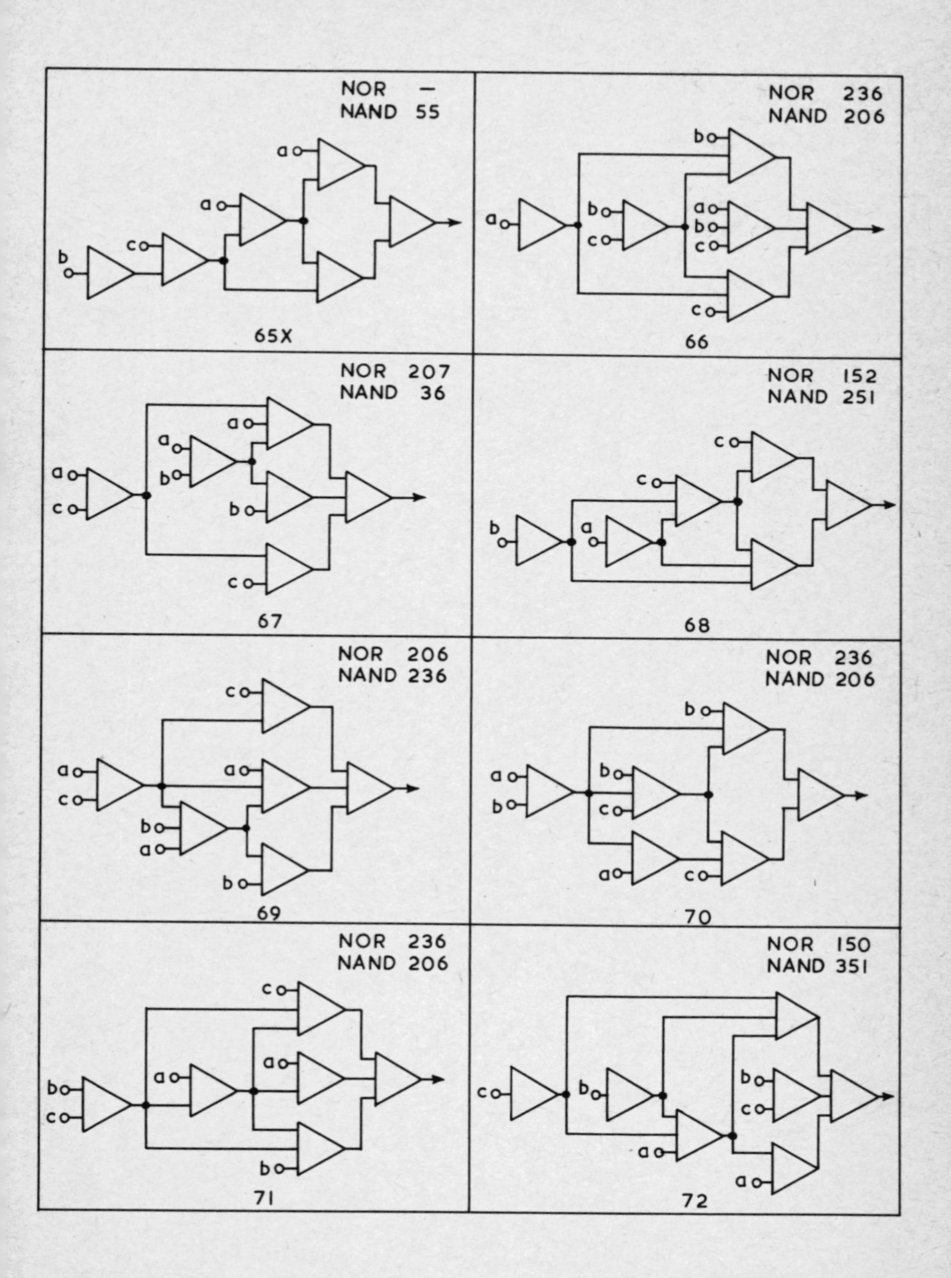
NOR —
NAND 55
65X
NOR 236
NAND 206
66
NOR 207
NAND 36
67
NOR 152
NAND 251
68
NOR 206
NAND 236
69
NOR 236
NAND 206
70
NOR 236
NAND 206
71
NOR 150
NAND 351
72

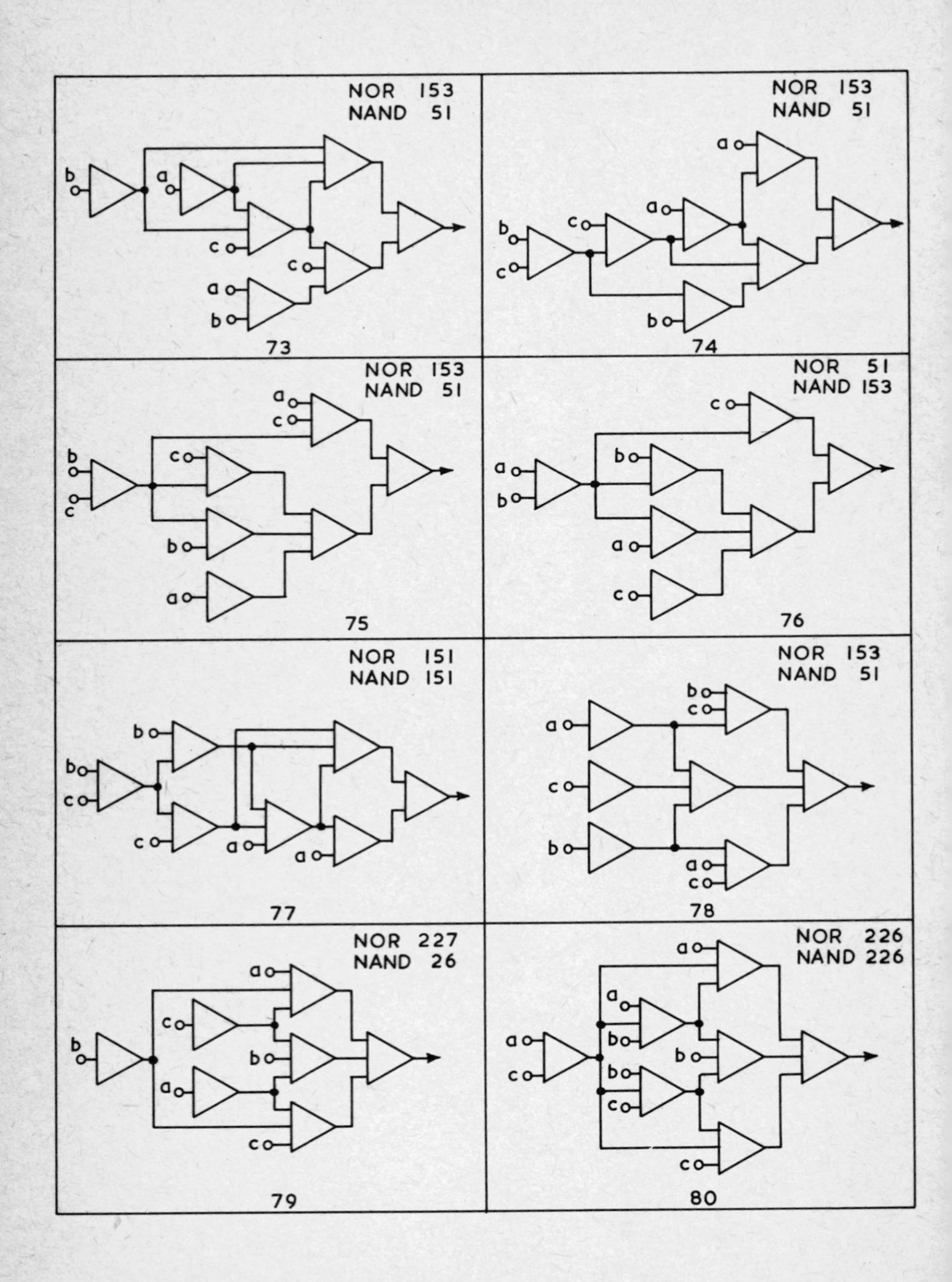
NOR 153
NAND 51
73
NOR 153
NAND 51
74
NOR 153
NAND 51
75
NOR 51
NAND 153
76
NOR 151
NAND 151
77
NOR 153
NAND 51
78
NOR 227
NAND 26
79
NOR 226
NAND 226
80

Appendix C

DERTOUZOS' TABLE OF THRESHOLD FUNCTIONS OF UP TO SIX VARIABLES†

Number		$\|b_i\|$						$\|a_i\|$					
$n \leqslant 3$	1	8	0	0	0			1	0	0	0		
	2	6	2	2	2			2	1	1	1		
	3	4	4	4	0			1	1	1	0		
$n \leqslant 4$	1	16	0	0	0	0		1	0	0	0	0	
	2	14	2	2	2	2		3	1	1	1	1	
	3	12	4	4	4	0		2	1	1	1	0	
	4	10	6	6	2	2		3	2	2	1	1	
	5	8	8	8	0	0		1	1	1	0	0	
	6	8	8	4	4	4		2	2	1	1	1	
	7	6	6	6	6	6		1	1	1	1	1	
$n \leqslant 5$	1	32	0	0	0	0	0	1	0	0	0	0	0
	2	30	2	2	2	2	2	4	1	1	1	1	1
	3	28	4	4	4	4	0	3	1	1	1	1	0
	4	26	6	6	6	2	2	5	2	2	2	1	1
	5	24	8	8	4	4	4	4	2	2	1	1	1
	6	24	8	8	8	0	0	2	1	1	1	0	0
	7	22	10	10	6	2	2	5	3	3	2	1	1
	8	22	10	6	6	6	6	3	2	1	1	1	1
	9	20	12	12	4	4	0	3	2	2	1	1	0
	10	20	12	8	8	4	4	4	3	2	2	1	1
	11	20	8	8	8	8	8	2	1	1	1	1	1
	12	18	14	14	2	2	2	4	3	3	1	1	1
	13	18	14	10	6	6	2	5	4	3	2	2	1
	14	18	10	10	10	6	6	3	2	2	2	1	1
	15	16	16	16	0	0	0	1	1	1	0	0	0
	16	16	16	12	4	4	4	3	3	2	1	1	1
	17	16	16	8	8	8	0	2	2	1	1	1	0
	18	16	12	12	8	8	4	4	3	3	2	2	1

(continued)

† Reproduced by permission, M. L. Dertouzos, *Threshold Logic: A Synthesis Approach* Research Monograph No.32, The MIT Press, Cambridge, Massachusetts, U.S.A., 1965.

Number		$\lvert b_i \rvert$							$\lvert a_i \rvert$						
	19	14	14	14	6	6	6		2	2	2	1	1	1	
	20	14	14	10	10	10	2		3	3	2	2	2	1	
	21	12	12	12	12	12	0		1	1	1	1	1	0	
$n \leqslant 6$	1	64	0	0	0	0	0	0	1	0	0	0	0	0	0
	2	62	2	2	2	2	2	2	5	1	1	1	1	1	1
	3	60	4	4	4	4	4	0	4	1	1	1	1	1	0
	4	58	6	6	6	6	2	2	7	2	2	2	2	1	1
	5	56	8	8	8	8	0	0	3	1	1	1	1	0	0
	6	56	8	8	8	4	4	4	6	2	2	2	1	1	1
	7	54	10	10	10	6	2	2	8	3	3	3	2	1	1
	8	54	10	10	6	6	6	6	5	2	2	1	1	1	1
	9	52	12	12	12	4	4	0	5	2	2	2	1	1	0
	10	52	12	12	8	8	4	4	7	3	3	2	2	1	1
	11	52	12	8	8	8	8	8	4	2	1	1	1	1	1
	12	50	14	14	14	2	2	2	7	3	3	3	1	1	1
	13	50	14	14	10	6	6	2	9	4	4	3	2	2	1
	14	50	14	10	10	10	6	6	6	3	2	2	2	1	1
	15	50	10	10	10	10	10	10	3	1	1	1	1	1	1
	16	48	16	16	16	0	0	0	2	1	1	1	0	0	0
	17	48	16	16	12	4	4	4	6	3	3	2	1	1	1
	18	48	16	16	8	8	8	0	4	2	2	1	1	1	0
	19	48	16	12	12	8	8	4	8	4	3	3	2	2	1
	20	48	12	12	12	12	8	8	5	2	2	2	2	1	1
	21	46	18	18	14	2	2	2	7	4	4	3	1	1	1
	22	46	18	18	10	6	6	2	9	5	5	3	2	2	1
	23	46	18	14	14	6	6	6	5	3	2	2	1	1	1
	24	46	18	14	10	10	10	2	7	4	3	2	2	2	1
	25	46	14	14	14	10	10	6	7	3	3	3	2	2	1
	26	44	20	20	12	4	4	0	5	3	3	2	1	1	0
	27	44	20	20	8	8	4	4	7	4	4	2	2	1	1
	28	44	20	16	16	4	4	4	6	4	3	3	1	1	1
	29	44	20	16	12	8	8	4	8	5	4	3	2	2	1
	30	44	20	12	12	12	12	0	3	2	1	1	1	1	0
	31	44	16	16	16	8	8	8	4	2	2	2	1	1	1
	32	44	16	16	12	12	12	4	6	3	3	2	2	2	1
	33	42	22	22	10	6	2	2	8	5	5	3	2	1	1
	34	42	22	22	6	6	6	6	5	3	3	1	1	1	1
	35	42	22	18	14	6	6	2	9	6	5	4	2	2	1
	36	42	22	18	10	10	6	6	6	4	3	2	2	1	1
	37	42	22	14	14	10	10	2	7	5	3	3	2	2	1
	38	42	18	18	18	6	6	6	5	3	3	3	1	1	1
	39	42	18	18	14	10	10	6	7	4	4	3	2	2	1
	40	42	18	14	14	14	14	2	5	3	2	2	2	2	1
	41	40	24	24	8	8	0	0	3	2	2	1	1	0	0
	42	40	24	24	8	4	4	4	6	4	4	2	1	1	1
	43	40	24	20	12	8	4	4	7	5	4	3	2	1	1
	44	40	24	20	8	8	8	8	4	3	2	1	1	1	1

(*continued*)

Number	$\lvert b_i \rvert$							$\lvert a_i \rvert$						
45	40	24	16	16	8	8	0	4	3	2	2	1	1	0
46	40	24	16	12	12	8	4	8	6	4	3	3	2	1
47	40	20	20	16	8	8	4	8	5	5	4	2	2	1
48	40	20	20	12	12	8	8	5	3	3	2	2	1	1
49	40	20	16	16	12	12	4	6	4	3	3	2	2	1
50	40	16	16	16	16	16	0	2	1	1	1	1	1	0
51	38	26	26	6	6	2	2	7	5	5	2	2	1	1
52	38	26	22	10	10	2	2	8	6	5	3	3	1	1
53	38	26	22	10	6	6	6	5	4	3	2	1	1	1
54	38	26	18	14	10	6	2	9	7	5	4	3	2	1
55	38	26	18	10	10	10	6	6	5	3	2	2	2	1
56	38	26	14	14	14	6	6	5	4	2	2	2	1	1
57	38	22	22	14	10	6	6	6	4	4	3	2	1	1
58	38	22	22	10	10	10	10	3	2	2	1	1	1	1
59	38	22	18	18	10	10	2	7	5	4	4	2	2	1
60	38	22	18	14	14	10	6	7	5	4	3	3	2	1
61	38	18	18	18	14	14	2	5	3	3	3	2	2	1
62	36	28	28	4	4	4	0	4	3	3	1	1	1	0
63	36	28	24	8	8	4	4	6	5	4	2	2	1	1
64	36	28	20	12	12	4	0	5	4	3	2	2	1	0
65	36	28	20	12	8	8	4	7	6	4	3	2	2	1
66	36	28	16	16	12	4	4	6	5	3	3	2	1	1
67	36	28	16	12	12	8	8	8	7	4	3	3	2	2
68	36	24	24	12	12	4	4	7	5	5	3	3	1	1
69	36	24	24	12	8	8	8	4	3	3	2	1	1	1
70	36	24	20	16	12	8	4	8	6	5	4	3	2	1
71	36	24	20	12	12	12	8	5	4	3	2	2	2	1
72	36	24	16	16	16	8	8	4	3	2	2	2	1	1
73	36	20	20	20	12	12	0	3	2	2	2	1	1	0
74	36	20	20	16	16	12	4	6	4	4	3	3	2	1
75	34	30	30	2	2	2	2	5	4	4	1	1	1	1
76	34	30	26	6	6	6	2	7	6	5	2	2	2	1
77	34	30	22	10	10	6	2	8	7	5	3	3	2	1
78	34	30	18	14	14	2	2	7	6	4	3	3	1	1
79	34	30	18	14	10	6	6	9	8	5	4	3	2	2
80	34	30	14	14	10	10	10	7	6	3	3	2	2	2
81	34	26	26	10	10	6	6	5	4	4	2	2	1	1
82	34	26	22	14	14	6	2	9	7	6	4	4	2	1
83	34	26	22	14	10	10	6	6	5	4	3	2	2	1
84	34	26	18	18	14	6	6	5	4	3	3	2	1	1
85	34	26	18	14	14	10	10	6	5	4	3	3	2	2
86	34	22	22	18	14	10	2	7	5	5	4	3	2	1
87	34	22	22	14	14	14	6	4	3	3	2	2	2	1
88	34	22	18	18	18	10	6	5	4	3	3	3	2	1
89	32	32	32	0	0	0	0	1	1	1	0	0	0	0
90	32	32	28	4	4	4	4	4	4	3	1	1	1	1
91	32	32	24	8	8	8	0	3	3	2	1	1	1	0
92	32	32	20	12	12	4	4	5	5	3	2	2	1	1

(*continued*)

Number	$\|b_i\|$							$\|a_i\|$						
93	32	32	16	16	16	0	0	2	2	1	1	1	0	0
94	32	32	16	16	8	8	8	4	4	2	2	1	1	1
95	32	32	12	12	12	12	12	3	3	1	1	1	1	1
96	32	28	28	8	8	8	4	6	5	5	2	2	2	1
97	32	28	24	12	12	8	4	7	6	5	3	3	2	1
98	32	28	20	16	16	4	4	6	5	4	3	3	1	1
99	32	28	20	16	12	8	8	7	6	5	4	3	2	2
100	32	28	16	16	12	12	12	5	4	3	3	2	2	2
101	32	24	24	16	16	8	0	4	3	3	2	2	1	0
102	32	24	24	16	12	12	4	5	4	4	3	2	2	1
103	32	24	20	20	16	8	4	6	5	4	4	3	2	1
104	32	24	20	16	16	12	8	7	6	5	4	4	3	2
105	32	20	20	20	20	8	8	3	2	2	2	2	1	1
106	30	30	30	6	6	6	6	3	3	3	1	1	1	1
107	30	30	26	10	10	10	2	5	5	4	2	2	2	1
108	30	30	22	14	14	6	6	4	4	3	2	2	1	1
109	30	30	18	18	18	2	2	5	5	3	3	3	1	1
110	30	30	18	18	10	10	10	3	3	2	2	1	1	1
111	30	30	14	14	14	14	14	2	2	1	1	1	1	1
112	30	26	26	14	14	10	2	6	5	5	3	3	2	1
113	30	26	22	18	18	6	2	7	6	5	4	4	2	1
114	30	26	22	18	14	10	6	8	7	6	5	4	3	2
115	30	26	18	18	14	14	10	6	5	4	4	3	3	2
116	30	22	22	22	18	6	6	4	3	3	3	2	1	1
117	30	22	22	18	18	10	10	5	4	4	3	3	2	2
118	28	28	28	12	12	12	0	2	2	2	1	1	1	0
119	28	28	24	16	16	8	4	5	5	4	3	3	2	1
120	28	28	20	20	20	4	0	3	3	2	2	2	1	0
121	28	28	20	20	12	12	8	4	4	3	3	2	2	1
122	28	28	16	16	16	16	12	3	3	2	2	2	2	1
123	28	24	24	20	20	4	4	5	4	4	3	3	1	1
124	28	24	24	20	16	8	8	6	5	5	4	3	2	2
125	28	24	20	20	16	12	12	7	6	5	5	4	3	3
126	26	26	26	18	18	6	6	3	3	3	2	2	1	1
127	26	26	22	22	22	2	2	4	4	3	3	3	1	1
128	26	26	22	22	14	10	10	5	5	4	4	3	2	2
129	26	26	18	18	18	14	14	4	4	3	3	3	2	2
130	26	22	22	22	14	14	14	4	3	3	3	2	2	2
131	24	24	24	24	24	0	0	1	1	1	1	1	0	0
132	24	24	24	24	12	12	12	2	2	2	2	1	1	1
133	24	24	20	20	16	16	16	5	5	4	4	3	3	3
134	22	22	22	18	18	18	18	3	3	3	2	2	2	2
135	20	20	20	20	20	20	20	1	1	1	1	1	1	1

Appendix D

IMPROVED TREE METHOD FOR THE STATE MINIMIZATION OF INCOMPLETELY SPECIFIED SEQUENTIAL MACHINES†

The implication tree method as described in Chapter 9 has one serious disadvantage inasmuch as all the secondary trees are to be constructed before a minimal cover of the machine can be determined. However if we are not interested in finding *all* solutions but are satisfied by obtaining *a* solution, then the method can be so modified that not all secondary trees need be constructed. Let the improved method be explained by working out the Machine M_3 of Fig.9-11.

At first the compatibility graph of the given ISSM is drawn. Then the MCs are determined and the implication trees from the MCs are constructed in the same way as has been described in Chapter 9. The various states which appear in a tree may or may not cover all the states of the ISSM. Whenever a tree does not cover all the states, the missing states are drawn to constitute a separate branch from the MC, so that each of these missing states forms a distinct node of this branch. This modified construction for Machine M_3, has been shown in Fig.D-1. (Compare this figure with Fig.9-15.)

Now we shall carry out bunching, which is more generalized and is defined as follows:

Definition D.1 The nodes $N_1, N_2, \ldots, N_n$ of a primary tree generated by an MC, MC_g can be *bunched* into a single node, N_0, where

$$N_0 = N_1 \cup N_2 \cup \ldots \cup N_n$$

if the following two conditions are satisfied:

1) There exists either a primary or a secondary tree with N_0 as the generating node.

2) The CPs (Compatible Pairs) contained in the tree of N_0 are also contained in the tree of MC_g.

It may be easily seen that the definition of bunching, as given in Definition 9.6.2, is a special case of the above definition, where the tree of N_0 does not contain any CP at all.

---

† N. N. Biswas and P. V. Acharya; Department of Electrical Communication Engineering, Indian Institute of Science, Bangalore, India, November, 1973.

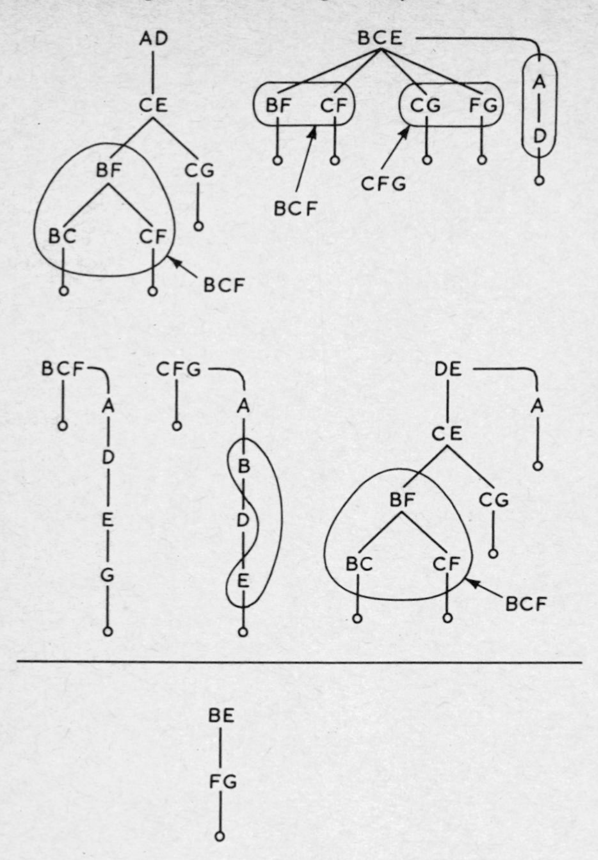

Figure D-1 Improved implication trees with generalized bunching for machine M_3 (cf Fig.9-15).

Applying this definition, the various bunchings that can be carried out are shown in Fig.D-1. The bunches BCF in the tree of AD; BCF, CFG and AD in the tree of BCE; and BCF in the tree of DE can be obtained from the primary trees themselves. In the tree of CFG, A and D cannot be bunched as this tree does not contain, say, the compatible pair BC which is contained in the tree of AD. To check if B and E can be bunched into BE, the secondary tree of BE is constructed. The tree of BE contains FG which is also contained in the tree of CFG (FG $\subset$ CFG). Hence B and E can be bunched. No bunching is possible in the tree of BCF. By construction, all the primary trees cover all the states of the ISSM. Hence, those trees with minimum number of compatibles are solutions. In Fig.D-1, each of the trees from AD, BCE, and CFG offers a solution with 4 states. These are

1) AD, CE, CG, BCF
2) BCE, BCF, CFG, AD
3) CFG, A, BE, D

(Compare these with the solutions obtained on page 260). The merit of this method is obvious. We had to construct only one secondary tree, *viz*. that of BE only. It should also be mentioned here, that the operations of substitution and augmentation are not required. Only the operation of bunching, as defined in D.1, is adequate.

Following this method the machine M_4, as depicted in Fig.9-21, can be minimized as shown in Fig.D-2. No bunching is possible in the trees of

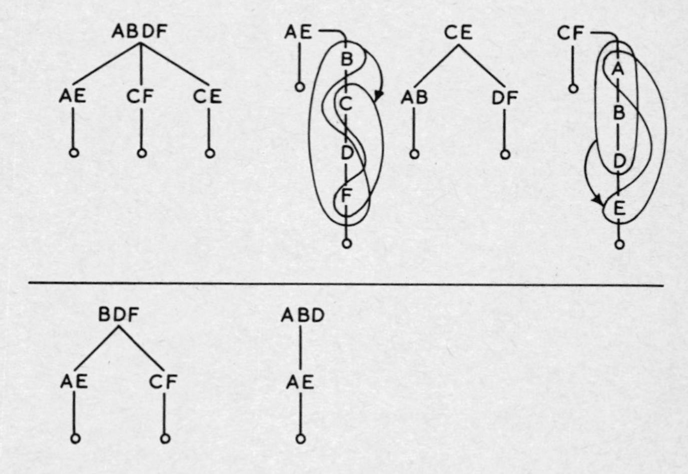

Figure D-2 Improved implication trees with generalized bunching for machine M_4 (cf Fig.9-23). The arrows indicate conditional bunching.

ABDF and CE. In the tree of AE, C and F can be bunched into CF. Now, to check if B, D, and F can be bunched, the secondary tree of BDF is constructed. It contains AE and CF which are also contained in the tree of AE (since C and F have been bunched into CF). So, B, D, and F are also bunched. Similarly in the tree of CF, A and E, and A, B, and D are bunched after the secondary tree of ABD is constructed. The trees of AE, CE, and CF each offers a 3-state solution which are as follows:

1) AE, BDF, CF
2) CE, AB, DF
3) CF, ABD, AE

(Compare these with the solutions obtained on page 264.)

Sometimes, a situation as shown in Fig. D-3 may arise. Here in the tree of CF, A, B, and D can be bunched if A and E can be bunched (as required by the secondary tree ABD). Again, as required by the tree of AEG, the states A, E, and G of the tree of CF can be bunched if A and D can be bunched. Hence the two bunches ABD and AEG are mutually dependent and one cannot exist without the other. The interdependence of such bunches is shown by arrows on the diagram. Following similar arguments, in Fig. D-2 the bunching of B, D, and F in the tree of AE is conditional on the bunching of C and F. This phenomenon of conditional bunching is also depicted by an arrow in the figure. It can be verified that the bunching of A, B, and D in the tree of CF is also dependent on the bunching of A and E.

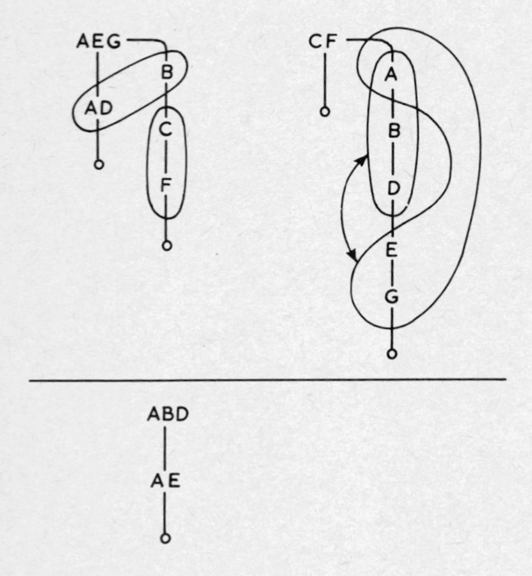

Figure D-3 A situation showing interdependence of two bunches.

This appendix shows the strength of the generalized bunching operation. However, before the procedure is terminated, it must be shown that for the *n*-state solution obtained, either *n* is the lower bound, or no solution with a number of states less than *n* exists.

Appendix E

ACRONYMS

BCD	Binary Coded Decimal
CCF	Conjunctive Canonical Form
CNF	Conjunctive Normal Form
CP	Compatible Pair
CSSM	Completely specified sequential machine
D	Delay
DCF	Disjunctive Canonical Form
DNF	Disjunctive Normal Form
ECC	Error Correcting Code
EDC	Error Detecting Code
EPI	Essential Prime Implicant
ESF	Elementary Symmetric Function
FF	Flip-Flop
IC	Integrated Circuit
ISSM	Incompletely Specified Sequential Machine
LS	Linearly Separable
LSD	Least Significant Digit
LSI	Large Scale Integration
MC	Maximal Compatible
MIRV	Minimum Integer Realization Vector
MPS	Minimal Product of Sums
MSD	Most Significant Digit
MSI	Medium Scale Integration
MSP	Minimal Sum of Products
NS	Next State
PI	Prime Implicant
POLS	Positive Ordered Linearly Separable
PS	Present State, Product of Sums
PSF	Partially Symmetric Function
QM	Quine–McCluskey

RPI	Redundant Prime Implicant
SA	State Assignment
SC	Sequential Circuit, Set Clear
SCT	Set Clear Trigger
SD	Self Dual
SM	Sequential Machine
SP	Sum of Products
SPI	Selective Prime Implicant
SR	Set Reset
T	Trigger
TF	Threshold Function
TG	Trigger Gate
TSF	Totally Symmetric Function
ULM	Universal Logic Module
VK	Veitch–Karnaugh

SUBJECT INDEX

NAME INDEX